Nanofluid Technologies and Thermal Convection Techniques

Ramesh Chand
Government Arya Degree College Nurpur (HP), India

A volume in the Advances in Mechatronics and Mechanical Engineering (AMME) Book Series

IGI GLOBAL
DISSEMINATOR OF KNOWLEDGE
www.igi-global.com

Published in the United States of America by
IGI Global
Engineering Science Reference (an imprint of IGI Global)
701 E. Chocolate Avenue
Hershey PA 17033
Tel: 717-533-8845
Fax: 717-533-8661
E-mail: cust@igi-global.com
Web site: http://www.igi-global.com

Library of Congress Cataloging-in-Publication Data

Library of Congress Cataloging-in-Publication Data

Names: Chand, Ramesh, 1973- author.
Title: Nanofluid technologies and thermal convection techniques / by Ramesh Chand.
Description: Hershey PA : CyberTech Publishing, [2017] | Includes bibliographical references.
Identifiers: LCCN 2016044397| ISBN 9781683180067 (hardcover) | ISBN 9781683180074 (ebook)
Subjects: LCSH: Nanofluids--Thermal properties. | Heat--Convection.
Classification: LCC TJ853.4.M53 C48 2017 | DDC 620.1/06--dc23 LC record available at https://lccn.loc.gov/2016044397

This book is published in the IGI Global book series Advances in Mechatronics and Mechanical Engineering (AMME) (ISSN: 2328-8205; eISSN: 2328-823X)

British Cataloguing in Publication Data
A Cataloguing in Publication record for this book is available from the British Library.

Advances in Mechatronics and Mechanical Engineering (AMME) Book Series

ISSN:2328-8205
EISSN:2328-823X

Editor-in-Chief: J. Paulo Davim, University of Aveiro, Portugal

Mission

With its aid in the creation of smartphones, cars, medical imaging devices, and manufacturing tools, the mechatronics engineering field is in high demand. Mechatronics aims to combine the principles of mechanical, computer, and electrical engineering together to bridge the gap of communication between the different disciplines.

The **Advances in Mechatronics and Mechanical Engineering (AMME) Book Series** provides innovative research and practical developments in the field of mechatronics and mechanical engineering. This series covers a wide variety of application areas in electrical engineering, mechanical engineering, computer and software engineering; essential for academics, practitioners, researchers, and industry leaders.

Coverage

- Computer-Based Manufacturing
- Vibration and acoustics
- Sustainable and green manufacturing
- Nanomaterials and nanomanufacturing
- Design and Manufacture
- Micro and nanomechanics
- Biologically Inspired Robotics
- Control Systems Modelling and Analysis
- Medical robotics
- Bioengineering Materials

IGI Global is currently accepting manuscripts for publication within this series. To submit a proposal for a volume in this series, please contact our Acquisition Editors at Acquisitions@igi-global.com or visit: http://www.igi-global.com/publish/.

The Advances in Mechatronics and Mechanical Engineering (AMME) Book Series (ISSN 2328-8205) is published by IGI Global, 701 E. Chocolate Avenue, Hershey, PA 17033-1240, USA, www.igi-global.com. This series is composed of titles available for purchase individually; each title is edited to be contextually exclusive from any other title within the series. For pricing and ordering information please visit http://www.igi-global.com/book-series/advances-mechatronics-mechanical-engineering/73808. Postmaster: Send all address changes to above address.

Titles in this Series

For a list of additional titles in this series, please visit:
http://www.igi-global.com/book-series/advances-mechatronics-mechanical-engineering/73808

Mathematical Concepts and Applications in Mechanical Engineering and Mechatronics
Mangey Ram (Graphic Era University, India) and J. Paulo Davim (University of Aveiro, Portugal)
Engineering Science Reference • ©2017 • 495pp • H/C (ISBN: 9781522516392) • US $215.00

Handbook of Research on Advances and Applications in Refrigeration Systems and Technologies
Pedro Dinis Gaspar (University of Beira Interior, Portugal) and Pedro Dinho da Silva (University of Beira Interior, Portugal)
Engineering Science Reference • ©2015 • 924pp • H/C (ISBN: 9781466683983) • US $545.00

Engineering Creative Design in Robotics and Mechatronics
Maki K. Habib (The American University in Cairo, Egypt) and J. Paulo Davim (University of Aveiro, Portugal)
Engineering Science Reference • ©2013 • 366pp • H/C (ISBN: 9781466642256) • US $195.00

Computational Methods for Optimizing Manufacturing Technology Models and Techniques
J. Paulo Davim (University of Aveiro, Portugal)
Engineering Science Reference • ©2012 • 395pp • H/C (ISBN: 9781466601284) • US $195.00

For an enitre list of titles in this series, please visit:
http://www.igi-global.com/book-series/advances-mechatronics-mechanical-engineering/73808

IGI GLOBAL
DISSEMINATOR OF KNOWLEDGE

www.igi-global.com

701 East Chocolate Avenue, Hershey, PA 17033, USA
Tel: 717-533-8845 x100 • Fax: 717-533-8661
E-Mail: cust@igi-global.com • www.igi-global.com

Dedicated to my daughters, Tanya Thakur and Pragya Thakur.
Thanks for being my strength and inspiration.

Table of Contents

Preface

Nanofluid refers to a fluid containing small amount of uniformly dispersed and suspended nanometer-sized particles in base fluid. When a small amount of nano-sized particles is added to the base fluid, the thermal conductivity of the fluid is enhanced and such a fluid is called nanofluid which was first coined by Choi (1995). Due to this property of the nanofluid they have wide range of industrial applications especially in the process where cooling is of primary interest. Buongiorno (2006) was the first researcher who dealt with convective transport in nanofluids. He noted that the nanoparticles absolute velocity can be viewed as the sum of the base fluid velocity and a relative (slip) velocity. He also discussed the effect of seven slip mechanisms: inertia, Brownian diffusion, thermophoresis, diffusiophoresis, Magnus effect, fluid drainage and gravity setting. He concludes that in the absence of turbulent eddies Brownian diffusion and thermophoresis dominate the other slip mechanisms. Tzou (2008a, 2008b) studied the on the onset of convection in a horizontal layer of nanofluid heated from below on the basis of Buongiorno's model. Nield and Kuznetsov (2009a, 2009b, 2011) and Kuznetsov and Nield (2010a, 2010b, 2010c) extended corresponding problem for porous medium. The above study deals with nanofluid as Newtonian nanofluid. There is growing importance of non-Newtonian visco-elastic fluids in geophysical fluid dynamics, chemical technology, petroleum, biological and material industries. The study of such type of non-Newtonian nanofluid is desirable. There are many visco-elastic fluids and one such class of visco-elastic fluid is Maxwellian visco-elastic fluid. Maxwellian visco-elastic fluid forms the basis for the manufacture of many important polymers and useful products. The work on visco-elastic fluid appears to be that of Herbert on plane coquette flow heated from below. He found a finite elastic stress in the undistributed state to be required for the elasticity to affect the stability. Using a three constants rheological model due to Oldroyd (1958), he demonstrated, for finite rate of strain, that the elasticity has a destabilizing effect, which results solely from the change in apparent viscosity. Vest and Arpaci

(1969) have studied the stability of a horizontal layer of Maxwellian visco-elastic fluid heated from below. Bhatia and Steiner (1972) studied the problem of thermal instability of a Maxwellian visco-elastic fluid in the presence of rotation and found that rotation has a destabilizing influence in contrast to its stabilizing effect on a viscous Newtonian fluid.

Although thermal instability problems in non-Newtonian nanofluid were studied by Nield (2010), Sheu (2011a, 2011b), Chand and Rana (2012, 2015a) and Rana, Thakur, and Kango (2014), Rana and Chand (2015a, 2015b) by taking different non-Newtonian fluids.

The choice of the boundary conditions imposed in all these studies on nanoparticles fraction is somewhat arbitrary, it could be argued that zero-flux for nanoparticles volume fraction is more realistic. Recently Nield and Kuznetsov (2014), Chand, Rana, and Hussein (2015), Chand and Rana (2014, 2015b), studied the thermal instability of nanofluid in a porous medium by taking normal component of the nanoparticle flux zero at boundary which is more physically realistic. Zero-flux for nanoparticles mean one could control the value of the nanoparticles fraction at the boundary in the same way as the temperature there could be controlled. The interest for investigations of visco-elastic nanofluids is also motivated by a wide range of engineering applications.

The objective of present work is to investigate theoretically the thermal instability in a horizontal layer of Maxwellian visco-elastic nanofluid on the basis of Buongiorno's model for more realistic boundary conditions. Galerkin weighted residuals method is used to find the solution of the eigen value problem. The stability criterions for stationary and oscillatory convection have been derived and graphs have been plotted to study the effects of various parameters on the stationary and oscillatory convection.

OVERVIEW OF THE BOOK

The Introduction provides the overview of the stability/instability, basic definition of stability/instability, Maxwellian visco-elastic fluids, nanofluids and its applications and other basic concepts which are necessary to understand the thermal instability in nanofluids.

Chapter 1 deals with theoretically investigation of thermal instability in horizontal layer Maxwellian visco-elastic nanofluid. A linear stability analysis based upon normal mode technique is used to find solution of the fluid layer. The onset criterion for stationary and oscillatory convection is derived analytically and graphically. The validity of 'Principle of Exchange of Stability' is examined and sufficient conditions for the non-existence of overstability are investigated.

Chapter 2 studies thermal convection in a horizontal layer Maxwellian visco-elastic nanofluid in presence of rotation. A linear stability theory based upon normal mode analysis is used to find the solution and eigen value problem is solved by Galerkin weighted residuals method. The effects of rotation, Lewis number, modified diffusivity ratio and concentration Rayleigh number on the stationary convection are investigated both analytically and graphically.

Chapter 3 studies thermal instability in a horizontal layer of Maxwellian visco-elastic nanofluid in the presence of uniform vertical magnetic field. The model used for the nanofluid describes the effects of Brownian motion and thermophoresis. Linear stability theory based upon normal mode analysis is employed to find expressions for Rayleigh number and critical Rayleigh number. The effects of magnetic field, Lewis number, modified diffusivity ratio and concentration Rayleigh number on the stationary convection are investigated both analytically and graphically. The influence of magnetic field on the stability is found to be stabilizes the fluid layer.

Chapter 4 deals with the study of thermal instability in a horizontal layer of Maxwellian visco-elastic nanofluid in the presence of both rotation and magnetic field. The model used incorporates the effect of Brownian diffusion, thermophoresis and magnetophoresis. The eigen value problem is solved by employing the Galerkin weighted residuals method. A linear stability theory based upon normal mode analysis is used to find expressions for Rayleigh number for a layer of Maxwellian visco-elastic nanofluid. The influence of rotation, magnetic field and other parameters on the stability is investigated both analytically and graphically. It is found that rotation and magnetic field both stabilizes fluid layer.

Chapter 5 studies thermal instability in a horizontal layer of Maxwellian visco-elastic nanofluid in porous medium on basis of Buongiorno's model. Darcy model has been used for porous medium. The normal mode technique is used to find the solution and the expression of Rayleigh number for stationary convection to find the effects porosity and other parameters for the problem. The results are presented both analytically and graphically.

Chapter 6 deals with study of thermal instability in a horizontal layer of Maxwellian visco-elastic nanofluid in porous medium. For porous medium, Brinkman-Darcy model is considered. A linear stability analysis based upon normal mode analysis is used to find solution of the fluid layer. The onset criterion for stationary and oscillatory convection is derived analytically and graphs have been plotted by giving numerical values to various parameters, to depict the stability characteristics. The effects of the Brinkman Darcy number and other parameters on the stability of the system are investigated. Regimes of oscillatory and non-oscillatory convection for various parameters are derived and discussed in detail.

Chapter 7 studies the effects of variable gravity on the thermal instability in a horizontal layer of Maxwellian visco-elastic nanofluid on basis of Buongiorno's model. The expression of Rayleigh number for stationary convection has been derived to find the effects of variable gravity and other parameters for the problem. The results are presented both analytically and graphically.

Chapter 8 deals with the study of thermal instability in a horizontal layer of Maxwellian visco-elastic nanofluid in the presence of 'Hall effect'. The model used incorporates the effect of Brownian diffusion, thermophoresis and magnetophoresis. The eigen value problem is solved by employing the Galerkin weighted residuals method. A linear stability theory based upon normal mode analysis is used to find expressions for Rayleigh number for a layer of nanofluid. The influence of 'Hall effect' and other parameters on the stability is investigated both analytically and graphically. It is found that 'Hall effect' destabilizes the stationary convection.

Chapter 9 studies the effects of internal heat source on the thermal instability in a horizontal layer of Maxwellian visco-elastic nanofluid. A linear stability analysis based upon normal mode analysis is used to find solution of the fluid layer. The onset criterion for stationary and oscillatory convection is derived analytically and graphs have been plotted by giving numerical values to various parameters, to depict the stability characteristics. The effects of internal heat source and other parameters on the stability of the system are investigated. The results are also presented graphically.

Chapter 10 deals with the study of double diffusive convection in a horizontal layer of Maxwellian visco-elastic nanofluid. The model used incorporates the effect of Brownian diffusion, thermophoresis and magnetophoresis. The eigen value problem is solved by employing the Galerkin weighted residuals method. A linear stability analysis based upon normal mode technique and perturbation method is used to find solution of the fluid layer. The influence of Dufour parameter, Soret parameter and other parameters on the stability is investigated both analytically and graphically.

Chapter 11 studies double diffusive convection in a horizontal layer of Maxwellian visco-elastic nanofluid in porous medium. Brinkman-Darcy model is used for porous medium. A linear stability analysis based upon normal mode technique is used to find solution of the fluid layer. The onset criterion for stationary convection is derived analytically and graphs have been plotted by giving numerical values to various parameters, to depict the stability characteristics.

REFERENCES

Bhatia, P. K., & Steiner, J. M. (1972). Convective instability in a rotating viscoelastic fluid. *Zeitschrift für Angewandte Mathematik und Mechanik, 52*(6), 321–330. doi:10.1002/zamm.19720520601

Buongiorno, J. (2006). Convective transport in nanofluids. *ASME Journal of Heat Transfer, 128*(3), 240–250. doi:10.1115/1.2150834

Chand, R., & Rana, G. C. (2012). Thermal instability of Rivlin-Ericksen elastico-viscous nanofluid saturated by a porous medium. *Journal of Fluids Engineering, 134*(12), 121203. doi:10.1115/1.4007901

Chand, R., & Rana, G. C. (2014). Thermal instability in a Brinkman porous medium saturated by nanofluid with no nanoparticle flux on boundaries. *Special Topics & Reviews in Porous Media: An International Journal, 5*(4), 277–286. doi:10.1615/SpecialTopicsRevPorousMedia.v5.i4.10

Chand, R., & Rana, G. C. (2015a). Instability of Walter's B' visco-elastic nanofluid layer heated from below. *Indian Journal of Pure & Applied Physics, 53*(11), 759–767.

Chand, R., & Rana, G. C. (2015b). Thermal instability in a horizontal layer of Walter's (Model B') visco-elastic nanofluid-a more realistic approach. *Applications and Applied Mathematics: An International Journal, 10*(2), 1027–1042.

Chand, R., Rana, G. C., & Hussein, A. K. (2015). Effect of suspended particles on the onset of thermal convection in a nanofluid layer for more realistic boundary conditions. *International Journal of Fluid Mechanics Research, 42*(5), 375–390. doi:10.1615/InterJFluidMechRes.v42.i5.10

Choi, S. (1995). Enhancing thermal conductivity of fluids with nanoparticles. In D. A. Siginer & H. P. Wang (Eds.), *Developments and applications of non-Newtonian flows* (Vol. 231, pp. 99–105). ASME FED.

Kuznetsov, A. V., & Nield, D. A. (2010a). Effect of local thermal non-equilibrium on the onset of convection in a porous medium layer saturated by a nanofluid. *Transport in Porous Media, 83*(2), 425–436. doi:10.1007/s11242-009-9452-8

Kuznetsov, A. V., & Nield, D. A. (2010b). Thermal instability in a porous medium layer saturated by a nanofluid: Brinkman Model. *Transport in Porous Media, 81*(3), 409–422. doi:10.1007/s11242-009-9413-2

Kuznetsov, A. V., & Nield, D. A. (2010c). The Onset of Double-Diffusive Nanofluid Convection in a Layer of a Saturated Porous Medium. *Transport in Porous Media*, *85*(3), 941–951. doi:10.1007/s11242-010-9600-1

Nield, D. A. (2010). A note on the onset of convection in a layer of a porous medium saturated by a non-Newtonian nanofluid of power-law type. *Transport in Porous Media*, *87*(1), 121–123. doi:10.1007/s11242-010-9671-z

Nield, D. A., & Kuznetsov, A. V. (2009a). Thermal instability in a porous medium layer saturated by a nanofluid. Int. *J. Heat Mass Transf.*, *52*(25-26), 5796–5801. doi:10.1016/j.ijheatmasstransfer.2009.07.023

Nield, D. A., & Kuznetsov, A. V. (2009b). The Cheng–Minkowycz problem for natural convective boundary-layer flow in a porous medium saturated by a nanofluid. *International Journal of Heat and Mass Transfer*, *52*(25-26), 5792–5795. doi:10.1016/j.ijheatmasstransfer.2009.07.024

Nield, D. A., & Kuznetsov, A. V. (2011). The effect of vertical through flow on thermal Instability in a porous medium layer saturated by a nanofluid. *Transport in Porous Media*, *87*(3), 765–775. doi:10.1007/s11242-011-9717-x

Nield, D. A., & Kuznetsov, A. V. (2014). Thermal instability in a porous medium layer saturated by a nanofluid: A revised model, Int. *J. Heat and Mass Transf.*, *68*, 211–214. doi:10.1016/j.ijheatmasstransfer.2013.09.026

Oldroyd, J. G. (1958). Non-Newtonian effects in steady motion of some idealized elastico- viscous liquids. *Proceedings of the Royal Society of London. Series A, Mathematical and Physical Sciences*, *245*(1241), 278–290. doi:10.1098/rspa.1958.0083

Rana, G. C., & Chand, R. (2015a). Rayleigh-Bénard convection in an elastic-viscous Walters' (model B') nanofluid layer. *Bulletin of the Polish Academy of Sciences*, *63*(1), 235–244.

Rana, G. C., & Chand, R. (2015b). Stability analysis of double-diffusive convection of Rivlin-Ericksen elastico-viscous nanofluid saturating a porous medium: A revised model. *Forsch Ingenieurwes*, *79*(1-2), 87–95. doi:10.1007/s10010-015-0190-5

Rana, G. C., Thakur, R. C., & Kango, S. K. (2014). On the onset of thermosolutal instability in a layer of an elastico-viscous nanofluid in porous medium. *FME Transactions*, *42*(1), 1–9. doi:10.5937/fmet1401001R

Sheu, L. J. (2011a). Thermal instability in a porous medium layer saturated with a visco-elastic nanofluid. *Transport in Porous Media*, *88*(3), 461–477. doi:10.1007/s11242-011-9749-2

Sheu, L. J. (2011b). Linear stability of convection in a in a visco elastic nanofluid layer. *World Acad. Sc. Engg. Tech.*, *58*, 289–295.

Tzou, D. Y. (2008a). Instability of nanofluids in natural convection. *ASME J. Heat Transf.*, *130*(7), 072401. doi:10.1115/1.2908427

Tzou, D. Y. (2008b). Thermal instability of nanofluids in natural convection. Int. *J. Heat Mass Transf.*, *51*(11-12), 2967–2979. doi:10.1016/j.ijheatmasstransfer.2007.09.014

Vest, C. M., & Arpaci, V. (1969). Overstability of visco-elastic fluid layer heated from below. *Fluid Mech.*, *36*(03), 613–623. doi:10.1017/S0022112069001881

Introduction

Fluid dynamics is the science which deals with the properties of the fluids in motion. Before dealing with fluid dynamics, we must know the terms 'fluid' and 'dynamics'. Dynamics is the branch of science which deals with the motion of bodies under the action of forces. Fluid is a substance that undergoes deformation under the action of forces. If the deformation in the material increases continually without limit under the action of shearing forces, however small, the material is called fluid. This continuous deformation under the action of forces compels the fluid to flow and this tendency is called "fluidity".

As we know that the matter exists in four forms, namely

- Solid
- Liquid
- Gas
- Plasma

Liquids and gases taken together are classified as fluids. It has been believed by the physicists for a long time that there is no clear dividing line between solids and fluids, since there are many materials which in some respect behave like a solid and in other respect like a fluid. For example, jelly, paint and pitch have dual character. However, a loose distinction can be made between solids and fluids. A solid mass has a definite shape of the container more or less instantaneously. The deformation in the piece of solid is small even under the action of large external forces, whereas in the case of fluids the deformation may be large under the suitably chosen forces however small in magnitude.

Fluids are classified as liquids and gases. As a result the distinction between liquids and gases is much less fundamental so far the dynamical studies are concerned. The most important difference between the mechanical properties of liquids and gases is the compressibility. Liquids have strong intermolecular forces whereas the gases experience weak intermolecular forces. As a result, the liquids are incompressible fluid while gases are compressible fluids. It should be mentioned that for velocities which are not comparable with the velocity of sound, the effect of compressibility on atmospheric air can be neglected and it may be considered to be a liquid and in this sense it is called incompressible air.

The fourth state of matter is called plasma. Plasma is essentially a highly ionized matter. Therefore, in plasma we have to take into account the charges on its particles and associated electromagnetic phenomena. We go to plasma state when we deal with Earth's molten core, ionosphere, stellar interiors and atmospheres.

Fluid dynamics has several sub disciplines and its study is important to physicist, whose main interest is to understand phenomena, study of fluid mechanics is important to engineer, whose interest is to solve industrial problems, aerospace engineer may be interested to designing the airplanes that have low resistance and at the same time have high 'lift' force to support the weight of the plane. Fluid dynamics has a wide range of applications, including calculating force and moments on aircraft, determining the mass flow rate of petroleum through pipelines, predicting weather patterns etc. Some of its principles are even used in traffic engineering, where traffic is treated as a continuous fluid.

Fluid dynamics offers a systematic structure that underlies these practical disciplines and that embraces empirical and semi-empirical laws derived from flow measurement, used to solve practical problems. The solution of a fluid dynamics problem typically involves calculation of various properties of the fluid, such as velocity, pressure, density and temperature as functions of space and time.

Fluid dynamics, like the study of any other branch of science, need mathematical analysis as well as experimentation. The analytical approach helps us finding the solution of certain idealized and simplified problems and understands the unity behind apparently dissimilar phenomena. The mathematical description of the state of a moving fluid is effected by means of function which give the distribution of the fluid velocity $v = v(x,y,z,t)$ and any two thermodynamic quantities pertaining to the fluid, for instance pressure $p = p(x,y,z,t)$ and density $\rho = \rho(x,y,z,t)$. All the thermodynamic quantities are determined by the values of any two of them, together with the equation of state; hence if we are given five quantities, namely three components of velocity v, the pressure p, density ρ, the state of moving fluid is completely determined.

It is very difficult to trace the origin of the science of fluid dynamics but the systematic study of fluid dynamics started only after the Euler's discovery of the equations of motion of an inviscid fluid. Later on, Lagrange gave the concept of velocity potential and stream function, Reynolds discovered the equation of turbulent motion, Prandtl put forward the boundary layer theory, the theories of turbulence and stabilities are the created by Taylor and Rayleigh respectively. The principle of resistance in flow in capillary tubes was given by Poiseuille, the credit for the equation of motion of viscous fluids goes to Navier and Stokes. Still later, some other good contribution were given by many more famous scientists, which include Bénard, Kutta, Prandtl, Lord Kelvin, Somemrfeld, Karman, Rayleigh and Zhukovski etc. Now a day, fluid dynamics has become a very vast subject and has given birth to other subjects like meteorology, Newtonian flows, non-Newtonian flows, gas dynamics and Magnetohydrodynamics (MHD) etc.

The foundational axioms of fluid dynamics are the laws of conservation of mass, conservation of momentum and conservation of energy. These are based on classical mechanics and are modified in relativistic mechanics. The central equations for fluid dynamics are non-linear differential equation describing the flow of a fluid whose stress depends linearly on velocity and pressure. In addition to the above, fluids are assumed to obey the continuum assumption.

STABILITY OF A HYDRODYNAMIC SYSTEM

For nearly century now, hydrodynamic stability has been recognized as one of the central problem of the fluid dynamics. It is concerned with how and when laminar flow breakdown, their subsequent development and eventual transition to turbulence. The system is stable with respect to small perturbations only when certain conditions are satisfied. When the disturbances can grow to finite amplitude and reach equilibrium and resulting in a new steady state, the new state may then become unstable to other types of disturbances, and may grow to another steady state, and so on. Finally, the system becomes a superposition of various large disturbances of random phases, and reaches a chaotic condition that is commonly described as turbulent.

Let us consider a hydrodynamic system, which in accordance with the equations governing it is in a stationary state. Let us consider a set of n parameters, which define the system such as pressure gradients, temperature gradients, rotation and magnetic fields and others parameters. If the system is disturbed slightly and the disturbances gradually die down, it is considered to be stable. If the disturbance grows in amplitude in such a way that the system progressively departs from the initial state a never reverts to it, the system is called unstable. A system cannot be considered as stable unless it is stable with respect to every possible disturbance to which the system is subjected.

Therefore, stability must imply that there exists no mode of disturbance for which it is unstable.

States of marginal stability can be one of the two kinds. If the amplitudes of a small disturbance can grow or be damped a periodically, the transition from stability to instability takes place via marginal state exhibiting a stationary pattern of motions. If the amplitude of a small disturbance can grow or be damped by oscillations of increasing or decreasing amplitude, the transition takes place via marginal state exhibiting oscillatory motion with a certain definite characteristic frequency. If at the onset of instability a stationary pattern of motion prevails, then one says that the 'Principle of Exchange of Stabilities' is valid and that instability sets in as stationary cellular convection, or secondary flow. On the other hand, if at the onset of instability oscillatory motions prevail then it is called the case of overstability.

The first major contribution on the study of hydrodynamics stability can be found in the theoretical paper of by Helmholtz (1868). Twelve year after the discovery of Helmholtz; Rayleigh (1880) developed a linear stability theory. Thereafter the combined efforts of Reynolds (1883) and Kelvin (1880, 1887), Rayleigh (1892a, 1892b) produced a rich harvest of knowledge on the subject. The first experimental investigations on thermal convection back date to Thomson (1882). Early in 20th century, the studies on hydrodynamics stability were connected with Bénard experiments on thermal instability in thin fluid layers. The experiments by Bénard (1900, 1901), in particular, have attracted great attention and are today considered classical fluid mechanics. The Bénard stability problem was first formulated and mathematically solved by Rayleigh (1916) for the idealized case of free boundaries with linear temperature gradient and further elaborated by Chandrasekhar (1961).

MAGNETOHYDRODYANMICS

Magnetohydrodynamics (MHD) is the union of the two fields of science, namely electromagnetic theory and fluid dynamics. Hydromagnetics is the science which deals with the motion of electrically conducting fluids in the presence of a magnetic field. The study of the interaction between magnetic field and electrically conducting moving fluids is currently receiving considerable interest. The field of MHD was initiated by Swedish physicist Hannes Alfvén, who received the Nobel Prize in Physics in 1970 for fundamental work and discoveries in Magnetohydrodynamics with fruitful applications in different parts of plasma.

Because of its tremendous importance in the quest for thermonuclear fusion, vast literature is available on the subject of hydromagnetic stability.

Fluid dynamics and electrodynamics theory were being developed independently of each other almost up to the first half of 20th century. The systematic study of MHD started only after 1942 when Alfven combined the two subjects by considering the motion of conducting fluids in the presence of magnetic field. This study has now come to be known as Magnetohydrodynamics or Hydromagnetics. It is concerned with physical systems specified by the equations that result from the fusion of those of hydrodynamics and electromagnetic theory. It is a well known fact that when a conductor moves in a magnetic field, electric currents are induced in it. These currents experience a mechanical force called the Lorentz force, due to the presence of magnetic field. This force tends to modify the initial motion of the conductor. Moreover, the induced currents generate their own magnetic field which is added on to the applied magnetic field. Thus there is coupling between the motion of the conductor and electromagnetic field, which is exhibited in a more pronounced form in liquid and gaseous conductors. Lorentz force is generally small unless inordinately high magnetic fields are applied. Thus this force is incapable of altering the motion as a whole considerably, but if it acts for a sufficiently long period, the molecules of gases and liquids may get accelerated considerably to alter initial state of these types of conductors.

Alfven (1942) proved his famous theorem that magnetic lines of force are glued to ideally conducting fluid. Every motion of the fluid perpendicular to the lines of force is forbidden because it can give infinite eddy currents. Thus the matter of the liquid is fastened to the lines of force. Alfven also discovered the simplest example of coupling between the mechanical forces and the magnetic lines of force in a highly conducting fluid moving in an external magnetic field and showed that this interaction would produce a new kind of wave which he called a Magnetohydrodynamic wave. The above discoveries of Alfven led to systematic study of Magnetohydrodynamic (MHD). The subject of MHD is, thus, comparatively of recent origin. It found its birth in attempts to explain certain phenomena in cosmic physics; for example, the generation and maintenance of original magnetic fields of Earth and Sun, the variability of magnetic stars and production of sunspots which are associated with magnetic fields.

Magnetic field introduces anisotropy, elasticity and lateral pressure in the fluid. Anisotropy results in the difference of electrical conductivity and diffusion coefficients along and perpendicular to the magnetic field. Elasticity and lateral pressure are responsible for the propagation of MHD waves. Attempts to show the existence of MHD waves in the laboratory were made by Lenhert (1954). Bullard (1949) and Batchelor (1950) pointed out that the magnetic field imparts to the fluid certain rigidity along with certain properties of elasticity which enables it to transmit disturbances by new modes of wave propagation. The experimental work of Lenhert (1954) concluded that the behavior of a conducting fluid is a very different in the absence and in the presence of magnetic field. For example, there is a tendency for all motions to become uniform along the magnetic field, or in other words, a tendency towards two-dimensional motion. These are some of the interesting properties associated with the magnetic field. Generally the magnetic field has a stabilizing effect on the instability. But a few exceptions are there. For example, Kent (1966) studied the effect of a horizontal magnetic field, which varies in the vertical direction, on the stability of parallel flows and showed that the systems is unstable under certain conditions, while in the absence of magnetic field the system is known to be stable.

Although the continuum approach is much simpler than the more rigorous gas kinematic one, it is not without difficulty. This is due to the reason that the hydrodynamic equations are basically non-linear even though the electrodynamics equations are linear. The coupling of the two systems of equations, hydrodynamic and electrodynamics causes the non-linear aspects to be carried over into the resulting MHD equations. The new phenomena which arise are interesting. For example, the coupling between longitudinal and transverse fields provides the possibility of energy transfer between the longitudinal and transverse modes of oscillations. This is of interest in astrophysics as well as in technology.

Initially the problem connected with the origin and maintenance of Earth's magnetic field created considerable interest in the study of hydro magnetic inertial waves. In the sequel we shall adopt a continuum picture. Plasma is essentially a highly ionized matter. The suns as well as the stars are in plasma state.

THE BASIC EQUATIONS OF STABILITY

For the mathematical description of hydrodynamics flow, we need some equations governing hydrodynamics flow of a viscous fluid of varying density and temperature. These equations based upon the basic of conservation laws of mass, momentum and energy together with the induction equation for the magnetic field which are given as follows:

- Equations of state (one)
- Equations of continuity (one)
- Equations of motion (three)
- Equations of energy (one)

These equations are mathematical expressions of basic physical laws. These are six in number and therefore, determine the six unknowns of the fluid motion viz., the three components of velocity **v**(u,v,w) the temperature T, the pressure p, and the density ρ, which are functions of both space coordinates and time.

EQUATION OF STATE

Variables that depend only upon the state of a system are called variables of state. The variables of state are pressure p, density ρ and temperature T. It is an experimental fact that a relationship between these three thermodynamic variables exists and can be written as F (p, ρ, T) = 0, which is commonly called the 'Equation of State'. For substances with which we shall be principally concerned, we can write the equation of state as

$$\rho = \rho_0\left[1 - \alpha\left(T - T_0\right)\right] \tag{1}$$

where α is the coefficient of volume expansion and T_0 is the temperature at density $\rho = \rho_0$.

EQUATION OF CONTINUITY

This equation expresses that the rate of generation of mass within a given volume is entirely due to the net inflow of mass through the surface enclosing the given volume (assuming that there are no internal sources). It amounts to the basic physical law that the matter is conserved; it is neither being created nor destroyed.

Assuming the fluid is incompressible, the equation of continuity is given by

$$\nabla \cdot v = 0, \tag{2}$$

where **v** is velocity of fluid.

EQUATION OF MOTION

The equations of motion are derived from Newton's second law of motions, which states that "Rate of change of linear momentum = Total force".

Equation of motion represents conservation of momentum. Thus equations of motion for a viscous, incompressible fluid is given by

$$\rho \frac{dv}{dt} = \rho X - \nabla p + \mu \nabla \cdot \mathbf{v}\,, \tag{3}$$

where X is the external force acting on fluid, μ is the viscosity of the fluid. Since external forces are of non-electromagnetic origin (gravity) only, then equation of motion can be written as

$$\rho \frac{dv}{dt} = -\nabla p + \rho \mathbf{g} + \mu \nabla \cdot \mathbf{v}\,. \tag{4}$$

EQUATION OF ENERGY

Equation of energy represents the law of conservation of energy. In the analysis of convective flows, a thermal energy balance is necessary to define the temperature field and the heat transport. Chandrasekhar (1961) the equation of energy for an incompressible fluid, takes the form

$$\rho c_v \frac{\partial T}{\partial t} + \rho c_v v \cdot \nabla T = k_m \nabla^2 T\,, \tag{5}$$

where c_v, k_m stand for specific heat at constant volume and thermal conductivity respectively. The viscous dissipation term, being very small in magnitude, has not been included in the equation (5).

HYDRODYNAMIC AND HYDROMAGNETIC STABILITY

Hydrodynamics as well as magnetohydrodynamics are governed by nonlinear partial differential equations. However, in spite of the complexity of the equations determining a fluid flow, some simple patterns of flow (such as between parallel planes, or rotating cylinders) are permitted as stationery solutions. However, these patterns of flow can be realized only for certain ranges of the parameters characterizing them, and cannot be realized outside these ranges. The reason being their inherent instability or in their inability to sustain themselves against small perturbations to which ever physical systems is subjected. Problems of hydrodynamics instability thus originated from the differentiation of the unstable flows from the stable patterns of permissible flows.

Stability can be defined as a quality of being immune to small disturbances. Thus, by stability we have permanent type of equilibrium state. An equilibrium state or steady flow, to be of permanent type, should not only satisfy the mechanical equation, but must be stable against arbitrary perturbations.

In recent years, the class of such problems of instability has been enlarged by the interest in hydrodynamic flow of electrically conducting fluids in the presence of magnetic fields. This is the domain of hydromagnetics, as we have discussed earlier, and there are problems of hydromagnetic stability even as there are problems of hydrodynamic stability.

Let us consider a hydrodynamic or hydromagnetic system in which the equations governing it are in stationery state. Let X_1, X_2, X_3, ..., X_j, be a set of parameters which define the system. These parameters include geometrical parameters such as the characteristic dimensions of the system; parameters characterizing the velocity field which may prevail in the system, magnitudes of the forces acting on the system such as pressure gradient, magnetic fields, rotation and others. While considering the stability of such a system, with a given set of parameters X_1, X_2, ..., X_j, we seek to determine the reaction of the system to small disturbances.

If the system is disturbed and the disturbance gradually dies down or if the system never departs appreciably from this stationery state, the system is said to be stable with respect to that particular disturbance. If the disturbance grows in amplitude in such a way that the system progressively departs from the initial state and never reverts to it, the system is called unstable with respect to that particular disturbance. A system may be considered as unstable even if there is only one special mode of disturbance with respect to which it is unstable. A system cannot be considered as stable unless it is stable with respect to every possible disturbance to which the system is subjected. Therefore stability must imply that there exists no mode of disturbance for which it is unstable.

If all the initial states are classified as stable, or unstable, according to the criterion stated, then the locus which separates the two classes of states defines the state of marginal stability of the system. By this definition, a marginal state is a state of neutral stability. The locus of the marginal state in the $(X_1, X_2, X_3, \ldots, X_j)$ space will be defined by an equation of the form $\Sigma(X_1, X_2, X_3, \ldots, X_j) = 0$. The determination of this locus is one of the prime objects of an investigation on hydrodynamic stability.

In thinking of the stability of a hydrodynamic system, it is often convenient to suppose that all parameters of the system, except one, are kept constant while the chosen one is continuously varied. We shall then pass from stable to unstable states when the particular parameter takes a critical value. We then say that instability sets in at this value of the chosen parameter when all the others have their pre-assigned values.

States of marginal stability can be one of two kinds. If the amplitude of a small disturbance can grow or be damped a periodically, the transition from stability to instability takes place via a marginal state exhibiting a stationary pattern of motions. If the amplitude of a small disturbance can grow or be damped by oscillations of increasing or decreasing amplitude, the transition takes place via a marginal state exhibiting oscillatory motions of definite characteristic frequency.

If at the onset of instability a stationary pattern of motions prevails, then one says that the 'Principle of the Exchange of Stabilities' is valid and that instability sets in as stationary cellular convection, or secondary flow. On the other hand, at the onset of instability oscillatory motions prevail, and then it is called the case of over stability.

THERMAL INSTABILITY

Convective phenomena are very common in nature and they hold a key role in many fields namely scientific fields, engineering, astrophysics and meteorology etc. Convection is of three types

1. Thermal Convection (Natural Convection)
2. Forced Convection
3. Mixed Convection

Thermal convection is a term that refers to heat transport by fluid flow generated by a thermal or solute concentration gradient. On the other hand, the term forced convection refers to convection phenomena generated by imposed pressure gradients. If both natural and forced convection occurs the term mixed convection is used.

Thermal instability often arises when fluid is heated from below. If the temperature difference between the two horizontal boundaries is sufficiently small, then the heat is transferred through the fluid by conduction alone. For greater temperature differences the conduction state becomes unstable and a convective motion is set up, or in other word when the temperature difference across the layer is great enough, the stabilizing effects of the viscosity and thermal conductivity are overcome by the destabilizing buoyancy, and an overturning instability ensures as thermal convection. We consider a horizontal layer of fluid in which an adverse temperature gradient is maintained by heating underside. Then the fluid at the bottom becomes lighter than the fluid at the top and thus it becomes a 'top heavy arrangement', which is potentially unstable. As a result of this there will be a natural tendency on the part of the fluid to redistribute itself to make up the deficiency in the arrangement. But this tendency is prevented to a certain extent by its own viscosity and therefore instability sets in only when the adverse temperature gradient exceeds certain critical value.

Convective instability seems to have been first described by the Thomson (1882), the elder brother of Lord Kelvin, but the first quantitative experiments were made by Bénard (1900). Simulated by the Bénard's experiments, Rayleigh (1916) formulated the theory of convective instability of layer of fluid between two horizontal planes. Rayleigh-Bénard convection originated from the experimental works of Bénard (1900) and theoretical analysis of Rayleigh (1916). Lord Rayleigh studied the dynamic origins of convective cells and proposed his theory on the buoyancy driven convection. He approached the analysis as a stability problem, searching for unstable modes and their growth rates. He used Euler's equation, the thermal energy balance and he worked within the Oberbeck-Boussinesq approximation. Moreover, he performed a linear stability analysis, neglecting all the non-linear terms, a hypothesis that allowed him to solve analytically the problem. He showed that instability would occur only when the adverse temperature gradient was so large that the dimensionless parameter $\mathrm{R} = \dfrac{\alpha\beta \mathrm{g} \mathrm{d}^4}{\nu\kappa}$ exceeds certain critical value. This parameter is called Rayleigh number. Here $\beta\left(=\dfrac{dT}{dz}\right)$ is the uniform temperature gradient, which is to be maintained, **g** is the acceleration due to gravity, d is depth of the layer, α is the coefficient of volume expansion, κ is the thermal diffusivity and ν is the kinematic viscosity.

Bénard instability first analyzed by Rayleigh (1916) and subsequently Jefferys (1928), Low (1929) and Pellow and Southwell (1940) extended the Rayleigh's work. Schmidt and Milverton (1935) verified experimentally the prediction of the onset of Bénard convection in fluids confined between horizontal isothermal solid surfaces.

Alfven (1942) proved this famous theorem that magnetic lines of force are glued to ideally conducting fluid. Every motion of the fluid perpendicular to the lines of force is forbidden because it can give infinite eddy currents. Thus the matter of the liquid is fastened to the lines of force. Alfven (1942) also discovered the simplest example of coupling between the mechanical forces and magnetic lines of force in a highly conducting fluid moving in an external magnetic field and showed that this interaction would produce a new kind of wave, which is called a Magneto Hydro Dynamic wave. The above discoveries of Alfven led to a systematic study of Magnetohydrodynamics (MHD).

For more detail and developments of the subject of stability one may refer to Joseph (1976), Chandrasekhar (1961), Sherman and Sutton (1962), Oberoi and Devanathan (1963), Joseph (1976), Drazen and Reid (1981), Zierep and Qertel (1982) and Banerjee and Gupta (1991).

THERMOSOLUTAL INSTABILITY

In classical thermal instability problems, it has been assumed that the driving density differences are produced by the spatial variation of single diffusing property (heat). It has been seen that a new phenomena occur when the simultaneous presence of two or more components with different diffusivities is considered. Such types of problems are known as 'thermosolutal instability problems or double-diffusive problems. The problem of thermosolutal instability in a layer of layer of fluid heated from below and subjected to a stable salinity gradient has been given by Veronis (1965), Nield (1968), Turner (1974), Knobloch (1980), Sunil et al. (2005), Motsa (2008), Wang and Tan (2008, 2011), Gupta and Sharma (2008), Nield and Kuznetsov (2011a), Kuznetsov and Nield (2010c), Malashetty and Kollur (2011), Malashettty et al. (2009b), Rana and Chand (2013b), Chand and Rana (2012e, 2014b), Goyal and Goyal (2015), Rana et al. (2014b, 2014c), Umavathi and Mohite (2014), Umavathi et al.(2015), Rana and Chand (2015c), Rana and Thakur (2016). Thermosolutal instability problems are of great importance because of its application to oceanography, astrophysics and various engineering disciplines and are well described in Turner (1973, 1974).

BOUSSINESQ APPROXIMATION

Boussinesq approximation has been used in the Rayleigh discussion, because, in solving the hydrodynamic equations, we have difficulties regarding their non-linear character and the variable nature of the various coefficients due to variations in temperature. The equations which were derived are, therefore of quite general validity. Due to these complications it is extremely difficult to solve these equations. So, there is a need for introducing some mathematical approximation to simplify the basic equations. However, as was first pointed out by Boussinesq (1903), there are many situations of practical occurrence in which the basic equation can be simplified considerably. These situations occur when variability in the density and in various coefficients is due to variations in the temperature of only moderate amounts. Boussinesq (1903) got rid of various coefficient variations by taking them to be constants by applying some approximations which are given below. However, non-linearity of equations still prevails under these approximations.

Boussinesq first pointed out that there are many situations of practical occurrence in which the basic equations can be simplified. These situations occur when the variation in the density and different coefficients is due to variations in temperature of only moderate amounts. The origin of simplification in these cases is due to the smallness of the coefficient of volume expansion α, whose range is 10^{-3} to 10^{-4}. For variations in temperature not exceeding 10^0C (say), the variations in density ρ are at most one percent. The variations in the other coefficients (consequent to the variations in density) must be of the same order. But there is one important exception that the variability of ρ in the term of external force in the equation of motion cannot be ignored. This is because the acceleration resulting from $(\delta\rho)$ Xi $=\alpha\Delta T$ Xi (where ΔT is a measure of the variations in temperature which occur) can be quite large. Accordingly, we may treat ρ as a constant in all terms in the equations of motion except the one in the external force. This is the 'Boussinesq approximation'. This approximation makes the mathematics simpler and it has also gained a wide recognition in other problems of non-homogeneous fluids.

Oberbeck and Boussinesq introduced an extremely useful hypothesis that allows a dramatic simplification in the approach to the convection studies. They assumed that all the physical properties of the fluid are temperature independent, except for the density in the gravitational body force term of the momentum balance equation. Here the density is considered as a linear function of the temperature which is given by

$$\rho = \rho_0\left[1-\alpha\left(T-T_0\right)\right].$$

CONTINUUM HYPOTHESIS

The continuum hypothesis is basically an approximation, in the same way planets are approximated by point particles when dealing with celestial mechanics, and therefore results in approximate solutions. Consequently, assumption of the continuum hypothesis can lead to results which are not of desired accuracy. Thus under the right circumstances, the continuum hypothesis produces extremely accurate results.

For fluids which are sufficiently dense to be a continuum, do not contain ionized species, and have velocities small in relation to the speed of light, the momentum equations for Newtonian fluids are the Navier-Stokes equations, which describe the flow of a fluid whose stress depends linearly on velocity gradients and pressure. The equations can be simplified in a number of ways, all of which make them easier to solve. Some of them allow appropriate fluid dynamics problems to be solved in closed form. In addition to the mass, momentum, and energy conservation equations, a thermodynamically equation of state giving the pressure as a function of other thermodynamic variables for the fluid is required to completely specify the problem.

NON-DIMENSIONAL PARAMETERS

Two fluids flows are said to be dynamically similar if the values of the dimensionless parameters in both the flows are remain the same. There are two general methods of obtaining the dimensionless parameters associated with a given problem:

1. Inspectional analysis
2. Dimensional analysis

The inspectional analysis method is possible only when we have a complete set of descriptive differential equations. The equations are made dimensionless and in this process certain dimensionless parameters appear as the co-efficient of the various terms in the equations. These co-efficient are called non-dimensional parameters.

In the dimensional analysis method, the non-dimensional parameters are found without knowledge of the governing differential equations. Instead, the relevant variables are collected and combined together to give the maximum number of independent dimensionless parameters. The complete set of relevant variables must be known and no extraneous variables can be introduced, otherwise the final set of non-dimensional parameters may be meaningless. The advantage of dimensional analysis is that the descriptive equations are not necessary and one is rewarded according to insight and cleverness. A useful role of thumb is that the number of independent non-dimensional parameters is equal to the total number of independent variables in the problem and the number of fundamental units, which may be taken as mass, length, time and the fundamental electrical unit, such as charge, making a total of four. The important dimensionless parameters are discussed below.

- Rayleigh number Ra,
- Concentration Rayleigh number Rn,
- Prandtl number Pr,
- Magnetic Prandtl number Pr_M,
- Lewis number Le,
- Taylor number Ta,
- Chandrasekhar number Q,
- Hall effect parameter M,
- Darcy number Da etc.

METHODS USED TO DETERMINING STABILITY

Following methods are used to determine the stability/instability of the system.

Perturbation Method

The perturbation method is the most suitable method to find the stability of the system. In this method, the hydrodynamic system whose instability one wish to establish, is supposed to undergo a specific small trail displacement and the effect of the additional forces brought into play is considered. If the forces thus produced tend to increase the displacement, thereby enhancing the deformation of the system still further, then system is unstable.

Energy Method

The more general method to discuss stability is the energy method. This is the oldest method of stability analysis which can accommodate finite disturbances also. It has also been applied with some success to oscillating flows between rotating cylinders.

We make use of an energy principle in this method. In a mechanical system for which there exists a potential energy function V', a stationery state of the system will be unstable or stable according as V' is a strict maximum or minimum. If T' the kinetic energy of the system, then in such a system $T' + V' = \text{Constant}$.

Let V' attains a strict minimum V'_0 for a stationary configuration, and when the system is disturbed slightly, then $V' > V'_0$ in a neighboring configuration.

Since we must have $T'+V'=T'_0+V'_0$, where T'_0, T $(T'_0>T')$ are the kinetic energies of the system in the stationary and disturbed states.

From above, we get

$$T' = T'_0 - (V'-V'_0),$$

$$< T'_0,$$

this shows that kinetic energy is decaying. Thus system therefore does not deviate further from the stationary configuration and remains in its proximity. Hence the system is stable.

On the other hand, if V'_0 denotes a strict maximum for a stationary configuration, then $T' > T'_0$ and the system will tend to depart more and more from the initial state, hence the system is unstable.

In the modern formulation of the energy method one considers the global energy of a difference motion. The global energy, boundary constraints and kinetic condition are use in two line of deduction. The first of these leads to a universal stabilities criterion, universal in the sense that specific details of the flow geometry and details of the basis motion need not be completely specified. A second line of deduction leads to the formulation of a maximum problem and achieves a sharper result by making more efficient use to known details of the basic flow; the procedure is elegantly developed by Serrin (1959) These result are of importance because they apply to a difference motion and so guarantee stability to finite disturbance. Serrin (1959) calculation of the stability limits for coquette flow between rotating cylinders shows that this method can give result which are not too conservative. Joseph (1965, 1966) extensions of the method to accommodate convective motions governed by the non-linear equations of Boussinesq demonstrate that even stronger results are possible. This method has been used Straughan (1991) in his book “The energy method, stability and nonlinear convection”.

Normal Mode Method

In the instability problems normal mode method is applied, because it gives the complete information of about the instability and rate of growth of instability. Stokes, Kelvin and Rayleigh adapted the method of normal mode to fluid dynamics. By giving an infinitesimal increment to various variables describing the flow, the linearized equations are obtained and these linearized perturbed equations are analyzed. Methods of analyzing the stability of flows were formulated in Reynolds's time. The method of normal modes for studying the oscillations and instability of a dynamics system of particles and rigid bodies was already highly developed. A known solution of Newton's or Lagrange's equations of motion for the system was perturbed. The equations were linearized by neglecting product of perturbations. It was further assumed that the perturbation of each quantity could be resolved into independent components or modes varying with time t like e^{nt} for some constant n, which is in general complex. The values of n for the modes were calculated from the linearnized equations. If the real part of n was found to be positive for any mode, the system was deemed unstable because general initial small perturbation of the system would grow exponentially in time until it was no longer small. The essential mathematical difference between fluid and particle dynamics is that the equations of motion are partial rather than ordinary differential equations. This difference leads to many technical difficulties in hydrodynamics stability, which have been overcome for only a few classes of flows with very simple configurations.

"Normal mode analysis method" is used to determine the stability of a stationary state of a hydrodynamic or hydrodynamic system. The beauty of the method is that it gives complete information about instability including the rate of growth of any unstable perturbation. This method has been used by Chandrasekhar (1961), while discussing the various instability problems.

Here we assume that the perturbations are infinitesimally small and use of the linear theory by retaining only linear terms in the equations governing perturbations. To study these equations we assume further that the perturbed quantities have time variation proportional to e^{nt}. The parameter 'n' is in general, function of the wave number 'a' and of other parameters defining the system. If the value of n (complex number) determined by the dispersion relation is:

1. Real and negative, the system is stable.
2. Real and positive, the system is unstable.
3. Complex, say $n = n_r + in_i$, where n_r and n_i are real and

3.1. $n_r < 0$, the system is stable;
3.2. $n_r > 0$, the system is unstable;
3.3. $n_r = 0$, the instability is oscillatory.

4. Further, if $n_r = 0$ implies that $n_i = 0$, then the stationary (cellular) pattern of flow prevails on the onset of instability. In other words, "Principle of Exchange of Stabilities" is valid.
5. If $n_r = 0$, does not imply that $n_i = 0$, then overstability occurs.
6. From this it follows that if n is real, then $n = 0$ will separate the stable and unstable modes and we will always have principle of exchange of stabilities.

Thus normal mode depends upon time exponentially with complex exponent. It is the real part of the complex solution, whose real and imaginary parts separately are the solution, because system is linear.

The main drawback of this method is that in some problems the dispersion relation becomes so much complicated that it is not possible to draw any meaningful conclusions from it. Finally, normal mode analysis is based on the liberalized stability theory and therefore, it has all the defects of linear theory. A review in this regards has been given by Drazin and Reid (1981).

Galerkin Method

Galerkin Method invented by Russian Mathematician Boris Grigoryevich Galerkin which is used to solve differential equation (second-order or fourth-order) which are difficult, even possible to solve analytically. Finlayson (1972) gave an extension of the Galerkin approximate method. The commonly used Galerkin method involves the trial functions of the vertical coordinate only for the onset of convection. It is one of the most popular and powerful numerical techniques that can be applied to many engineering problems which are governed by a differential equations with boundary conditions.

BOUNDARY CONDITIONS

The fluid is confined between planes $z = 0$ and $z = d$. Regardless of the nature of these boundary surface, we must require

$\theta = 0$ and $w = 0$ at $z=0$ and $z = d$.

There are two further boundary conditions which, however, depend upon the nature of the surface at z = 0 and z = d. Now if the surface is rigid then there is no slip occurs on this surface implies that not only w, but horizontal component of velocity u and v vanish. Thus u = 0 and v = 0.

Since the condition must be satisfied for al x and y on the surface, it follows from the equation of continuity

$$\frac{\partial u}{\partial x}+\frac{\partial v}{\partial y}+\frac{\partial w}{\partial z}=0,$$

thus $\frac{\partial w}{\partial z}=0$ on the rigid surface.

For the free surface both the normal stress and shearing stress are zero.

Thus $\frac{\partial u}{\partial z}=\frac{\partial v}{\partial z}=0$ on the free surfaces.

Also from equation of continuity, we have $\frac{\partial^2 w}{\partial x^2}=0$.

The boundary condition on the normal component of vorticity

$\varsigma = 0$ on the rigid surface,

$\frac{\partial \varsigma}{\partial z}=0$ on free surface.

MORE REALISTIC BOUNDARY CONDITION FOR NANOFLUID

Nield and Kuznetsov (2014) pointed out that this type of boundary condition on volume fraction of nanoparticles is physically not realistic as it is difficult to control the nanoparticles volume fraction on the boundaries, and suggested an alternative boundary condition that is, the flux of volume fraction of nanoparticles is zero on the boundaries

Thus boundary conditions when the flux of volume fraction of nanoparticles is zero on the boundaries

$$w=0,\ T=T_0,\ D_B\frac{\partial \phi}{\partial z}+\frac{D_T}{T_0}\frac{\partial T}{\partial z}=0 \ \text{at } z=0$$

and

$$w = 0, T = T_1, \; D_B \frac{\partial \phi}{\partial z} + \frac{D_T}{T_1} \frac{\partial T}{\partial z} = 0 \text{ at } z = d.$$

NON-NEWTONIAN FLUIDS

Many materials such as polymer solutions or melts, drilling mud, clastomers, certain oils and greases and many other emulsions and gel are classified as non-Newtonian fluids. Most of these fluids used in industries. For these kinds of fluids, the commonly accepted assumption of a linear relationship between the stress and the rate of strain does not hold or the fluid that show distinct deviation from "Newtonian hypothesis" (stress on fluid is linearly proportional to strain rate of fluid) is called non-Newtonian fluids. Non-Newtonian fluids are those in which viscosity at a given pressure and temperature is a function of velocity gradient. The non-Newtonian fluids have been modeled by constitutive equations which vary greatly in complexity. The non-Newtonian fluid considered described by the constitutive relations

$$p'_{ik} = p_{ik} - {}'_{ij} p'',$$

$$\left(1 + \text{»} \frac{d}{dt}\right) p_{ik} = 2\text{¼}\left(1 + \text{»}_0 \frac{d}{dt}\right) e_{ik},$$

$$e_{ik} = \frac{1}{2}\left(\frac{\partial u_i}{\partial x_k} + \frac{\partial u_k}{\partial x_j}\right), \tag{6}$$

where $p'_{ik}, p_{ik}, e_{ik}, {}'_{ik}$ and p'' denote respectively the normal stress tensor, shear stress tensor, rate of strain tensor, Kronecker delta and scalar pressure. Here $\frac{d}{dt}$ is the convection derivative, λ the relaxation time and, λ_0 $(<\lambda)$ is the retardation time. If $\lambda_0 = 0$ the fluid is Maxwellian visco-elastic fluid; while for $\lambda_0 \neq 0$, the fluid is referred as Oldroydian visco-elastic fluid and for $\lambda = \lambda_0 = 0$, the fluid is known as Newtonian viscous fluid. Non-Newtonian fluids help us understand the wide variety of fluids that exist in the physical world. Plastic solids, power-law fluids, visco-elastic fluids, and time-dependent viscosity fluids are others that exhibit complex and counterintuitive relationships between shear stress and viscosity/elasticity. These fluids help us understand the wide variety of fluids that exist in the physical world and characterized by power-law model. The visco-elastic fluid is one of many models that have proposed to describe the non-Newtonian behavior of such fluids. The first visco-elastic rate type model, widely used all over is due to Maxwell (1866). Maxwell fluid is viscous in nature and has a great storage of energy. The work on visco-elastic fluid appears to be that of Herbert on plane coquette flow heated from below. He found a finite elastic stress in the undistributed state to be required for the elasticity to affect the stability. Using a three constants rheological model due to Oldroyd (1958, 1965), he demonstrated, for finite rate of strain, that the elasticity has a destabilizing effect, which results solely from the change in apparent viscosity. There are many visco-elastic fluids which cannot be characterized by Maxwell or Oldroyd's constitutive relations e.g. Rivlin-Ericksen and Water's B' elastic-viscous fluids. Ericksen (1953), Rivlin and Ericksen (1955) proposed a model for Rivlin-Ericksen elastic-viscous fluid and Walters (1960, 1962) for Water's B' elastic-viscous fluids. The principles and applications of rheology were given by Fredricksen (1964). A detailed study of convection in Maxwell visco-elastic fluid is given by Tom and Strawbridge (1953), Green (1968), Vest and Arpaci (1969), Seerin (1959), Bhatia and Steiner (1972, 1973), Srivastava (1971), Sokolov and Tanner (1972), Sharma (1975), Rosenblat (1986), Rudraiah et al. (1989), Sharma and Sharma (1990), Martinez-Mardones (1990, 2000, 2002, 2003), Larson (1992), Sharma and Kumari (1993), Sharma and Sunil (1994), Shenoy (1994), Sharma and Kumar (1996a, 1996b, 1997), Prakash and Kumar (1999a, 1999b), Sharma et al. (1999, 2000,) Prakash and Chand (1999, 2000, 2002), Sharma and Rana (2001), Sharma et al. (2001), Sharma and Kishore (2001), Kaloni and Lou (2002), Kim et al. (2003),Yoon et al. (2004), Sunil et al. (2005), Kumar and Singh (2006), Malashetty et al. (2006, 2009a, 2009b), Sharma et al. (2006), Laroze et al. (2007), Wang and Tan (2008, 2011), Gupta and Sharma (2008), Zhang et al. (2008), Malashetty et al. (2009a, 2009b), Awad et al. (2010), Agarwal (2010), Nield (2010), Chand (2010, 2011, 2013c, 2015a), Chand and Kango (2011), Fakhar and Anwar (2012), Gupta and Agarwal (2011), Kang et al. (2011, 2014), Thakur and Rana (2013), Gupta et al. (2012), Khan et al. (2012), Singh and Mehta (2013), Chand and Rana (2012d, 2012e, 2014b), Rana and Thakur (2012b, 2012c, 2013a, 2013b), Agarwal and Verma (2014), Kango et al. (2013), Rana and Chand (2012, 2013a), Kumar and Kumar (2013), Chand et al. (2015c) and Rana et al. (2012, 2014a, 2014b, 2015, 2016a, 2016b), Umavathi et al. (2015), Umavathi and Mohite (2016), Umavathi and Kumar (2016) etc.

NANOFLUIDS AND NANOFLUIDS TECHNOLOGIES

Nanofluid technology, a new interdisciplinary field of huge importance which comprises nanotechnology, nanoscience and thermal engineering, has been developed largely over the past few years due to the need of heat transfer enhancement in a variety of applications in transportation, electronics, nuclear, medical, food and space. It is now possible to develop high-performance coolants whose thermal properties are drastically different from those of the conventional heat transfer fluids (water, ethylene glycol) because in the nanoscale range, fundamental properties of nanofluids depend strongly on particle shape, size and the surface/interface area. These fluids are colloidal suspensions of solid nanoparticles into a carrier liquid. The presence of nanoparticles in the carrier fluid increases the effective thermal conductivity and thus enhances the heat transfer rate. Suspensions of solids in fluids are used to improve thermal properties for more than century. In the 19th century Maxwell (1881) proposed model for thermal conductivity enhancement in suspensions. However, application of micro- and larger particles is connected with many disadvantages, e.g. high concentrations of solid phase are needed to achieve satisfying enhancement of thermal properties.

The flow of nanofluid is of great interest in many area of modern science, engineering and technology, chemical and nuclear industries and biomechanics. The term nanofluids are colloidal suspensions of nanoparticles of size 100 nm in a base fluid. Nanoparticles taken as oxide ceramics (Al_2O_3, CuO), metal carbides (SiC), nitrides (AlN, SiN) or metals (Al, Cu) etc. and base fluids mostly used in the preparation of nanofluids are the common working fluids such as, water, ethylene or tri-ethylene- glycols, oil and other lubricants, bio-fluids, polymer solutions, other common fluids. In order to improve the stability of nanoparticles inside the base fluid, some additives are added to the mixture in small amounts. The term 'nanofluid' was first coined by Choi (1995) of the Argonne National Laboratory, USA in 1995. Nanofluids are not naturally occurring but they are synthesized in the laboratory. The choice of base fluid and nanoparticles depends on the application for which the nanofluid is intended. In the presence of a mere few percents of nanoparticles, a significant increase of the effective thermal conductivity. Common fluids have limited heat transfer capabilities while some of the nanoparticles (metals) have very high thermal conductivity in comparison to these common fluids. Presence of these nanoparticles in the base fluids may increase the thermal conductivity of the fluids by 15-40%.

PREPARATION METHODS FOR NANOFLUIDS

Preparation of nanofluids is of very much important in the area of nanofluid research due to its wide range of applications. Research papers and review articles involving the preparation methods for nanofluids were published by authors Williams et al. (2006), Yu and Xie (2012), Drzazga et al. (2012), Taylor et al. (2013), Mukherjee and Paria (2013), Manimaran et al. (2014), Quddoos et al. (2014) and Prakash et al. (2015). Based upon these studies it is observed that nanofluids can be prepared by two-step or one-step method.

Two-Step Method

Two-step method is the most widely used method for preparing nanofluids. Nanoparticles used in this method are first produced as dry powders by physical or chemical methods e.g. grinding, laser ablation, sol-gel processing, etc. and then, the nanosized powder will be suspended in base fluid Paul et al. (2010) in the second processing step with the help of intensive magnetic force agitation, ultrasonic agitation, high-shear mixing, homogenizing, and ball milling. This method is the most economic method to produce nanofluids in large scale, because nanopowder synthesis techniques have already been scaled up to industrial production levels. Due to the high surface area and surface activity, nanoparticles have the tendency to aggregate. The important technique to enhance the stability of nanoparticles in fluids is the use of surfactants. However, the functionality of the surfactants under high temperature is also a big concern, especially for high-temperature applications. Due to the difficulty in preparing stable nanofluids by two-step method, several advanced techniques are developed to produce nanofluids, including one-step method.

Eastman et al. (1997), Lee et al. (1999), Wang et al. (1999) used two-step method to produce alumina nanofluids. Murshed et al. (2005) prepared TiO_2-water nanosuspension by the same method. Two-step method can also be used for synthesis of carbon nanotube based nanofluids.

In spite of such disadvantages this process is still popular as the most economic process for nanofluids production.

Though this method is economic, the problem of drying, storage, and transportation exist. Also, the problem of agglomeration and clogging leads to reduced thermal conductivity of nanofluids.

One-Step Method

To get a more stable nanofluid single step preparation process is preferred, as name indicates it is synthesisized in only one step. In the single step method, the nanoparticles preparation and nanofluid preparation are carried out simultaneously. The nanoparticles are directly prepared by a physical vapor deposition technique or a liquid chemical method. In this method, the processes of drying, storage, transportation, and dispersion of nanoparticles are avoided, so the agglomeration of nanoparticles is minimized, and the stability of fluids is increased. The method avoids the undesired particle aggregation fairly well. However, this method only applicable for small scale production and, at current stage, it is almost impossible to scale up to industrial scale. Furthermore, this method is only applicable for low vapor pressure base fluid which limits its application.

Eastman et al. (2001) developed a one-step physical vapor condensation method to prepare Cu/ethylene glycol nanofluids. Zhu et al. (2004) presented a single-step chemical process for the preparation of Cu nanofluids. Lo et al. (2005) developed vacuum based submerged arc nanoparticle synthesis to prepare CuO, Cu_2O and Cu based nanofluids with different dielectric liquids. Jwo et al. (2007) prepared CuO–water nanofluid using temperature arc method and indicated about the improvement of thermal conductivity of nanofluid by 9.6%.

This method also proved to be a good way to produce mineral oil based silver nanofluids. One-step physical method cannot synthesize nanofluids in large scale, and the cost is also high, so the one-step chemical method is developing rapidly.

However, there are some disadvantages for one-step method. The most important one is that the residual reactants are left in the nanofluids due to incomplete reaction or stabilization. It is difficult to elucidate the nanoparticle effect without eliminating this impurity effect.

An advantage of one-step synthesis method is that nanoparticle agglomeration is minimized. But prime problem is that only low vapor pressure fluids are compatible with such a process.

Other Methods

There are some other literatures where different approaches (other than these two procedures) were expressed. Feng et al. (2006) used aqueous organic phase transfer method for preparation of gold, silver, platinum nanoparticles. Phase transfer method can also be applied to prepare kerosene based Fe_3O_4

nanofluids which do not show time dependent thermal conductivity. Grafting of oleic acid onto the surface Fe_3O_4 makes it compatible with kerosene. Wei et al. (2010) established a continuous flow microfluidic microreactor to synthesize copper nanofluids. The microstructure and properties of nanofluids can appropriately be varied by adjusting parameters such as concentration, flow rate, additives. Moreover a novel preparation of aqueous CuO nanofluid can be done through novel precursor transformation method with the help of ultrasonic and microwave irradiation. Here the precursor, $Cu(OH)_2$ is completely converted to CuO in water under that process. The use of ammonium citrate is to prevent the growth and aggregation of nanoparticles, resulting in a stable CuO aqueous nanofluid with higher thermal conductivity than those prepared by other dispersion methods.

POTENTIAL FEATURES OF NANOFLUIDS

Nanofluids have some special features that make them very special for various engineering and industrial applications. Some of the special qualities of nanofluids are:

- Rise in thermal conductivity beyond exception and much higher than theoretical predictions,
- Ultrafast heat transfer ability,
- Better stability than other colloids,
- Reduction of erosion and clogging in micro channels,
- Reduction in pumping power,
- Reduce friction coefficient,
- Better lubrication,
- Very high rise in viscosity of base fluid,
- Increased thermal conductivity (TC) at low nanoparticles concentrations,
- Strong temperature-dependent TC,
- Non-linear increase in TC with nanoparticles concentration,
- Increase in boiling critical heat flux (CHF).

These characteristics, Heat transfer and other key features of fluid flow and heat transfer behaviors of nanofluid was reported by researchers Masuda et al. (1993), Choi (1995, 1999), Wang et al. (1999), Xuan and Lee (2000, 2003), Das (2003), Xie et al. (2001, 2002), Patel et al. (2003), Eastman et al. (2001, 2004), Kim et al. (2004), Das et al. (2003, 2006, 2008), Wang and

Majumdar (2006), Wang (2007), Cheng et al. (2008), Garg et al. (2008), Abu-Nada et al. (2008), Kakac and Pramuanjaroenkij (2009), Hwang, et al. (2009), Shima et al. (2009), Abu-Nada (2011), Parekh and Lee (2011), Mohammed et al. (2011), Paul et al.(2011), Patel (2012), Prabhat et al. (2012), Adil et al.(2014), Nerella et al. (2014).

These features make of nanofluid useful in many applications.

APPLICATIONS OF NANOFLUIDS

Nanofluids have been extensively used in a wide range of applications and can be considered the next-generation heat transfer fluids as they offer exciting new possibilities to enhance heat transfer performance compared to pure fluids. Nanofluids applications in various fields are as follow

Heat Transfer Applications

- Industrial Cooling Applications
- Smart Fluids
- Nuclear Systems Cooling
- Extraction of Geothermal Power and Other Energy Sources
- Heating Buildings and Reducing Pollution
- Space and Defense

Automotive Applications

- Nanofluid Coolant
- Nanofluid in Fuel
- Brake and Other Vehicular Nanofluids

Electronic Applications

- Cooling of Microchips
- Microscale Fluidic Applications

Biomedical Applications

- Nanodrug Delivery
- Cancer Theraupetics
- Cryopreservation
- Nanocryosurgery
- Sensing and Imaging
- Nanofluid Detergent
- Antibacterial Activity

Energy Applications

- Energy Storage
- Solar Absorption

Mechanical Applications

- Friction Reduction
- Magnetic Sealing

Other Applications

- Intensify Microreactors
- Nanofluids as Vehicular Brake Fluids
- Nanofluids-Based Microbial Fuel Cell
- Mass Transfer Enhancement.

Philip and Shima (2012) studied the thermal properties while Keblinski et al. (2009) and Minsta et al. (2009) studied the thermal conductivity of nanofluids and it was found that nanofluid exhibit enhanced thermal properties. The novel features of nanofluids make them potentially useful in many applications ranging from use in the automotive industry to the medical arena to use in power plant cooling systems as well as computers. Manna (2000) described the detailed applications of nanofluid in automotive radiators, lubrication, additives for fuels, shock absorbers replacing or along with the traditional materials used for similar purposes. Routbort et al. (2008) that employed nanofluids for industrial cooling that could result in great energy savings and resulting emissions reductions. Nguyen et al. (2006) suggest the use of nanofluids in cooling of micro-electronic components. Donzelli et al. (2009) showed that a particular class of nanofluids can be used as a smart material working as a heart valve to control the flow of heat. Buongiorno and Hu (2005) suggested the possibility of nanofluid in advanced nuclear system while Kleinstreuer et al. (2008) in the delivery of nano-drug system, Tsaia et al. (2004) in electronics cooling, Hone (2004) in carbon nanotubes, Tzeng et al. (2005), Buongiorno et al. (2008) safety of nuclear reactors, Wong and Leon (2010) in radiators, Kole and Dey (2010) as car engine coolant, Gupta et al. (2012a) toward green environment, Faiz and Zahir (2014) in filter operation, Kasaeian et al. (2015) and Khanafer and Vafai (2013) in solar energy system, Javadi et al. (2013) and Zhang et al. (2014) in solar collector etc.

Kim et al. (2007) assess the feasibility of nanofluids in nuclear applications by improving the performance of any water-cooled nuclear system that is heat removal limited. Possible applications include pressurized water reactor (PWR) primary coolant, standby safety systems, accelerator targets, plasma diverters, and so forth. The use of nanofluids in nuclear power plants seems like a potential future application. Engine oils, automatic transmission fluids, coolants, lubricants, and other synthetic high-temperature heat transfer fluids found in conventional truck thermal systems-radiators, engines, heating, ventilation and air-conditioning (HVAC)-have inherently poor heat transfer properties. These could benefit from the high thermal conductivity offered by nanofluids that resulted from addition of nanoparticles Yu et al. (2008). In the nanofluid research applied to the cooling of automatic transmissions, Tzeng et al. (2005) dispersed CuO and Al_2O_3 nanoparticles into engine transmission oil. The experimental setup was the transmission of a four-wheel-drive vehicle. The transmission had an advanced rotary blade coupling, where high local temperatures occurred at high rotating speeds. Chaudhari and Walke (2014) reported the application of nanofluid in solar energy while Chieruzzi et al. (2013) studied the heat capacity of nanofluids for thermal energy storage. Applications of nanotechnology to improve the performance of solar collectors were studied by Hussein et al. (2016a) and Hussein (2016). Magnetic nanoparticles have recently got wide interest in many fields. Nakano et al. (2008), Lai et al. (2009), Singh and Lillard (2009), Zhang et al. (2010) reported the biomedical applications of nanofluids in drug delivery and anticancer drugs system and Mahendran and Philip (2012) used nanofluid based optical sensor for rapid visual inspection of defects in structures such as rail tracks and pipelines.

The latest developments and various applications of nanofluids also reported in detail by Wong and Leon (2010), Li et al. (2011), Saidur et al. (2011), Yu and Xie (2012), Aybar et al. (2014), Bourantas et al. (2014), Sheikholeslami et al. (2012, 2013, 2015a, 2015b, 2015c, 2015d, 2015f), Rao et al. (2014), Ravikumar and Goud (2014), Rashid et al. (2014), Sadique and Verma (2014), Verma and Tiwari (2015), Sandeep et al. (2015), Sharma et al. (2015).

CONVECTION IN NANOFLUIDS

A comprehensive study of convective transport in nanofluids was made by Buongiorno (2006). He dealt with almost all aspects of convective transport in nanofluids. He noted that the nanoparticles absolute velocity can be viewed as the sum of the base fluid velocity and a relative (slip) velocity. He took seven slip mechanisms that show relative velocity between the nanoparticles and the base fluid including inertia, Brownian diffusion, thermophoresis, diffisiophoresis, the magnus effect, fluid drainage and gravity. He pointed out that in the absence of turbulent effect, only Brownian diffusion; thermophoresis are important mechanisms in nanofluids. Hence based on these two effects, he proceeded to write down conservation equations.

Alloui et al. (2010) studied the natural convection of nanofluids in a shallow cavity heated from below. They observed that the presence of nanoparticles in a fluid is found to reduce the strength of flow field, this behavior being more pronounced at low Rayleigh number. Also the temperatures on the solid boundaries are reduced (enhanced) by the presence of the nanoparticles. For completeness, it is to mention that a substantially different treatment of Bénard problem for nanofluid is given by Kim et al. (2007). These authors simply modified three quantities namely the thermal expansion coefficients, the thermal diffusivity, and kinematic viscosity that appear in the definition of Rayleigh number. Convection in nanofluids based on Buongiorno's model received attention of several researchers. Tzou (2008a, 2008b) studied the Bénard problem for a nanofluid on the basis of Buongiorno's model and observed that nanofluid is less stable than regular fluid. Later, many authors this problem was revisited by Nield and Kuznetsov (2010a) by taking different types of non-dimensional parameters. An extension to the problem Nield and Kuznetsov (2010a) has been was made by many authors. Agarwal (2014), Agarwal and Bhaduria (2011, 2014), Agarwal et al. (2011 2012, 2014), Bhadauria and Agarwal (2011a, 2011b, 2012), Bhadauria et al. (2011), Chand (2013a, 2013b, 2015b), Chand et al. (2013a, 2013b, 2015a, 2015d, 2015f, 2016a), Chand and Rana (2012a, 2012b, 2012c, 2014a, 2014c, 2014d, 2014e), Dhananjay et al. (2010), Gupta et al. (2013, 2015, 2016), Kuznetsov (2011a, 2011b), Kuznetsov and Nield (2010a, 2010b, 2010c, 2011), Nield and Kuznetsov (2009a, 2009b, 2010a, 2010b, 2011a, 2011b, 2012, 2013), Rana and Agarwal (2015), Rana et al. (2014b, 2014c), Umavathi (2015), Yadav (2014), Yadav and Kim (2015a, 2015b), Yadav and Lee (2015a, 2015b), Yadav et al. (2011, 2012a, 2012b, 2013a, 2013b, 2013c, 2014a, 2015, 2016a, 2016b), Kiran et al. (2016), Mahajan and Arora (2013), Khan et al. (2013), Shivakumara and Dhananjaya (2014), Kumar and Awasthi (2016), Rana and Thakur (2016).

The above references mentioned deal with nanofluids as Newtonian nanofluids.

Many investigators such as Chen et al. (2007a, 2007b, 2009), and Schmidt et al. (2008) also dealt with Non-Newtonian rheological behavior of nanofluids.

Many investigations such as Chen et al. (2007, 2009), Shivakumara et al. (2006, 2015), Schmidt et al. (2008), Sheu et al. (2008), Sheu (2011a, 2011b), Chand and Rana (2012c, 2015b, 2015c), Rana and Chand (2015a, 2015c), Rana et al. (2014a, 2014b, 2015), Umavathi and Mohite (2016), Yadav et al. (2014b) also deal with non-Newtonian rheological behavior of nanofluid. The convection of non-Newtonian fluids in a porous medium has a wide spectrum of applications; such as oil recovery, food processing, and the spread of contaminants in the environment, and in various processes in the chemical and material industries.

All the study based upon the assumption that the value of the nanoparticle fraction at the boundary could manage in the same way as the temperature. But in due course, it turned out that physically the above boundary conditions may be difficult to establish and the boundary condition for nanoparticles needs to be more realistic.

Recently, Nield and Kuznetsov (2014), Chand and Rana (2015a, 2015d), Chand et al. (2014, 2015b, 2015e), Rana and Chand (2015b, 2015c) suggested that the value of the temperature can be imposed on the boundaries, but the nanoparticle fraction adjusts so that the nanoparticles flux is zero on the boundaries. In this respect, this model is more realistic physically than that employed by previous authors. Under the circumstances, it is desirable to investigate convective instability problems by utilizing these boundary conditions to get meaningful insight in to the problems.

GOVERNING EQUATIONS FOR NANOFLUIDS

Buongiorno (2006) dealt with almost all aspects of convective transport in nanofluids. The convective transport in nanofluids and proposed a model incorporating the effects of Brownian diffusion and thermophoresis. On the basis of the transport equations of Buongiorno (2006), the stability of problems of the onset of convection in nanofluid layer has been studied by many authors. The governing equations for the convection in nanofluids based upon Buongiorno's model are as follows

EQUATION OF STATE FOR NANOFLUID

Density ρ of nanofluids can be determined by using Buongiorno's model Buongiorno (2006) is given as

$$\rho = \phi\rho_p + \left(1-\phi\right)\rho_f, \tag{7}$$

where φ is the volume fraction of the nanoparticles, ρ_p density of nanoparticles and ρf density of base fluid.

Taking the density of the nanofluid as that of the base fluid, the equation of state can be written as

$$\rho \cong (\varphi\rho_\rho + (1-\varphi)\{\rho(1-\alpha(T-T_0))\}), \tag{8}$$

where ρ is the coefficient of thermal expansion.

EQUATION OF CONTINUITY FOR NANOFLUID AND NANOPARTICLES

The equation of continuity for nanofluid is given by

$$\nabla \cdot v = 0 . \tag{9}$$

The equation of continuity for the nanoparticles is

$$\frac{\partial \phi}{\partial t} + v \cdot \nabla \phi = D_B \nabla^2 \phi + \frac{D_T}{T_1} \nabla^2 T , \tag{10}$$

where D_B is the Brownian diffusion coefficient and D_T is the thermoporetic diffusion coefficient of the nanoparticles and T_1 is reference temperature.

EQUATION OF MOTION FOR NANOFLUID

Combing the equations (4) and equation (8), the equation of motion for nanofluid is given by

$$\rho \frac{dv}{dt} = -\nabla p + \left(\phi \rho_p + (1-\phi) \left\{ \rho \left(1 - \alpha \left(T - T_0 \right) \right) \right\} \right) g + \mu \nabla^2 v . \tag{11}$$

EQUATION OF ENERGY FOR NANOFLUID

Equation of energy for nanofluid is given by

$$(\rho c) \frac{\partial T}{\partial t} + (\rho c)_f \, \mathbf{v} \cdot \nabla T = \mathrm{k_m} \nabla^2 \mathrm{T} + (\text{Ác})_{\mathrm{p}} \left(\mathrm{D_B} \nabla \text{Æ} \cdot \nabla \mathrm{T} + \frac{\mathrm{D_T}}{\mathrm{T_1}} \nabla \mathrm{T} \cdot \nabla \mathrm{T} \right), \tag{12}$$

where (ρc) is heat capacity of fluid and $(\rho c)_p$ is heat capacity of nanoparticles and km is thermal conductivity.

EFFECT OF VARIOUS FACTORS ON STABILITY

The following factors effects the onset of convection in a horizontal layer of Maxwellian visco-elastic nanofluid heated from below.

Rotation

The rotation has a significant effect on the onset of thermal instability. It introduces a number of new elements in fluid dynamics, and some of its consequences are unexpected, for example, the role of viscosity is inverted. The origin of this and other consequences of rotation can be attributed to certain theorems relating to vorticity, in the dynamics of rotating fluids.

The steady of rotating fluid for Bénard problem was experimentally demonstrated by Rossby (1969) for wide range of Taylor number. Convection implies that motions which, occur have necessarily a three dimensional character, but Taylor-Proudman theorem doesn't allow it for an inviscid fluid as long as the non-linear terms in the equations of motion are neglected. Therefore, in contrast to non-rotating fluids, an inviscid fluid in rotation is expected to be thermally stable for all adverse temperature gradients. In fact, thermal instability can arise only in the presence of viscosity, while Taylor-Proudman theorem forbids any variations of the velocity in the direction of the angular velocity of rotation. Chandrasekhar (1961) has obtained the following conclusions on the effects of rotation on the onset of thermal instability:

1. The onset of the instability is a stationary convection as long as the Prandtl number p_1 exceeds a certain critical value p_1^*.The precise value of p_1^* depends on the nature of the boundary surfaces.
2. Rotation inhibits the onset instability and extend of inhibition depends on the Taylor number $\mathrm{Ta} = \dfrac{4\Omega^2 d^4}{\nu^2}$ and the Prandtl number $\mathrm{Pr} = \dfrac{\nu}{\kappa}$.
3. It $p_1 < p_1^*$, then two care arises
   3.1. The onset of instability will be as overstable oscillations of the Taylor number exceeds a certain limits $T_A^{(p_1)}$ depending on p_1.
   3.2. The onset of instability will be as stationary convection of $T_A \leq T_A^{(p_1)}$.

The above results can be formulated as the following general principle: "Thermal instability as stationary convection will set in at the minimum (adverse) temperature gradient which is necessary to maintain a balance between the rate of dissipation of energy by viscosity and the rate of liberation of the thermodynamically available energy by the buoyancy force acting on the fluid. Likewise, the onset of thermal instability will be as overstable oscillation if it is possible to balance in a synchronous manner the periodically varying amounts of kinetic energy with similarly varying amounts of dissipation and liberation of energy".

The inhibiting effect of rotation on the instability of a fluid layer heated from below has been recognized as a phenomenon of major importance in Bénard convection as convection in a rotating system relevant to many industrial and geophysical applications. Thermal instability problem in a rotating micropolar fluid was studied by Qin and Kaloni (1992).

Magnetic Field

Bullard (1949) and Batchelor (1950) pointed out that the magnetic field imparts to the fluid in certain rigidity. The experimental work of Lehnert (1954) conducted the behavior of conducting fluids very different in the absence and in the presence of magnetic field. For example, there is tendency for all motions to become uniform along the magnetic field, or other words, a tendency towards two dimensional motions. These are some interesting properties associated with magnetic field. Generally the magnetic field has a stabilizing effect on the instability. But a few exception are there for example Kent (1966) studied the effect of a horizontal magnetic field, which varies in vertical direction, on the stability of parallel flows and showed that system is unstable under certain conditions, while in the absence of magnetic field the system is known to be stable.

We consider a fluid having electrical conductivity and under the influence of a magnetic field. The electrical conductivity and the prevalence of magnetic field contribute to effects of two kinds. First, by the motion of the electrically conducting fluid across the magnetic lines of force, electric currents are generated and the associated magnetic field contribute to changes in the existing fields, and second, the fact that the fluid elements carrying currents transverse magnetic lines of force contribute to additional forces acting on the fluid elements. It is this two-fold interaction between the motions and the fields that is responsible for patterns of behaviors, which are often unexpected and striking.

The interaction between the fluid motions and the magnetic fields are contained in Maxwell's equations and in the equations of hydrodynamics suitably modified. Since we shall not be concerned with effects which are related in any way to the propagation of electromagnetic waves, we can neglect the displacement currents in Maxwell's equation, closely related to this is the further possibility of avoiding any explicit reference to the charge density. The reason for this is not that it is small in itself, but rather than its variations affect the equation expressing the conservation of charge only by terms of order $\mathbf{v}^2/c^2$, where $\mathbf{v}$ is the fluid velocity and c is the velocity of light. The terms of this order can be ignored. The interaction of the magnetic field in hydrodynamics can be interpreted as follows:

Magnetic field has the effect of inhibiting the onset of instability by convection. When a magnetic field is impressed on an electrically conducting fluid, there will also be the dissipation of energy by Joule heating, in addition to the dissipation by viscosity. As was to be expected, the effect of inhibiting the onset of instability becomes more pronounced as the strength of magnetic field increases and elongates the cells which appear at the marginal stability for certain ranges of the concerned parameters. In the case of an inviscid fluid of zero resistivity when external forces are derivable from the potential function v, the analogue of the Taylor-Proudman theorem states that "all steady slow motions in the presence of a uniform magnetic field are necessarily two-dimensional" It forbids variation of motion in the direction of the magnetic field. Convection implies that motions, which occur, have necessarily a three-dimensional character. But such motions are forbidden for fluids of zero resistivity as slow steady two-dimensional motions are only allowed. A fluid of zero resistivity should therefore be therefore being thermally stable for all adverse temperature gradients. In a steady state, the energy released by buoyancy force acting on the fluid must balance the energy dissipated by both means and this can be achieved at higher temperature gradient that are sufficient in the absence of Joule heating, showing thereby the stabilizing effect of the magnetic field. In the presence of a magnetic field, disturbances can be propagated as Alfven waves.

Chandrasekhar (1961) analysis predicted another interesting point that the marginal state could either be stationary or oscillatory in character. It was found in agreement with the experimental results of Nakagawa (1955, 1957) and others. It was shown that, when magnetic field is present and the magnetic Prandtl number $p_2 = \frac{\nu}{\eta}$ is less than the Prandtl number $p_1 = \frac{\nu}{\kappa}$, which is a requirement met by a large margin under most terrestrial conditions, overstability cannot occur and the principle of exchange of stabilities is valid.

Equation of motion for nanofluid in the presence of magnetic field are given as

$$\rho \frac{d\mathbf{v}}{dt} = -\nabla p + \left(\text{ÆÁ}_{p} + \left(1 - \text{Æ}\right)\left\{\text{Á}_{f}\left(1 - \pm\left(T - T_{0}\right)\right)\right\}\right)\mathbf{g} + \mu\nabla^{2}\mathbf{v} + \frac{\mu_{e}}{4\pi}\left(\mathbf{H}\cdot\nabla\right)\mathbf{H}, \tag{13}$$

where H is magnetic field and μ_e is the magnetic permeability.

Maxwell equations are

$$\frac{d\mathbf{H}}{dt} = \left(\mathbf{H}\cdot\nabla\right)\mathbf{v} + \eta\nabla^{2}\mathbf{v}, \tag{14}$$

$$\nabla\cdot\mathbf{H} = 0, \tag{15}$$

where η_1 is the electrical resitivity.

The theoretical investigations of Bénard problem in the presence of magnetic field is given by Rainbow (1948), Alchaar et al. (1995), Ghasemi and et al. (2011), Mahmoudi et al. (2012), Islam (2012), Chand (2013b), Mahajan and Arora (2013), Yadav et al. (2013c), Gupta et al. (2013, 2015), Wakif et al. (2016) etc.

Hall Effect

If an electric field is applied right angle to magnetic field, the whole current will not flow along the electric field. The tendency of the electric current of flow across an electric field in the presence of magnetic field is called 'Hall effect'. The Hall Effect is likely to be important in geophysical and astrophysical situation. The study of MHD flows with Hall currents has important engineering applications in MHD generators, Hall accelerators, refrigeration coils, electric transformers etc. Recently, the study of Magnetohydrodynamics (MHD) became important in engineering applications, such as in designing cooling system with liquid metals, MHD generator and other devices in the petroleum industry, materials processing, Plasma studies, nuclear reactors, geophysics, astrophysics, aeronautics and aerodynamics. It serves as the basis for understanding some of the important phenomena occurring in heat exchanger devices. The study of MHD flows with Hall currents has important engineering applications in flight MHD, MHD generators, Hall accelerators, refrigeration coils, electric transformers etc. The theoretical investigations of Bénard problem with Hall effect is given by Gupta (1967), Sharma and Gupta (1993), Sharma and Kumar (1996a, 1996b),Prakash and Chand (1999), Sharma et al. (2000), Sunil and Singh (2000), Rani and Tomar

(2010), and El-Aziz (2013), Chand and Rana (2014a), Gupta et al. (2016), Yadav and Lee (2015b).

Maxwell equations in the presence of Hall effect are given by

$$\frac{d\mathbf{H}}{dt} = \left(\mathbf{H} \cdot \nabla\right) v + \eta_1 \nabla^2 H - \frac{C}{4\pi Ne} \nabla \times \left(\left(\nabla \times \mathbf{H}\right) \times \mathbf{H}\right),, \tag{16}$$

$$\nabla \cdot H = 0 \tag{17}$$

where C, N, e, stand for the speed of light, electron number density and charge of electron respectively.

Porous Medium

Flows through porous medium are very much prevalent in nature, and therefore, the study of flow through porous media has attracted the attention of large number of scholars because of its scientific and engineering applications. Thermal instability in a porous medium is a phenomenon related to various fields. It has many applications in geophysics, food processing, petroleum industry, bio-mechanics, oil reservoir modeling, hydrology, chemical engineering, building of thermal insulations and nuclear reactors. Many researchers have investigated thermal instability problems by taking different types of fluids. Horton and Rogers (1945), Lapwood (1948), Wooding (1960), Nield (1968), Patil and Rudraiah (1973), Vafai and Tien (1981), Palm and Tyvand (1984), Scheidegger (1990), Sharma and Sharma (1991), Kaviany (1995), Ingham and Pop (1981, 2002), Vadasz (1997, 1998a, 1998b, 2000, 2006), Vafai (2000), Vafai and Hadim (2000), Bejan et al. (2004), Leong and Lai (2004), Basak et al. (2006), Nield (1968, 1987, 2008), Straughan (2008), Storesletten (1998, 2004), Nield and Bejan (2013), Rana and Thakur (2012a), Sulochana et al. (2015), studied in detail the convective flow in a porous medium.

1. Porous Media

Porous media may be defining as solid bodies that contain 'pores'. 'Pores' are void spaces, which must be distributed more or less frequently through the material if it is to be called porous. Extremely small voids in a solid are called 'molecular interstices' and very large ones are called 'caverns'. Pores are void spaces intermediate in size between caverns and molecular interstices. Therefore the limitation of their size is intuitive and rather indefinite. The pore in porous medium may be interconnect or non-interconnect. Flow of interstitial fluid is possible only if at least part of the pore space is interconnected. The interconnect part of pore system is called the effective pore space. Many natural substances such as rocks, soils, biological tissues, and manmade materials such as cements, foams and ceramics can be considered as porous media. Flows through porous media are very much prevalent in nature, and therefore, the study of the flows through porous media has attracted the attention of a number of scholars because of its scientific and engineering application. Such flows has shown their great importance in petroleum engineering to study the movement of natural gas, oil and water through the oil reservoirs; in chemical engineering for filtration and purification processes. Further, in the fields of agriculture engineering to study the underground resources, seepage of water in riverbeds, one needs to investigate the flows of fluids through porous media. Two macroscopic properties of non-ideal porous media, which may be used to describe fluid flow, are porosity and medium permeability. As the fluid flow through a porous medium the gross effect is represented by Darcy's law. According to which the usually viscous term $\mu\nabla^2 v$ is replaced by the resistance term $-\left(\frac{\mu}{k_1}\right)\mathbf{q}$ in the equation of motion, where μ is viscosity, **v** is fluid velocity and **q** is Darcian (filter) velocity of fluid. The fluid velocity **v** and Darcian velocity **q** are connected by relation $\mathbf{v} = \frac{\mathbf{q}}{\varepsilon}$, where ε is porosity of porous medium.

3. Porosity

Porosity is defined as the ratio of volume of void to the total volume. It is denoted by ε ($0 < \varepsilon < 1$). If the calculation of porosity is based upon the interconnected pore space instead of on the total pore space, the resulting quantity is termed as effective porosity, which is given by $e = \frac{\varepsilon}{1-\varepsilon}$. Porosity macroscopically characterizes the effective pores volume of the medium.

Natural media have porosity usually less than 0.6, the sandstone lies in the range 0.08 – 0.38, the soil between 0.43 – 0.54 and the leather between 0.56 – 0.59. Artificial materials like fiberglass, mineral wool may have porosity slightly less than the value 1 that corresponds to the limit of a clear fluid.

3. Permeability

The conductance of the medium is defined with direct reference of Darcy's law as the seepage velocity of the percolating water per unit drop of hydraulic head. The permeability is related to pore-size distribution since the distribution of the size of entrances exists, and lengths of the pore walls make up the conductance of the given pore-structure. The dimensions of permeability are length squared. In oil industry, it is measured in 'Darcy' with 1 Darcy = 9.87×10^{-9} cm^2. The permeability and porosity are related, since of the porosity is zero, the permeability is zero.

4. Darcy's Law

Darcy's law named after a French scientist and engineer Henry P.G. Darcy. Based on his pioneering work Darcy (1856) he formulated what is today the well known Darcy's law. He based his law on the results of experiments on the flow of water through layers of sand. Darcy's law refers to the case of laminar flow in the porous medium and, moreover, it refers to a tight packed solid with a fluid owing in very small pores. This low porosity regime allows one to obtain a momentum balance equation that is dramatically simpler than the Navier-Stokes equation. When fluid slowly percolates through the pores of the rock, the gross effect is represented by the Darcy's law. According to which the resistance term $-\frac{\mu}{k_1}\mathbf{q}$ will replace the usual viscous term in the equation of motion, μ is viscosity of the fluid, k_1 is permeability of medium and **q** is the seepage velocity of fluid.

Thus momentum equation for porous medium is

$$-\frac{\mu}{k_1}\mathbf{q} = \nabla p. \tag{18}$$

A detailed account of the Darcy-Bénard problem in porous medium has been studied by Horton and Rogers (1945), Lapwood (1948), Wooding (1960), Ingham and Pop (1981), Poulikakos and Bejan (1985), Rees (2000), Vafai and Hadim (2000) and Nield and Bejan (2013), Borujerdi et al. (2008) etc.

5. Brinkman's Model

Darcy's model is suitable for a relatively low porosity. Moreover it cannot allow the assignment of no-slip conditions on a boundary surface. On increasing the porosity of the medium, Brinkman (1947a, 1947b, 1949, 1952) proposed a momentum equation adding, with respect to the Darcy equation, a Laplacian term that appears in the equation of motion, namely

$$-\frac{\mu}{k_1}\mathbf{q} = \nabla p + \tilde{\mu}\nabla^2\mathbf{q}, \tag{19}$$

where $\tilde{\mu}$ is known as the effective viscosity.

A good account of convection problems in Brinkman porous medium were given by Tong and Subramanian (1985), Kuznetsov and Nield (2010b), Chand and Rana (2012b, 2012d), Mahajan and Sharma (2012), Rana and Thakur (2012a, 2012b, 2012c, 2013a), Rana and Chand (2012), Bala and Chand (2014), Sharma and Gupta (2015) etc.

6. Anisotropic Porous Medium

In geothermal system with a ground structure composed of many strata of different permeabilities, the overall horizontal permeability may be up to ten times as large as the vertical component. Processes such as sedimentation, compaction, frost action, and reorientation of the solid matrix are responsible for the creation of anisotropic natural porous media. Anisotropy can also be a characteristic of artificial porous material like pelleting used in chemical engineering process and fiber materials used in insulating purposes. The review of research on convective flow through anisotropic porous medium has been given by McKibbin (1985, 1992) and Storeletten (1998, 2004). Nield and Bejan (2013), Malashetty and Basavaraja (2002, 2003), Malashetty and Kollur (2011), Malashetty et al. (2009b), Agarwal et al. (2011), Chand et al. (2013a), Shivakumara and Dhananjaya (2014), Chand et al. (2013a, 2014), Bala and Chand (2015b) studied the effect of anisotropy on the onset of convection in a horizontal layer of fluid.

EFFECT OF INTERNAL HEAT

Convection induced by internal heat sources has wide range of applications in geophysics, astrophysics, fire and combustion modelling, thermal ignition, miniaturization of electronic components etc. In such flows the buoyancy force is incremented due to heat source resulting in modification of heat/mass transfer characteristics. The onset of convection in a Darcy porous layer with a uniform heat source was studied by Borujerdi et al. (2008) and found that a smooth monotonic variation in the critical Rayleigh number. Char and Chiang (1994) studied the effect of a quadratic basic state temperature profile caused by constant internal heat generation for Bénard-Marangoni convection. Shivakumara and Suma (2000) have investigated the effect of through flow and constant internal heat generation on the onset of convection using rigid and perfectly conducting boundaries. Heat source at the bottom of a nanofluid-filled enclosure was studied by Aminossadati and Ghasemi (2009). The effect of uniform internal heat generation on the onset of Brinkman-Bénard convection in a ferrofluid was studied by Nanjundappa et al. (2011) and found that the effect of increase in the value of internal heat source, magnetic Rayleigh number and nonlinearity of the magnetization parameter is to hasten, while increase in the value of Biot number, the ratio of viscosities and reciprocal of Darcy number is to delay the onset of thermo-magnetic convection in a ferrofluid saturated porous layer. Yadav et al. (2012) studied the internal heat source effects on the onset of Darcy-Brinkman convection in a porous layer saturated by nanofluid and found that both temperature and volumetric fraction of nanoparticle distributions are having a destabilizing factor to make the system more unstable.

VARIABLE GRAVITY

The idealization of uniform gravity assumed in theoretical investigations, although valid for laboratory purposes can scarcely be justified for large-scale convection phenomena occurring in atmosphere, the ocean or mantle of the Earth. It then becomes imperative to consider gravity as variable quantity varying with distance from surface or reference point. Pradhan and Samal (1987), Pradhan et al. (1989) studied the thermal instability of a fluid layer in a variable gravitational field while Alex and Prabhamani (2001), Alex et al. (2001) studied the variable gravity effects on the thermal instability in a porous medium with internal heat source and inclined temperature gradient. The other theoretical investigations of Bénard problem in presence of variable gravity were given by Rionero and Straughan (1991), Prakash and Chand (1999, 2002), Prakash and Kumar (1999a, 1999b), Sharma and Rana (2001), Chand (2010, 2011, 2013c), Chand et al. (2013a, 2015d), Chand and Kumar (2013) and Kango et al. (2013), Bala and Chand (2015a), Harfash and Alshara (2015) and found that variable gravity play important role instability of fluid layer.

REFERENCES

Abu-Nada, E. (2011). Rayleigh-Bénard convection in nanofluids: Effect of temperature dependent properties. *International Journal of Thermal Sciences*, *50*(9), 1720–1730. doi:10.1016/j.ijthermalsci.2011.04.003

Abu-Nada, E., Masoud, Z., & Hijazi, A. (2008). Natural convection heat transfer enhancement in horizontal concentric annuli using nanofluids. *International Communications in Heat and Mass Transfer*, *35*(5), 657–665. doi:10.1016/j.icheatmasstransfer.2007.11.004

Adil, A., Gupta, S., & Ghosh, P. (2014). Numerical prediction of heat transfer characteristics of nanofluids in a mini channel flow. *Journal of Energy*, *2014*, 1–7. doi:10.1155/2014/307520

Alfven, H. (1942). On the existence of electromagnetic-hydrodynamic waves. *Nature*, *150*(3805), 405–415. doi:10.1038/150405d0

Aggarwal, A. K. (2010). Effect of rotation on thermosolutal convection in a Rivlin–Ericksen fluid permeated with suspended particles in porous medium. *Adv. Theor. Appl. Mech.*, *3*(4), 177–188.

Agarwal, S. (2014). Natural convection in a nanofluid-saturated rotating porous layer: A more realistic approach. *Transport in Porous Media*, *104*(3), 581–592. doi:10.1007/s11242-014-0351-2

Aggarwal, A. K., & Verma, A. (2014). The effect of compressibility, rotation and magnetic field on thermal instability of Walters fluid permeated with suspended particles in porous medium. *Thermal Science*, *18*(2suppl.2), 539–550. doi:10.2298/TSCI110805087A

Agarwal, S., & Bhadauria, B. S. (2011). Natural convection in a nanofluid saturated rotating porous layer with thermal non equilibrium model. *Transport in Porous Media*, *90*(2), 627–654. doi:10.1007/s11242-011-9807-9

Agarwal, S., & Bhadauria, B. S. (2014). Unsteady heat and mass transfer in a rotating nanofluid layer. *Continuum Mechanics and Thermodynamics*, *26*(4), 437–445. doi:10.1007/s00161-013-0309-6

Agarwal, S., Bhadauria, B. S., & Siddheshwar, P. G. (2011). Thermal instability of a nanofluid saturating a rotating anisotropic porous medium. *Special Top. Rev. Porous Media*, *2*(1), 53–64. doi:10.1615/SpecialTopicsRevPorousMedia.v2.i1.60

Agarwal, S., Rana, P., & Bhadauria, B. S. (2014). Rayleigh–Bénard convection in a nanofluid layer using a thermal nonequilibrium model. *Journal of Heat Transfer*, *136*(12), 122501–12514. doi:10.1115/1.4028491

Agarwal, S., Sacheti, N. C., Chandran, P., Bhadauria, B. S., & Singh, A. K. (2012). Non-linear convective transport in a binary nanofluid saturated porous layer. *Transport in Porous Media*, *93*(1), 29–49. doi:10.1007/s11242-012-9942-y

Alchaar, S., Vasseur, P., & Bilgen, E. (1995). Effect of a magnetic field on the onset of convection in a porous medium. *Heat and Mass Transfer*, *30*(4), 259–267. doi:10.1007/BF01602772

Alex, S. M., & Prabhamani, R. P. (2001). Effect of variable gravity field on Soret driven thermosolutal convection in a porous medium. *International Communications in Heat and Mass Transfer*, *28*(4), 509–518. doi:10.1016/S0735-1933(01)00255-X

Alex, S. M., Prabhamani, R. P., & Vankatakrishan, K. S. (2001). Variable gravity effects on thermal instability in a porous medium with internal heat source and inclined temperature gradient. *Fluid Dynamics Research*, *29*(1), 1–6. doi:10.1016/S0169-5983(01)00016-8

Alloui, Z., Vasseur, P., & Reggio, M. (2010). Natural convection of nanofluids in a shallow cavity heated from below. *International Journal of Thermal Sciences*, *50*(3), 385–393. doi:10.1016/j.ijthermalsci.2010.04.006

Aminossadati, S. M., & Ghasemi, B. (2009). Natural convection cooling of a localised heat source at the bottom of a nanofluid-filled enclosure. *European Journal of Mechanics - B/Fluids*, *28*(5), 630–640. doi:10.1016/j.euromechflu.2009.05.006

Awad, F. G., Sibanda, P., & Motsa, S. S. (2010). On the linear stability analysis of a Maxwell fluid with double-diffusive convection. *Applied Mathematical Modelling*, *34*(11), 3509–3517. doi:10.1016/j.apm.2010.02.038

Aybar, H., Sharifpur, M., Azizian, M. R., Mehrabi, M., & Meyer, J. P. (2014). A review of thermal conductivity models for nanofluids. *Heat Transfer Engineering*. doi:10.1080/01457632.2015.987586

Bala, A., & Chand, R. (2014). Thermal instability in a horizontal layer of Ferrofluid in Brinkman porous medium. *Journal of Scientific and Engineering Research*, *1*(2), 25–34.

Bala, A., & Chand, R. (2015a). Variable gravity effect on the thermal instability of Ferrofluid in a Brinkman porous medium. International Journal of Astronomy. *Astrophysics and Space Science*, *2*(5), 39–44.

Bala, A., & Chand, R. (2015b). Thermal instability in a horizontal layer of ferrofluid in anisotropic porous medium. *Open Science Journal of Mathematics and Application*, *3*(6), 176–180.

Banerjee, M. B., & Gupta, J. R. (1991). Studies in Hydrodynamic and Hydromagnetic Stability. Silver Line Publications.

Basak, T., Roy, S., Paul, T., & Pop, I. (2006). Natural convection in a square cavity filled with a porous medium: Effects of various thermal boundary conditions. *International Journal of Heat and Mass Transfer*, *49*(7-8), 1430–1441. doi:10.1016/j.ijheatmasstransfer.2005.09.018

Batchelor, G. K. (1950). On the spontaneous magnetic field in a conducting liquid in turbulent motion. *Proceedings of the Royal Society of London*, *A201*(1066), 405–416. doi:10.1098/rspa.1950.0069

Bejan, A., Dincer, I., Lorente, S., Miguel, A. F., & Reis, A. H. (2004). *Porous and complex flow structures in modern technologies*. New York: Springer. doi:10.1007/978-1-4757-4221-3

Bénard, H. (1900). Les tourbillons cellularies dans une nappe liquid. Revue Generale des Sciences Pures et. *Appliqués*, *11*, 1261–1271.

Bénard, H. (1901). Les tourbillons cellularies dans une nappe liquidetransportant de la chaleur par convection en regime permanent. *Ann. Chimie (Paris)*, *23*, 62–71.

Bhadauria, B. S., & Agarwal, S. (2011a). Natural convection in a nanofluid saturated rotating porous layer: A nonlinear study. *Transport in Porous Media*, *87*(2), 585–602. doi:10.1007/s11242-010-9702-9

Bhadauria, B. S., & Agarwal, S. (2011b). Convective transport in a nanofluid saturated porous layer with thermal non equilibrium model. *Transport in Porous Media*, *88*(1), 107–131. doi:10.1007/s11242-011-9727-8

Bhadauria, B. S., & Agarwal, S. (2012). Natural convection in a rotating nanofluid layer. In MATEC Web of Conferences. EDP Sciences. doi:10.1051/matecconf/20120106001

Bhadauria, B. S., Agarwal, S., & Kumar, A. (2011). Non-linear two-dimensional convection in a nanofluid saturated porous medium. *Transport in Porous Media*, *90*(2), 605–625. doi:10.1007/s11242-011-9806-x

Bhatia, P. K., & Steiner, J. M. (1972). Convective instability in a rotating viscoelastic fluid. *Zeitschrift für Angewandte Mathematik und Mechanik, 52*(6), 321–330. doi:10.1002/zamm.19720520601

Bhatia, P. K., & Steiner, J. M. (1973). Thermal instability in a viscoelastic fluid layer in hydromagnetic. *Journal of Mathematical Analysis and Applications, 41*(2), 271–280. doi:10.1016/0022-247X(73)90201-1

Borujerdi, A. N., Noghrehabadi, A. R., & Rees, D. A. S. (2008). Influence of Darcy number on the onset of convection in a porous layer with a uniform heat source. *International Journal of Thermal Sciences, 47*(8), 1020–1025. doi:10.1016/j.ijthermalsci.2007.07.014

Bourantas, G. C., Skouras, E. D., Loukopoulos, V. C., & Burganos, V. N. (2014). Heat transfer and natural convection of nanofluids in porous media. *European Journal of Mechanics - B/Fluids, 43*, 45–56. doi:10.1016/j.euromechflu.2013.06.013

Boussinesq, J. (1903). *Théorie Analytique de la chaleur.* Gauthier-Villars Paris, Vol.-II.

Brinkman, H. C. (1947a). A calculation of the viscous force exerted by a flowing fluid on a dense swarm of particles. *Applied Scientific Research, 1A*, 27–34.

Brinkman, H. C. (1947b). On the permeability of media consisting of closely packed porous particles. *Applied Scientific Research, A1*, 81–86.

Brinkman, H. C. (1949). Problem of fluid flow through swarms of particles and through macromolecules in solution. *Research London, 2*, 190–194. PMID:18126669

Brinkman, H. C. (1952). The viscosity of concentrated suspensions and solutions. *The Journal of Chemical Physics, 20*(4), 571–577. doi:10.1063/1.1700493

Bullard, E. C. (1949). The magnetic field within the earth. *Proc. Roy. Soc. (Lon.) A, 197*, 433-453.

Buongiorno, J. (2006). Convective transport in nanofluids. *ASME Journal of Heat Transfer, 128*(3), 240–250. doi:10.1115/1.2150834

Buongiorno, J., & Hu, W. (2005). Nanofluid coolants for advanced nuclear power plants. In *Proceedings of ICAPP'05.*

Buongiorno, J., Hu, L., Kim, S. J., Hannink, R., Truong, B., & Forrest, E. (2008). Nanofluids for enhanced economics and safety of nuclear reactors: An evaluation of the potential features, issues and research gaps. *Nuclear Technology*, *162*, 80–91.

Chand, R. (2010). Gravitational effect on thermal instability of Maxwell visco-elastic fluid in porous medium. *Ganita Sandesh*, *24*(2), 166–170.

Chand, R. (2011). Effect of suspended particles on thermal instability of Maxwell visco-elastic fluid with variable gravity in porous medium. *Antarctica Journal of Mathematics*, *8*(6), 487–497.

Chand, R. (2013a). Thermal instability of rotating nanofluid. *Int. J. Appl. Math and Mech.*, *9*(3), 70–90.

Chand, R. (2013b). On the onset of Rayleigh-Bénard convection in a layer of nanofluid in Hydromagnetics. *Int. J. of Nanoscience*, *12*(6), 1350038. doi:10.1142/S0219581X13500385

Chand, R. (2013c). Thermal Instability of rotating Maxwell visco-elastic fluid with variable gravity in porous medium. *Journal of the Indian Math. Soc.*, *80*(2), 23-31.

Chand, R. (2015a). On the onset of thermal convection in a layer of Oldroydian visco-elastic fluid saturated by Brinkman–Darcy porous medium. *Studia Geotechnica et Mechanica*, *37*(4), 3–10. doi:10.1515/sgem-2015-0039

Chand, R. (2015b). Electro-thermal convection in a Brinkman porous medium saturated by nanofluid. *Ain Shams Engineering Journal.* doi.org/10.1016/j.asej.%202015.10.008

Chand, R., & Kango, S. K. (2011). Thermosolutal instability of dusty rotating Maxwell visco-elastic fluid in porous medium. *Advances in Applied Science Research*, *2*(6), 541–553.

Chand, R., & Rana, G. C. (2012a). Oscillating convection of nanofluid in porous medium. *Transport in Porous Media*, *95*(2), 269–284. doi:10.1007/s11242-012-0042-9

Chand, R., & Rana, G. C. (2012b). On the onset of thermal convection in rotating nanofluid layer saturating a Darcy-Brinkman porous medium. *International Journal of Heat and Mass Transfer*, *55*(21-22), 5417–5424. doi:10.1016/j.ijheatmasstransfer.2012.04.043

Chand, R., & Rana, G. C. (2012c). Thermal instability of Rivlin-Ericksen elastico-viscous nanofluid saturated by a porous medium. *Journal of Fluids Engineering*, *134*(12), 121203. doi:10.1115/1.4007901

Chand, R., & Rana, G. C. (2012d). Effect of rotation on thermal instability of Oldroydian visco-elastic fluid saturated by Brinkman Darcy porous medium. *Research J. Engineering and Tech.*, *3*(2), 92–99.

Chand, R., & Rana, G. C. (2012e). Dufour and Soret effects on the thermosolutal instability of Rivlin-Ericksen elastico-viscous fluid in porous medium. *Z. Naturforsch.*, *67a*, 685–691.

Chand, R., & Rana, G. C. (2014a). Hall effect on the thermal instability in a horizontal layer of nanofluid. *Journal of Nanofluids*, *3*(3), 247–253. doi:10.1166/jon.2014.1100

Chand, R., & Rana, G. C. (2014b). Double diffusive convection in a layer of Maxwell viscoelastic fluid in porous medium in the presence of Soret and Dufour effects.Journal of Fluids, Article ID 479107.

Chand, R., & Rana, G. C. (2014c). The effects of radiation on the onset of thermal instability in a layer of nanofluid layer in a porous medium. *Int. J. of Appl. Math and Mech.*, *10*(12), 76–91.

Chand, R., & Rana, G. C. (2014d). Thermal instability in a Brinkman porous medium saturated by nanofluid with no nanoparticle flux on boundaries. *Special Topics & Reviews in Porous Media: An International Journal*, *5*(4), 277–286. doi:10.1615/SpecialTopicsRevPorousMedia.v5.i4.10

Chand, R., & Rana, G. C. (2014e). Hall effect on thermal instability in a horizontal layer of nanofluid saturated in a porous medium. Int. J. *Theoretical and Applied Multiscale Mechanics*, *3*(1), 58–73. doi:10.1504/IJTAMM.2014.069455

Chand, R., & Rana, G. C. (2015a). Magneto convection in a layer of nanofluid in porous medium- a more realistic approach. *Journal of Nanofluids*, *4*(2), 196–202. doi:10.1166/jon.2015.1142

Chand, R., & Rana, G. C. (2015b). Magneto convection in a layer of nanofluid with Soret effect. *Acta Mechanica et Automatica*, *9*(2), 63–69. doi:10.1515/ama-2015-0011

Chand, R., & Rana, G. C. (2015c). Instability of Walter's B' visco-elastic nanofluid layer heated from below. *Indian Journal of Pure & Applied Physics*, *53*(11), 759–767.

Chand, R., & Rana, G. C. (2015d). Thermal instability in a horizontal layer of Walter's (Model B') visco-elastic nanofluid-a more realistic approach. *Applications and Applied Mathematics: An International Journal, 10*(2), 1027–1042.

Chand, R., Kango, S. K., & Rana, G. C. (2014). Thermal instability in anisotropic porous medium saturated by a nanofluid- a realistic approach. *NSNTAIJ, 8*(12), 445–453.

Chand, R., Rana, G. C., & Hussein, A. K. (2015a). On the onset of thermal instability in a low Prandtl number nanofluid layer in a porous medium. *Journal of Applied Fluid Mechanics, 8*(2), 265–272.

Chand, R., Rana, G. C., & Hussein, A. K. (2015b). Effect of suspended particles on the onset of thermal convection in a nanofluid layer for more realistic boundary conditions. *International Journal of Fluid Mechanics Research, 42*(5), 375–390. doi:10.1615/InterJFluidMechRes.v42.i5.10

Chand, R., Kango, S. K., & Singh, V. (2015c). Megneto-convection in a layer of Maxwell visco-elastic fluid in a porous medium with Soret effect. *Research J. of Engineering and Tech., 6*(7), 23–30. doi:10.5958/2321-581X.2015.00005.7

Chand, R., Rana, G. C., & Kango, S. K. (2015d). Effect of variable gravity on thermal instability of rotating nanofluid in porous medium. *FME Transactions, 43*(1), 62–69. doi:10.5937/fmet1501062c

Chand, R., Rana, G. C., & Kumar, S. (2013a). Variable gravity effects on thermal instability of Nanofluid in anisotropic porous medium. Int. *J. of Applied Mechanics and Engineering, 18*(3), 631–642.

Chand, R., Rana, G. C., & Singh, K. (2015e). Thermal instability in a Rivlin-Ericksen elastico-viscous nanofluid in a porous medium: A revised model. *International Journal of Nanoscience and Nanoengineering, 2*(1), 1–5.

Chand, R., Rana, G. C., & Yadav, D. (2016a). Electrothermo convection in a porous medium saturated by nanofluid. *Journal of Applied Fluid Mechanics, 9*(3), 1081–1088.

Chand, R., Yadav, D., & Rana, G. C. (2015f). Electro thermo convection in a horizontal layer of rotating nanofluid. *Int. Journal of Nanoparticles, 8*(3-4), 241–261. doi:10.1504/IJNP.2015.073726

Chand, R., Rana, G. C., Kumar, A., & Sharma, V. (2013b). Thermal instability in a layer of nanofluid subjected to rotation and suspended particles. *Research Journal Science and Tech., 5*(1), 32–40.

Chandrasekhar, S. (1961). *Hydrodynamic and Hydromagnetic Stability*. New York: Dover Publication.

Char, M., & Chiang, K. (1994). Stability analysis of Bénard Marangoni convection in fluids with internal heat source. *Journal of Physics. D, Applied Physics*, *27*(4), 748–755. doi:10.1088/0022-3727/27/4/012

Chaudhari, K., & Walke, P. (2014). Applications of nanofluid in solar energy–a review. *Int. J. Eng. Res. Technol.*, *3*, 460–463.

Chen, H. S., Ding, Y. L., & Lapkin, A. (2009). Rheological behaviour of nanofluids containing tube/rod-like nanoparticles. *Powder Technology*, *194*(1-2), 132–141. doi:10.1016/j.powtec.2009.03.038

Chen, H. S., Ding, Y. L., He, Y. R., & Tan, C. Q. (2007a). Rheological behavior of ethylene glycol based titania nanofluids. *Chemical Physics Letters*, *444*(4-6), 333–337. doi:10.1016/j.cplett.2007.07.046

Chen, H. S., Ding, Y. L., He, Y. R., & Tan, C. Q. (2007b). Rheological behaviour of ethylene glycol based titania nanofluids. *Chemical Physics Letters*, *444*(4-6), 333–337. doi:10.1016/j.cplett.2007.07.046

Cheng, L., Bandarra, F. E. P., & Thome, J. R. (2008). Nanofluid two-phase flow and thermal physics: A new research frontier of nanotechnology and its challenges. *Journal of Nanoscience and Nanotechnology*, *8*(7), 3315–3332. doi:10.1166/jnn.2008.413 PMID:19051876

Chieruzzi, M., Cerritelli, G., Miliozzi, A., & Kenny, J. (2013). Effect of nanoparticles on heat capacity of nanofluids based on molten salts as PCM for thermal energy storage. *Nanoscale Research Letters*, *8*(1), 448–456. doi:10.1186/1556-276X-8-448 PMID:24168168

Choi, S. (1995). Enhancing thermal conductivity of fluids with nanoparticles. In D. A. Siginer & H. P. Wang (Eds.), *Developments and applications of non-Newtonian flows* (Vol. 231, pp. 99–105). MD: ASME FED.

Choi, S. (1999). *Nanofluid technology: current status and future research*. Argonne: Energy Technology Division, Argonne National Laboratory.

Darcy, H. (1856). *Les Fontaines Publiques de la Ville de Dijon*. Paris: Dalmont.

Das, S., Putra, N., Thiesen, P., & Roetzel, W. (2003). Temperature dependence of thermal conductivity enhancement for nanofluids. *Journal of Heat Transfer*, *125*(4), 567–574. doi:10.1115/1.1571080

Das, S. K., Choi, S. U. S., & Patel, H. (2006). Heat transfer in nanofluids- a review. *Heat Transfer Engineering*, *27*(10), 3–19. doi:10.1080/01457630600904593

Das, S. K., Choi, S. U. S., Yu, W., & Pradeep, T. (2008). *Nanofluids Science and Technology*. New Jersey: Wiley.

Das, S. K., Putra, N., Thiesen, P., & Roetzel, W. (2003). Temperature dependence of thermal conductivity enhancement of nanofluids. *ASME J. Heat Transfer*, *125*(4), 567–574. doi:10.1115/1.1571080

Dhananjay, Agrawal, G. S. & Bhargava, R. (2010). Rayleigh Bénard convection in nanofluid. *Int. J. of Appl. Math and Mech.*, *7*(2), 61–76.

Donzelli, G., Cerbino, R., & Vailati, A. (2009). Bistable heat transfer in a nanofluid. *Physical Review Letters*, *102*(10), 104503. doi:10.1103/PhysRevLett.102.104503 PMID:19392118

Drzazga, M., Lemanowicz, M., Dzido, G., & Gierczycki, A. (2012). Preparation of metal oxide-water nanofluids by two-step method. Prosimy cytować jako: Inż. *Ap. Chem.*, *51*(5), 213–215.

Drazin, P. G., & Reid, W. H. (1981). *Hydrodynamic Stability*. Cambridge, UK: Cambridge University Press.

Eastman, J. A., Choi, S. U. S., Li, S., Yu, W., & Thompson, L. J. (2001). Anomalously increased effective thermal conductivities of ethylene glycol-based nanofluids containing copper nanoparticles. *Applied Physics Letters*, *78*(6), 718–720. doi:10.1063/1.1341218

Eastman, J. A., Choi, S. U. S., Yu, W., & Thompson, L. J. (2004). Thermal transport in nanofluids. *Annual Review of Materials Research*, *34*(1), 219–246. doi:10.1146/annurev.matsci.34.052803.090621

Eastman, J. A., Choi, S. U. S., Li, S., Thompson, L. J., & Lee, S. (1997). Enhanced thermal conductivity through the development of nanofluids. *Materials Research Society Symposium-Proceedings*. Materials Research Society.

El-Aziz, M. A. (2013). Effects of hall current on the flow and heat transfer of a nanofluid over a stretching sheet with partial slip. *International Journal of Modern Physics C*, *24*(7), 1350044–21. doi:10.1142/S0129183113500447

Ericksen, J. L. (1953). Characteristic surfaces of equations of motion of non-Newtonian fluids. *Zeitschrift fur angewandte Mathematik und Physik*, *4*, 260-267.

Faiz, F., & Zahir, E. (2014). A comparative study of nanofluids for tune able filter operation. *International Journal of Engine Research*, *3*(1), 9–12. doi:10.17950/ijer/v3s1/103

Feng, X., Ma, H., Huang, S., Pan, W., Zhang, X., Tian, F., & Luo, J. et al. (2006). Aqueous-organic phase-transfer of highly stable gold, silver, and platinum nanoparticles and new route for fabrication of gold nanofilms at the oil/water interface and on solid supports. *The Journal of Physical Chemistry B*, *110*(25), 12311–12317. doi:10.1021/jp0609885 PMID:16800553

Finlayson, B. A. (1972). *The method of weighted residuals and variational principles*. New York: Academics Press.

Fredricksen, A. G. (1964). *Principles and applications of Rheology*. Prentice-Hall Inc.

Garg, J., Poudel, B., Chiesa, M., Gordon, J. B., Ma, J. J., Wang, J. B., & Chen, G. et al. (2008). Enhanced thermal conductivity and viscosity of copper nanoparticles in ethylene glycol nanofluid. *Journal of Applied Physics*, *103*(7), 074301. doi:10.1063/1.2902483

Ghasemi, B., Aminossadati, S. M., & Raisi, A. (2011). Magnetic field effect on natural convection in a nanofluid-filled square enclosure. *International Journal of Thermal Sciences*, *50*(9), 1748–1756. doi:10.1016/j.ijthermalsci.2011.04.010

Goyal, M. R., & Goyal, A. (2015). Bound for the complex growth rate in Soret driven double diffusive convection in Rivlin-Ericksen viscoelastic fluid in porous medium. Int. *Journal of Mathematical Archive*, *6*(11), 195–205.

Green, T. (1968). Oscillating convection in an elasticoviscous liquid. *Physics of Fluids*, *11*(7), 1410–1412. doi:10.1063/1.1692123

Gupta, A. S. (1967). Hall effects on thermal instability. *Rev. Roumaine Math. Pures Appl.*, *12*, 665–677.

Gupta, U., & Aggarwal, P. (2011). Thermal instability of compressible Walters (Model B) fluid in the presence of Hall currents and suspended particles. *Therm. Sci.*, *15*(2), 487–500. doi:10.2298/TSCI1102487G

Gupta, U., & Sharma, G. (2008). Thermosolutal instability of a compressible Rivlin–Ericksen fluid in the presence of rotation and Hall currents saturating a porous medium. *Applied Mathematics and Computation*, *196*(1), 158–173. doi:10.1016/j.amc.2007.05.069

Gupta, H., Agrawal, G., & Mathur, J. (2012). An over view of nanofluids: A new media towards green environment. *Int. J. Environ. Sci.*, *3*, 433–440.

Gupta, U., Aggarwal, P., & Wanchoo, R. K. (2012). Thermal convection of dusty compressible Rivlin–Ericksen viscoelastic fluid with Hall currents. *Therm. Sci.*, *16*(1), 177–191. doi:10.2298/TSCI110128113G

Gupta, U., Ahuja, J., & Wanchoo, R. K. (2013). Magneto convection in a nanofluid layer. *International Journal of Heat and Mass Transfer*, *64*, 1163–1171. doi:10.1016/j.ijheatmasstransfer.2013.05.035

Gupta, U., Ahuja, J., Kumar, R. & Wanchoo, R. K. (2016). On Hydromagnetic stability of a horizontal nanofluid layer with Hall currents. *Materials Physics and Mechanics, 27*, 9-21.

Gupta, U., Sharma, J., & Sharma, V. (2015). Instability of binary nanofluid with magnetic field. *Applied Mathematics and Mechanics*, *36*(6), 693–706. doi:10.1007/s10483-015-1941-6

Harfash, A. J., & Alshara, A. K. (2015). Chemical reaction effect on double convection in porous media with magnetic and variable gravity effects. *Kerean J. of Chemical Engg.*, *32*(6), 1046–1059. doi:10.1007/s11814-014-0327-5

Helmholtz, H. (1868). Uber discountinuirliche Flussigkeitsbewegungen. *Wissenschaftliche Abhandlungen*, *21*, 129–166.

Hone, J. (2004). Carbon nanotubes: thermalproperties. In Dekker encyclopedia of nanoscience and nanotechnology. doi:10.1201/9781439834398.ch26

Horton, C. W., & Rogers, F. T. (1945). Convection currents in porous medium. *Journal of Applied Physics*, *16*(6), 367–370. doi:10.1063/1.1707601

Hussein, A. K. (2016). Applications of nanotechnology to improve the performance of solar collectors–Recent advances and overview. *Renewable & Sustainable Energy Reviews*, *62*, 767–792. doi:10.1016/j.rser.2016.04.050

Hussein, A., Walunj, A., & Kolsi, L. (2016a). Applications of nanotechnology to enhance the performance of the direct absorption solar collectors. *J. Thermal Eng.*, *2*, 529–540.

Hwang, K. S., Jang, S. P., & Choi, S. U. S. (2009). Flow and convective heat transfer characteristics of water-based Al_2O_3 nanofluids in fully developed laminar flow regime. Int. *J. Heat Mass Transf.*, *52*(1-2), 193–199. doi:10.1016/j.ijheatmasstransfer.2008.06.032

Ingham, D. B., & Pop, I. (1981). *Transport Phenomena in Porous Media*. New York: Elsvier.

Ingham, D. B., & Pop, I. (Eds.). (2002). Transport Phenomena in Porous Media. Pergamon.

Islam, T. (2012). The magnetoviscous–thermal instability. *The Astrophysical Journal*, *746*(8), 1–12.

Javadi, F., Saidur, R., & Kamalisarvestani, M. (2013). Investigating performance improvement of solar collectors by using nanofluids. *Renewable & Sustainable Energy Reviews*, *28*, 232–245. doi:10.1016/j.rser.2013.06.053

Jefferys, H. (1928). Some cases of instabilities in fluid motion. *Proceedings of the Royal Society of London. Series A, Containing Papers of a Mathematical and Physical Character*, *188*(779), 195–208. doi:10.1098/rspa.1928.0045

Joseph, D. D. (1965). On The stability of the Boussinesq equations. *Archive for Rational Mechanics and Analysis*, *20*(1), 59–70. doi:10.1007/BF00250190

Joseph, D. D. (1966). Nonlinear stability of the Boussinesq equations by method of energy. *Archive for Rational Mechanics and Analysis*, *20*, 59–70.

Joseph, D. D. (1976). *Stability in Fluid Motion* (Vols. 1-2). Berlin: Springer.

Jwo, C. S., Teng, T. P., & Chang, J. (2007). A simple model to estimate thermal conductivity of fluid with acicular nanoparticles. *Journal of Alloys and Compounds*, *434-435*, 569–571. doi:10.1016/j.jallcom.2006.08.229

Kakac, S., & Pramuanjaroenkij, A. (2009). Review of convective heat transfer enhancement with nanofluids. *International Journal of Heat and Mass Transfer*, *52*(13-14), 3187–3196. doi:10.1016/j.ijheatmasstransfer.2009.02.006

Kaloni, P. N., & Lou, J. X. (2002). Stability of Hadley circulations in a Maxwell fluid. *Journal of Non-Newtonian Fluid Mechanics*, *103*(2-3), 167–186. doi:10.1016/S0377-0257(01)00197-5

Kang, J., Fu, C., & Tan, W. (2011). Thermal convective instability of viscoelastic fluids in a rotating porous layer heated from below. *Journal of Non-Newtonian Fluid Mechanics*, *166*(1-2), 93–101. doi:10.1016/j.jnnfm.2010.10.008

Kang, J., Zhou, F., Tan, W., & Xia, T. (2014). Thermal instability of a nonhomogeneous power-law nanofluid in a porous layer with horizontal through flow. *Journal of Non-Newtonian Fluid Mechanics*, *213*, 50–56. doi:10.1016/j.jnnfm.2014.09.006

Kango, S. K., Rana, G. C., & Chand, R. (2013). Triple-diffusive convection in Walter's (Model B') fluid with varying gravity field saturating a porous medium. *Studia Geotechnica et Mechanica*, *XXXV*(3), 45–56.

Kasaeian, A., Eshghi, A., & Sameti, M. (2015). A review on the applications of nanofluids in solar energy systems. *Renewable & Sustainable Energy Reviews*, *43*, 584–598. doi:10.1016/j.rser.2014.11.020

Kaviany, S. (1995). *Principles of Heat Transfer in Porous Media*. New York: Springer-Verlag. doi:10.1007/978-1-4612-4254-3

Keblinski, P., Hu, L. W., & Alvarado, J. L. (2009). A benchmark study on thermal conductivity of nanofluid. *Journal of Applied Physics*, *106*(9), 094312. doi:10.1063/1.3245330

Kelvin, L. (1880). On a disturbing infinity in Lord Rayleighs solution for waves in a plane vortex stratum. *Nature*, *23*(576), 45–46. doi:10.1038/023045a0

Kelvin, L. (1887). On the stability of steady and of the periodic fluid motion. *Philosophical Magazine*, *23*(144), 459–464. doi:10.1080/14786448708628034

Kent, A. (1966). Instability of laminar flow of a magnetofluid. *Physics of Fluids*, *9*(7), 1286–1289. doi:10.1063/1.1761842

Khan, I., Fakhar, K., & Anwar, M. I. (2012). Hydromagnetic rotating flows of an Oldroyd-B fluid in a porous medium. *Special Topics & Reviews in Porous Media-An International Journal*, *3*(1), 89–95. doi:10.1615/SpecialTopicsRevPorousMedia.v3.i1.80

Khan, Z. H., Khan, W. A., & Pop, I. (2013). Triple diffusive free convection along a horizontal plate in porous media saturated by a nanofluid with convective boundary condition. *International Journal of Heat and Mass Transfer*, *66*, 603–612. doi:10.1016/j.ijheatmasstransfer.2013.07.074

Khanafer, K., & Vafai, K. (2013). Applications of nanomaterials in solar energy and desalination sectors. *Advances in Heat Transfer*, *45*, 303–329. doi:10.1016/B978-0-12-407819-2.00005-0

Kim, J., Kang, Y. T., & Choi, C. K. (2004). Analysis of convective instability and heat transfer characteristics of nanofluids. *Physics of Fluids*, *16*(7), 2395–2401. doi:10.1063/1.1739247

Kim, M. C., Lee, S. B., & Chung, B. J. (2003). Thermal instability of viscoelastic fluids in porous media. *International Journal of Heat and Mass Transfer*, *46*(26), 5065–5072. doi:10.1016/S0017-9310(03)00363-6

Kim, S. J., Bang, I. C., Buongiorno, J., & Hu, L. W. (2007). Study of pool boiling and critical heat flux enhancement in nanofluids. *Bulletin of the Polish Academy of Sciences-Technical Sciences*, *55*(2), 211–216.

Kiran, P., Bhadauria, B. S., & Kumar, V. (2016). Thermal convection in a nanofluid saturated porous medium with internal heating and gravity modulation. *Journal of Nanofluids*, *5*(3), 1–12. doi:10.1166/jon.2016.1220

Kleinstreuer, C., Li, J., & Koo, J. (2008). Microfluidics of nano-drug delivery. Int. *J. Heat Mass Transf.*, *51*(23-24), 5590–5597. doi:10.1016/j.ijheatmasstransfer.2008.04.043

Knobloch, E. (1980). Convection in binary fluids. *Physics of Fluids*, *23*(9), 1918–1927. doi:10.1063/1.863220

Kole, M., & Dey, T. K. (2010). Thermal conductivity and viscosity of Al_2O_3 nanofluid based on car engine coolant. *Journal of Physics. D, Applied Physics*, *43*(31), 315501. doi:10.1088/0022-3727/43/31/315501

Kumar, V., & Awasthi, M. K. (2016). Onset of triple-diffusive convection in a nanofluid layer. *Journal of Nanofluids*, *5*(2), 284–291. doi:10.1166/jon.2016.1217

Kumar, V., & Kumar, P. (2013). Thermal convection in a (Kuvshiniski-type) viscoelastic rotating fluid in the presence of magnetic field through porous medium. *IJE Trans A: Basics*, *26*, 753–760.

Kumar, P., & Singh, M. (2006). On a viscoelastic fluid heated from below in a porous medium. *J. Non-Equilib Thermodyn*, *31*(2), 189–203. doi:10.1515/JNETDY.2006.009

Kuznetsov, A. V. (2011a). Non-oscillatory and oscillatory nanofluid bio-thermal convection in a horizontal layer of finite depth. *European Journal of Mechanics - B/Fluids*, *30*(2), 156–165. doi:10.1016/j.euromechflu.2010.10.007

Kuznetsov, A. V. (2011b). Nanofluid bioconvection in water-based suspensions containing nanoparticles and oxytactic microorganisms: oscillatory instability. *Nanoscale Research Letters, 6*. doi:X-6-10010.1186/1556-276

Kuznetsov, A. V., & Nield, D. A. (2010a). Effect of local thermal non-equilibrium on the onset of convection in a porous medium layer saturated by a nanofluid. *Transport in Porous Media*, *83*(2), 425–436. doi:10.1007/s11242-009-9452-8

Kuznetsov, A. V., & Nield, D. A. (2010b). Thermal instability in a porous medium layer saturated by a nanofluid: Brinkman Model. *Transport in Porous Media, 81*(3), 409–422. doi:10.1007/s11242-009-9413-2

Kuznetsov, A. V., & Nield, D. A. (2010c). The onset of double-diffusive nanofluid convection in a layer of a saturated porous medium. *Transport in Porous Media, 85*(3), 941–951. doi:10.1007/s11242-010-9600-1

Kuznetsov, A. V., & Nield, D. A. (2011). The effect of local thermal non-equilibrium on the onset of convection in porous medium layer saturated by a nanofluid: Brinkman model. *Journal of Porous Media, 14*(4), 285–293. doi:10.1615/JPorMedia.v14.i4.10

Lai, J., Nelson, V. M., Nash, A., Hoffman, A. S., Yager, P., & Stayton, P. (2009). Dynamic bio processing and microfluidic transport control with smart magnetic nanoparticles in laminar-flow devices. *Lab on a Chip, 9*(14), 1997–2002. doi:10.1039/b817754f PMID:19568666

Lapwood, E. R. (1948). Convection of a fluid in porous medium. *Proceedings of the Cambridge Philosophical Society, 44*(04), 508–519. doi:10.1017/S030500410002452X

Laroze, D., Martinez-Mardones, J., & Bragard, J. (2007). Thermal convection in a rotating binary viscoelastic liquid mixture. *The European Physical Journal. Special Topics, 146*(1), 291–300. doi:10.1140/epjst/e2007-00187-6

Larson, R. G. (1992). Instabilities in viscoelastic flows. *Rheologica Acta, 31*(3), 213–221. doi:10.1007/BF00366504

Lee, S., Choi, S. U. S., Li, S., & Eastman, J. A. (1999). Measuring thermal conductivity of fluids containing oxide nanoparticles. *Journal of Heat Transfer, 121*(2), 280–289. doi:10.1115/1.2825978

Lehnert, B. (1954). Magneto-hydrodynamic waves in liquid sodium. *Physical Review, 94*(4), 815–824. doi:10.1103/PhysRev.94.815

Leong, J. C., & Lai, F. C. (2004). Natural convection in rectangular layers porous cavities. J. Thermophys. *Heat Transfer, 18*, 457–463.

Li, Y., Xie, H., Yu, W., & Li, J. (2011). Investigation on heat transfer performances of nanofluids in solar collector. *Materials Science Forum, 694*, 33–36. doi:10.4028/www.scientific.net/MSF.694.33

Lo, C. H., Tsung, T. T., Chen, L. C., Su, C. H., & Lin, H. M. (2005). Fabrication of Copper Oxide Nanofluid Using Submerged Arc Nanoparticle Synthesis System (SANSS). *Journal of Nanoparticle Research, 7*(2-3), 313–320. doi:10.1007/s11051-004-7770-x

Low, A. R. (1929). On the criterion for stability of a layer of viscous fluid heated from below. *Proceedings of the Royal Society of London. Series A, Containing Papers of a Mathematical and Physical Character, 125*(796), 180–195. doi:10.1098/rspa.1929.0160

Mahajan, A., & Arora, M. (2013). Convection in rotating magnetic nanofluids. *Applied Mathematics and Computation, 219*(11), 6284–6296. doi:10.1016/j.amc.2012.12.012

Mahajan, A., & Sharma, M. K. (2012). Brinkman flow in Couple-stress fluids. Special Topics & Reviews in Porous Media-. *International Journal (Toronto, Ont.), 3*(3), 215–219.

Mahendran, V., & Philip, J. (2012). Nanofluid based optical sensor for rapid visual inspection of defects in ferromagnetic materials. *Applied Physics Letters, 100*(7), 073104. doi:10.1063/1.3684969

Mahmoudi, A. H., Pop, I., & Shahi, M. (2012). Effect of magnetic field on natural convection in a triangular enclosure filled with nanofluid. *International Journal of Thermal Sciences, 59*, 126–140. doi:10.1016/j.ijthermalsci.2012.04.006

Malashetty, M. S., & Basavaraja, D. (2002). Rayleigh-Bénard convection subject to time dependent wall temperature/gravity in a fluid-saturated anisotropic porous medium. *Heat and Mass Transfer, 38*(7-8), 551–563. doi:10.1007/s002310100245

Malashetty, M. S., & Basavaraja, D. (2003). The effect of thermal/gravity modulation on the onset of convection in a horizontal anisotropic porous layer. *Int. J. Appl. Mech. Eng., 8*(3), 425–439.

Malashetty, M. S., & Kollur, P. (2011). The onset of double diffusive convection in a couple stress fluid saturated anisotropic porous layer. *Transport in Porous Media, 86*(2), 435–459. doi:10.1007/s11242-010-9630-8

Malashetty, M. S., Shivakumara, I. S., Kulkarni, S., & Swamy, M. (2006). Convective instability of Oldroyd B fluid saturated porous layer heated from below using a thermal nonequilibrium model. *Transport in Porous Media, 64*(1), 123–139. doi:10.1007/s11242-005-1893-0

Malashetty, M. S., Swamy, M., & Heera, R. (2009a). The onset of convection in a binary viscoelastic fluid saturated porous layer. *Zeitschrift für Angewandte Mathematik und Mechanik*, *89*(5), 356–365. doi:10.1002/zamm.200800199

Malashetty, M. S., Wenchang, T., & Swamy, M. (2009b). The onset of double convection in a binary viscoelastic fluid saturated anisotropic porous layer. *Physics of Fluids*, *21*(8), 084101. doi:10.1063/1.3194288

Manimaran, R., Palaniradja, K., Alagumurthi, N., Sendhilnathan, S., & Hussain, J. (2014). Preparation and characterization of copper oxide nanofluid for heat transfer applications. *Appl. Nanosci.*, *4*(2), 163–167. doi:10.1007/s13204-012-0184-7

Manna, I. (2000). Synthesis, characterization and application of nanofluid: An overview. *Journal of the Indian Institute of Science*, *89*, 21–33.

Martinez-Mardones, J., & Perez-Garcia, C. (1990). Linear instability in viscoelastic fluid convection. *Journal of Physics Condensed Matter*, *2*(5), 1281–1290. doi:10.1088/0953-8984/2/5/019

Martinez-Mardones, J., Tiemann, R., & Walgraef, D. (2000). Rayleigh-Bénard convection in binary viscoelastic fluid. *Physica A*, *283*(1-2), 233–236. doi:10.1016/S0378-4371(00)00159-X

Martınez-Mardones, J., Tiemann, R., & Walgraef, D. (2002). Convection in binary viscoelastic fluid. *Revista Mexicana de Física*, *48*(Suplemento), 103–105.

Martinez-Mardones, J., Tiemann, R., & Walgraef, D. (2003). Amplitude equation for stationary convection in a binary viscoelastic fluid. *Physica A*, *327*(1-2), 29–33. doi:10.1016/S0378-4371(03)00433-3

Masuda, H., Ebata, A., Teramae, K., & Hishinuma, N. (1993). Alternation of thermal conductivity and viscosity of liquid by dispersing ultra-fine particles (dispersion of-Al_2O_3,SiO_2 and TiO_2 ultra-fine particles). *Netsu Bussei (Japan)*, *4*(4), 227–233. doi:10.2963/jjtp.7.227

Maxwell, J. C. (1881). *A Treatise on Electricity and Magnetism*. Oxford, UK: Clarendon Press.

Maxwell, J. C. (1866). On dynamical theory of gases. *Philos. Trans. R. Soc. London Sec. A*, *157*, 2678–2690.

Minsta, H. A., Roy, G., Nguyen, C. T., & Doucet, D. (2009). New temperature dependent thermal conductivity data for water-based nanofluids. *International Journal of Thermal Sciences*, *48*(2), 363–371. doi:10.1016/j.ijthermalsci.2008.03.009

Mohammed, H. A., Al-Aswadi, A. A., Shuaib, N. H., & Saidur, R. (2011). Convective heat transfer and fluid flow study over a step using nanofluids: A review. *Renewable & Sustainable Energy Reviews*, *15*(6), 2921–2939. doi:10.1016/j.rser.2011.02.019

Motsa, S. S. (2008). On the onset of convection in a porous layer in the presence of Dufour and Soret effects. *SJPAM*, *3*, 58–65.

Mukherjee, S., & Paria, S. (2013). Preparation and stability of nanofluids-a review. *IOSR Journal of Mechanical and Civil Engineering*, *9*(2), 63–69. doi:10.9790/1684-0926369

Murshed, S. M. S., Leong, K. C., & Yang, C. (2005). Enhanced thermal conductivity of TiO_2 -water based nanofluids. *International Journal of Thermal Sciences*, *44*(4), 367–373. doi:10.1016/j.ijthermalsci.2004.12.005

Nakagawa, Y. (1955). An experiment on the inhibition of thermal convection by magnetic field. *Nature*, *175*(4453), 417–419. doi:10.1038/175417b0

Nakagawa, Y. (1957). Experiments on the inhibition of thermal convection by magnetic field. *Proceedings of the Royal Society of London. Series A, Mathematical and Physical Sciences*, *240*(1220), 108–113. doi:10.1098/rspa.1957.0070

Nakano, M., Matsuura, H., & Ju, D. (2008). Drug delivery system using nano-magnetic fluid. In *Proceedings of the 3rd International Conference on Innovative Computing, Information and Control (ICICIC '08)*.

Nanjundappa, C. E., Shivakumara, I. S., Lee, J., & Ravisha, M. (2011). Effect of internal heat generation on the onset of Brinkman-Bénard convection in a ferrofluid saturated porous layer. *International Journal of Thermal Sciences*, *50*(2), 160–168. doi:10.1016/j.ijthermalsci.2010.10.003

Nerella, S., Sudheer, N., & Bhramara, P. (2014). Enhancement of heat transfer by nanofluids in solar collectors. *Int. J. Innov. Eng.Technol.*, *3*, 115–120.

Nield, D. A. (1968). Onset of the thermohaline convection in porous medium. *Water Resources Research*, *5*(3), 553–560. doi:10.1029/WR004i003p00553

Nield, D. A. (1987). Convective heat transfer in porous media with columnar structures. *Transport in Porous Media*, *2*(2), 177–185. doi:10.1007/BF00142658

Nield, D. A. (2008). General heterogeneity effects on the onset of convection in a porous medium. In P. Vadaz (Ed.), *Emerging Topics in Heat and Mass Transfer in Porous Media*. New York: Springer. doi:10.1007/978-1-4020-8178-1_3

Nield, D. A. (2010). A note on the onset of convection in a layer of a porous medium saturated by a non-Newtonian nanofluid of power-law type. *Transport in Porous Media*, *87*(1), 121–123. doi:10.1007/s11242-010-9671-z

Nield, D. A., & Bejan, A. (2013). *Convection in porous media*. New York: Springer-Verlag. doi:10.1007/978-1-4614-5541-7

Nield, D. A., & Kuznetsov, A. V. (2009a). Thermal instability in a porous medium layer saturated by a nanofluid. Int. *J. Heat Mass Transf.*, *52*(25-26), 5796–5801. doi:10.1016/j.ijheatmasstransfer.2009.07.023

Nield, D. A., & Kuznetsov, A. V. (2009b). The Cheng–Minkowycz problem for natural convective boundary-layer flow in a porous medium saturated by a nanofluid. *International Journal of Heat and Mass Transfer*, *52*(25-26), 5792–5795. doi:10.1016/j.ijheatmasstransfer.2009.07.024

Nield, D. A., & Kuznetsov, A. V. (2010a). The onset of convection in a horizontal nanofluid layer of finite depth. *European Journal of Mechanics - B/Fluids*, *29*(3), 217–223. doi:10.1016/j.euromechflu.2010.02.003

Nield, D. A., & Kuznetsov, A. V. (2010b). The onset of convection in a layer of cellular porous material: Effect of temperature-dependent conductivity arising from radiative transfer. *Journal of Heat Transfer*, *132*(7), 074503. doi:10.1115/1.4001125

Nield, D. A., & Kuznetsov, A. V. (2011a). The onset of double-diffusive convection in a nanofluid layer. *International Journal of Heat and Fluid Flow*, *32*(4), 771–776. doi:10.1016/j.ijheatfluidflow.2011.03.010

Nield, D. A., & Kuznetsov, A. V. (2011b). The effect of vertical through flow on thermal Instability in a porous medium layer saturated by a nanofluid. *Transport in Porous Media*, *87*(3), 765–775. doi:10.1007/s11242-011-9717-x

Nield, D. A., & Kuznetsov, A. V. (2012). The onset of convection in a layer of a porous medium saturated by a nanofluid: Effects of conductivity and viscosity variation and cross diffusion. *Transport in Porous Media*, *92*(3), 837–846. doi:10.1007/s11242-011-9935-2

Nield, D. A., & Kuznetsov, A. V. (2013). Onset of convection with internal heating in a porous medium saturated by a nanofluid. *Transport in Porous Media*, *99*(1), 73–83. doi:10.1007/s11242-013-0174-6

Nield, D. A., & Kuznetsov, A. V. (2014). Thermal instability in a porous medium layer saturated by a nanofluid: A revised model, Int. *J. Heat and Mass Transf.*, *68*, 211–214. doi:10.1016/j.ijheatmasstransfer.2013.09.026

Nguyen, C. T., Roy, G., Galanis, N., & Suiro, S. (2006). Heat transfer enhancement by using Al2O3-water nanofluid in a liquid cooling system for microprocessors. In *Proceedings of the 4th WSEAS International Conference on Heat Transfer. Thermal Engineering and Environment*, (pp. 103-108).

Oberoi, C., & Devanathan, C. (1963). *Proc. Summer Seminar in Magnetohydrodynamics*. IIT Bangalore.

Oldroyd, J. G. (1958). Non-Newtonian effects in steady motion of some idealized elastico- viscous liquids. *Proceedings of the Royal Society of London. Series A, Mathematical and Physical Sciences*, *245*(1241), 278–290. doi:10.1098/rspa.1958.0083

Oldroyd, J. G. (1965). Some steady flows of a general elastic-viscous liquids. *Proceedings of the Royal Society of London*, *283*(1392), 115–133. doi:10.1098/rspa.1965.0010

Palm, E., & Tyvand, P. A. (1984). Thermal convection in a rotating porous layer. *Zeitschrift für Angewandte Mathematik und Physik*, *35*(1), 122–133. doi:10.1007/BF00945182

Parekh, K., & Lee, H. S. (2011). Experimental investigation of thermal conductivity of magnetic nanofluids. In *Proceedings of the 56th DAE Solid State Physics Symposium 2011,AIP Conference Proceedings*, (vol. 1447, pp. 385–386).

Patel, R. (2012). Effective viscosity of magnetic nanofluids through capillaries. *Physical Review E: Statistical, Nonlinear, and Soft Matter Physics*, *85*(2), 026316. doi:10.1103/PhysRevE.85.026316 PMID:22463326

Patel, H. E., Das, S. K., Sundararajan, T., Nair, A. S., George, B., & Pardeepa, T. (2003). Thermal conductivities of naked and monolayer protected metal nanoparticles based nanofluids: Manifestation of anomalous enhancement and chemical effects. *Applied Physics Letters*, *83*(14), 2931–2933. doi:10.1063/1.1602578

Patil, R. P., & Rudraiah, N. (1973). Stability of Hydromagnetic thermo convective flow through porous medium. Trans. ASME. *Journal of Applied Mechanics, 40*(4), 879–884. doi:10.1115/1.3423181

Paul, G., Chopkar, M., Manna, I., & Das, P. K. (2010). Techniques for measuring the thermal conductivity of nanofluids: A review. *Renewable & Sustainable Energy Reviews, 14*(7), 1913–1920. doi:10.1016/j.rser.2010.03.017

Paul, G., Philip, J., Raj, B., Das, P. K., & Manna, I. (2011). Synthesis, characterization, and thermal property measurement of nano-$Al_{95}Zn0_5$ dispersed nanofluid prepared by a two-step process. Int. *J. Heat Mass Tran., 54*(15-16), 3783–3788. doi:10.1016/j.ijheatmasstransfer.2011.02.044

Pellow, A., & Southwell, R. V. (1940). On maintained convection motion in a fluid heated from below. *Proceedings of the Royal Society of London. Series A, Mathematical and Physical Sciences, 176*(966), 312–343. doi:10.1098/rspa.1940.0092

Philip, J., & Shima, P. D. (2012). Thermal properties of nanofluids. Adv. *Coll. and Interface Science, 183-184*(15), 30–45. doi:10.1016/j.cis.2012.08.001 PMID:22921845

Poulikakos, D., & Bejan, A. (1985). The departure from Darcy flow in natural convection in a vertical porous layer. *Physics of Fluids, 28*(12), 3477–3484. doi:10.1063/1.865301

Prabhat, N., Buongiorno, J., & Lin-Wen, H. (2012). Convective heat transfer enhancement in nanofluids: Real anomaly or analysis artifact? *Journal of Nanofluids, 1*(1), 55–62. doi:10.1166/jon.2012.1003

Pradhan, G. K., & Samal, P. C. (1987). Thermal stability of a fluid layer under variable body forces. *Journal of Mathematical Analysis and Applications, 122*(2), 487–495. doi:10.1016/0022-247X(87)90280-0

Pradhan, G. K., Samal, P. C., & Tripathy, U. K. (1989). Thermal stability of a fluid layer in a variable gravitational field. *Indian J. Pure Appl. Math., 20*(7), 736–745.

Prakash, K., & Chand, R. (1999). Thermosolutal instability of Maxwell visco-elastic fluid with Hall Current, suspended particles and variable gravity in porous medium. *Ganita Sandesh, 13*(1), 1–12.

Prakash, K., & Chand, R. (2000). Effect of kinematic visco-elasticity and rotation on thermal instability of Rivlin- Ericksen elastico- viscous fluid in porous medium. *Ganita Sandesh, 14*(1), 1–8.

Prakash, K., & Chand, R. (2002). Thermal instability of Oldroydian visco-elastic fluid in the presence of Finite Larmor Radius rotation and variable gravity in porous medium. Proc. Natl. Acad. Sci. India Sect. A Phys. *Sci.*, *72*(4), 373–386.

Prakash, K., & Kumar, N. (1999a). Thermal instability in Rivlin-Ericksen elastic-viscous fluid in the presence of Finite Larmor Radius and variable gravity in porous medium. *Journal of the Physical Society of Japan*, *68*(4), 1168–1172. doi:10.1143/JPSJ.68.1168

Prakash, K., & Kumar, N. (1999b). Effects of suspended particles, rotation and variable gravity field on the thermal instability of Rivlin-Ericksen elastic-viscous fluid in porous medium. *Indian Journal of Pure and Applied Math.*, *30*(11), 1157–1166.

Prakash, S. B., Kotin, K. N., & Kumar, P. (2015). Preparation and characterization of nanofluid (CuO-Water, TiO_2-Water). *International Journal of Science and Engineering*, *1*(3), 14–20.

Qin, Y., & Kaloni, P. N. (1992). A thermal instability problem in a rotating micropolar fluid. *International Journal of Engineering Science*, *30*(9), 1117–1126. doi:10.1016/0020-7225(92)90061-K

Quddoos, A., Anand, A., Mishra, G. K., & Nag, P. (2014). Nanofluids: Introduction, preparation, stability analysis and stability enhancement techniques. *International Journal in Physical & Applied Sciences*, *1*(3), 31–36.

Rainbow, J. (1948). The magnetic fluid clutch. *Transactions of the American Institute of Electrical Engineers*, *67*(2), 1308–1315. doi:10.1109/T-AIEE.1948.5059821

Rana, P., & Agarwal, S. (2015). Convection in a binary nanofluid saturated rotating porous layer. *Journal of Nanofluids*, *4*(1), 59–65. doi:10.1166/jon.2015.1123

Rana, G. C., & Chand, R. (2012). Effect of rotation on the onset of compressible Rivlin-Ericksen fluid heated from below saturating a Darcy-Brinkman porous medium. *Research J. Engineering and Tech.*, *3*(2), 76–81.

Rana, G. C., & Chand, R. (2013a). Double-diffusive convection in compressible Rivlin-Ericksen fluid permeated with suspended particles in a Brinkman porous medium. *Int. J. of Appl. Math and Mech.*, *9*(10), 58–73.

Rana, G. C., & Chand, R. (2013b). Combined effect of suspended particles and rotation on thermosolutal convection in a visco elastic fluid saturating a Darcy-Brinkman porous medium. *Iranian Journal of Science & Technology*, *37*(A1), 319–325.

Rana, G. C., & Chand, R. (2015a). Rayleigh-Bénard convection in an elastic-viscous Walters' (model B') nanofluid layer. *Bulletin of the Polish Academy of Sciences*, *63*(1), 235–244.

Rana, G. C., & Chand, R. (2015b). Onset of thermal convection in a rotating nanofluid layer saturating a Darcy-Brinkman porous medium: A more realistic model. *Journal of Porous Media*, *18*(6), 629–635. doi:10.1615/JPorMedia.v18.i6.60

Rana, G. C., & Chand, R. (2015c). Stability analysis of double-diffusive convection of Rivlin-Ericksen elastico-viscous nanofluid saturating a porous medium: A revised model. *Forsch Ingenieurwes*, *79*(1-2), 87–95. doi:10.1007/s10010-015-0190-5

Rana, G. C., & Thakur, R. C. (2012a). A mathematical theorem on the onset of Couple-stress fluid permeated with suspended dust particles saturating a porous medium. *Int. Journal of Multiphysics*, *6*(1), 61–72. doi:10.1260/1750-9548.6.1.61

Rana, G. C., & Thakur, R. C. (2012b). Effect of suspended particles on thermal convection in Rivlin-Ericksen fluid in a Darcy-Brinkman porous medium. *Journal of Mechanical Engineering Science*, *2*, 162–171. doi:10.15282/jmes.2.2012.3.0014

Rana, G. C., & Thakur, R. C. (2012c). Effect of suspended particles on the onset of thermal convection in compressible elastico-viscous fluid in a Brinkman porous medium. *Caspian Journal of Applied Sciences Research*, *1*(10), 65–73.

Rana, G. C., & Thakur, R. C. (2013a). Combined effect of suspended particles and rotation on double-diffusive convection in a visco-elastic fluid saturated by a Darcy-Brinkman porous medium. *The Journal of Computational Multiphase Flows*, *5*(2), 101–113. doi:10.1260/1757-482X.5.2.101

Rana, G. C., & Thakur, R. C. (2013b). Effect of suspended particles on the onset of thermal convection in a compressible visco-elastic fluid in a Darcy-Brinkman porous medium. *FDMP*, *9*(3), 251–265.

Rana, G. C., & Thakur, R. C. (2016). The onset of double-diffusive convection in a layer of nanofluid under rotation[Thermal Engineering]. *Engenharia Térmica*, *15*(1), 88–95.

Rana, G. C., Chand, R., & Sharma, V. (2016a). The effect of rotation on the onset of electrohydrodynamic instability of an elastico-viscous dielectric fluid layer. *Bulletin of the Polish Academy of Sciences Technical Sciences*, *64*(1), 143–149. doi:10.1515/bpasts-2016-0016

Rana, G. C., Chand, R., & Sharma, V. (2016b). Onset of electrohydrodynamic instability of a rotating viscoelastic fluid layer saturating a porous medium. *Acta Technica*, *61*, 31–44.

Rana, G. C., Chand, R., & Yadav, D. (2015). The onset of electrohydrodynamic instability of an elastico-viscous Walters (Model B) dielectric fluid layer. *FME Transactions*, *43*(2), 154–160. doi:10.5937/fmet1502154r

Rana, G. C., Thakur, R. C., & Jamwal, H. S. (2014a). The onset of thermal instability of viscoelastic rotating fluid permeated with suspended particles in porous medium. *Structural Integrity and Life*, *14*(3), 193–198.

Rana, G. C., Thakur, R. C., & Kango, S. K. (2014b). On the onset of thermosolutal instability in a layer of an elastico-viscous nanofluid in porous medium. *FME Transactions*, *42*(1), 1–9. doi:10.5937/fmet1401001R

Rana, G. C., Thakur, R. C., & Kango, S. K. (2014c). On the onset of double-diffusive convection in a layer of nanofluid under rotation saturating a porous medium. *Journal of Porous Media*, *17*(8), 657–667. doi:10.1615/JPorMedia.v17.i8.10

Rana, G. C., Thakur, R. C., & Kumar, S. (2012). Thermosolutal convection in compressible Walters (model B) fluid permeated with suspended particles in a Brinkman porous medium. *The Journal of Computational Multiphase Flows*, *4*(2), 211–224. doi:10.1260/1757-482X.4.2.211

Rani, N., & Tomar, S. K. (2010). Thermal convection problem of micropolar fluid subjected to Hall current. *Applied Mathematical Modelling*, *34*(2), 508–519. doi:10.1016/j.apm.2009.06.007

Rao, N., Gahane, L. & Ranganayakulu, S. (2014). Synthesis, applications and challenges of nanofluids – review. *IOSR J. Appl. Phys.*, 21– 28.

Rashid, F., Dawood, K., & Hashim, A. (2014). Maximizing of solar absorption by (TiO_2– water) nanofluid with glass mixture. *Int. J. Res. Eng. Technol.*, *2*, 87–90.

Ravikumar, J. & Goud, P. (2014). Nanofluids: a promising future. *J. Chem. Pharm. Sci.*, 57–61.

Rayleigh, L. (1880). On the stability or instability of certain fluid motions. *Proceedings of the London Mathematical Society, 11*, 57–70.

Rayleigh, L. (1892a). On the question of stability of flow of fluids. *Philosophical Magazine, 34*(206), 59–70. doi:10.1080/14786449208620167

Rayleigh, L. (1892b). On the instability of cylindrical fluid surface. *Philosophical Magazine, 34*(207), 177–180. doi:10.1080/14786449208620304

Rayleigh, L. (1916). On convective currents in a horizontal layer of fluid when the higher temperature is on the underside. *Philosophical Magazine, 32*(192), 529–546. doi:10.1080/14786441608635602

Rees, D. A. S. (2000). *The stability of Darcy-Bénard convection, Hand book of porous media. 12* (K. Vafai & H. Haddim, Eds.). Dekker.

Reynolds, O. (1883). On the experiment investigation of the circumstances which determine whether the motion of the water shall be direct or sinuous, and of law of resistance in parallel channels. *Philosophical Transactions of the Royal Society of London, 174A*(0), 935–982. doi:10.1098/rstl.1883.0029

Rionero, S., & Straughan, B. (1990). Convection in a porous medium with internal heat source and variable gravity effects. *International Journal of Engineering Science, 28*(6), 497–503. doi:10.1016/0020-7225(90)90052-K

Rivlin, R. S., & Ericksen, J. L. (1955). Stress-deformation relations for isotropic materials. *J. Rational Mech. Anal., 1*, 323–334.

Rosenblat, S. (1986). Thermal convection in a viscoelastic liquid. *Journal of Non-Newtonian Fluid Mechanics, 21*(2), 201–223. doi:10.1016/0377-0257(86)80036-2

Rossby, H. T. (1969). A steady of Bénard convection with and without rotation. *Journal of Fluid Mechanics, 36*(02), 309–335. doi:10.1017/S0022112069001674

Routbort, J. (2008). *Argonne National Lab, Michellin North America, St. Gobain Corp*. Michellin, North America.

Rudraiah, N., Kaloni, P. N., & Radhadevi, P. V. (1989). Oscillatory convection in a viscoelastic fluid through a porous layer heated from below. *Rheologica Acta, 28*(1), 48–52. doi:10.1007/BF01354768

Sadique, M., & Verma, A. (2014). Nanofluid-based receivers for increasing efficiency of solar panels. *Int. J. Adv. Mech. Eng., 4*, 77–82.

Saidur, R., Leong, K., & Mohammad, H. (2011). A review on applications and challenges of nanofluids. *Renewable & Sustainable Energy Reviews, 15*(3), 1646–1668. doi:10.1016/j.rser.2010.11.035

Sandeep, N., Sulochana, C., & Kumar, B. R. (2015). MHD boundary layer flow and heat transfer past a stretching/shrinking sheet in a nanofluid. *Journal of Nanofluids, 4*(4), 1–6. doi:10.1166/jon.2015.1181

Scheidegger, A. E. (1960). *The Physics through porous media.* University Toronto Press.

Schmielt, R. J., & Milverton, S. W. (1935). On the stability of a fluid when heated from below. *Proceedings of the Royal Society of London. Series A, Mathematical and Physical Sciences, 152*(877), 586–594. doi:10.1098/rspa.1935.0209

Schmidt, A. J., Chiesa, M., Torchinsky, D. H., Johnson, J. A., Boustani, A., McKinley, G. H., & Chen, G. et al. (2008). Experimental investigation of nanofluid shear and longitudinal viscosities. *Applied Physics Letters, 92*(24), 244107. doi:10.1063/1.2945799

Seerin, J. (1959). On the stability of viscous fluid motions. *Archive for Rational Mechanics and Analysis, 3*(3), 1–9. doi:10.1007/BF00284160

Sharma, R. C. (1975). Effect of rotation on thermal instability of a viscoelastic fluid. *Acta Physiologica Hungarica, 40*, 11–17.

Sharma, J., & Gupta, U. (2015). Double–diffusive nanofluid convection in porous medium with rotation: Darcy Brinkman model. *Procedia Engineering, 127*, 783–790. doi:10.1016/j.proeng.2015.11.413

Sharma, R. C., & Gupta, U. (1993). Thermal instability of compressible fluids with hall currents and suspended particles in porous medium. *International Journal of Engineering Science, 31*(7), 1053–1060. doi:10.1016/0020-7225(93)90113-9

Sharma, V., & Kishore, K. (2001). Hall effect on thermosolutal instability of Rivlin–Ericksen fluid with varying gravity field in porous medium. *International Journal of Pure and Applied Mathematics, 32*(11), 1643–1657.

Sharma, R. C., & Kumar, P. (1996a). Effect of rotation on thermal instability in Rivlin-Ericksen elastico-viscous fluid. *Z. Naturforch., 51a*, 821–824.

Sharma, R. C., & Kumar, P. (1996b). Hall effect on thermosolutal instability in Maxwellian visco-elastic fluid in porous medium. *Arch. Mech.*, *48*(1), 199–209.

Sharma, R. C., & Kumar, P. (1997). Thermal instability in Rivlin-Ericksen elastico-viscous fluid in hydromagnetics. *Z. Naturforch.*, *52a*, 369–371.

Sharma, R. C., & Kumari, V. (1993). Thermosolutal convection in a Maxwellian viscoelastic fluid in porous medium. *Czechoslovak Journal of Physics*, *43*(1), 31–40. doi:10.1007/BF01589582

Sharma, V., & Rana, G. C. (2001). Thermal instability of a Walters B elastic-viscous fluid in the presence of a variable gravity field and rotation in a porous medium. *J. Non Equilib. Thermodyn.*, *26*(1), 31–40. doi:10.1515/JNETDY.2001.003

Sharma, R. C., & Sharma, V. (1991). Stability of stratified fluid in porous medium in the presence of suspended particles and variable magnetic field. *Czechoslovak Journal of Physics*, *41*(5), 450–462. doi:10.1007/BF01597948

Sharma, R. C., & Sharma, Y. D. (1990). Thermal instability in a Maxwellian viscoelastic fluid in porous medium. *Journal of Mathematical and Physical Sciences*, *24*(2), 115–123.

Sharma, R. C., & Sunil, . (1994). Thermal instability of an Oldroydian fluid with suspended particles in hydromagnetics in porous medium. *Polymer-Plastics Technology and Engineering*, *33*(3), 323–339. doi:10.1080/03602559408013096

Sharma, R. C., Sunil & Chand, S. (1999). Thermosolutal instability of Walters' rotating fluid (model B') in porous medium. *Arch. Mech.*, *51*(2), 181–191.

Sharma, R. C., Sunil, & Chand, S. (2000). Hall effect on thermal instability of Rivlin-Ericksen elastic viscous fluid. *Indian J. Pure Appl. Math.*, *31*(4), 49-59.

Sharma, V., Sunil, & Gupta, U. (2006). Stability of stratified elastico-viscous Walters (model B) fluid in the presence of horizontal magnetic field and rotation in porous medium. *Arch. Mech.*, *58*(2), 187–197.

Sharma, R. C., Sunil, & Pal, M. (2001). Thermosolutal convection in Rivlin-Ericksen rotating fluid in porous medium in hydromagnetics. *Indian J. Pure Appl. Math.*, *32*(1), 143–156.

Sharma, A. K., Tiwari, A. K., & Dixit, A. R. (2015). Progress of nanofluid application in machining: A review. *Materials and Manufacturing Processes*, *30*(7), 813–828. doi:10.1080/10426914.2014.973583

Sheikholeslami, M., & Abelman, S. (2015a). Two phase simulation of nanofluid flow and heat transfer in an annulus in the presence of an axial magnetic field. *IEEE Transactions on Nanotechnology*, *14*(56), 1–9.

Sheikholeslami, M., & Ellahi, R. (2015b). Three dimensional mesoscopic simulation of magnetic field effect on natural convection of nanofluid. *International Journal of Heat and Mass Transfer*, *89*, 799–808. doi:10.1016/j.ijheatmasstransfer.2015.05.110

Sheikholeslami, M., & Ganji, D. D. (2015c). Nanofluid flow and heat transfer between parallel plates considering Brownian motion using DTM. *Computer Methods in Applied Mechanics and Engineering*, *283*, 651–663. doi:10.1016/j.cma.2014.09.038

Sheikholeslami, M. & Rashidi, M. M. (2015d). Effect of space dependent magnetic field on free convection of Fe_3O_4 water nanofluid. *J. Taiwan Inst. Chem. Eng*. 10.1016/jtice.2015.03.035

Sheikholeslami, M., Ganji, D. D., Javed, Y., & Ellahi, R. (2015e). Effect of thermal radiation on magnetohydrodynamics nanofluid flow and heat transfer by means of two phase model. *Journal of Magnetism and Magnetic Materials*, *374*, 36–43. doi:10.1016/j.jmmm.2014.08.021

Sheikholeslami, M., Rashidi, M. M., & Ganji, D. D. (2015f). Effect of non-uniform magnetic field on forced convection heat transfer of Fe_3O_4 water nanofluid. *Computer Methods in Applied Mechanics and Engineering*, *294*, 299–312. doi:10.1016/j.cma.2015.06.010

Sheikholeslami, M., Gorji-Bandpay, M., & Ganji, D. D. (2012). Magnetic field effects on natural convection around a horizontal circular cylinder inside a square enclosure filled with nanofluid. *International Communications in Heat and Mass Transfer*, *39*(7), 978–986. doi:10.1016/j.icheatmasstransfer.2012.05.020

Sheikholeslami, M., Gorji-Bandpay, M., & Soleimani, S. (2013). Two phase simulation of nanofluid flow and heat transfer using heat line analysis. *International Communications in Heat and Mass Transfer*, *47*, 73–81. doi:10.1016/j.icheatmasstransfer.2013.07.006

Shenoy, A. V. (1994). Non-Newtonian fluid heat transfer in porous media. *Advances in Heat Transfer, 24*, 90–101. doi:10.1016/S0065-2717(08)70233-8

Sherman, A., & Sutton, G. W. (1962). *Magnetohydrodynamics*. North-Western University Press, and Evanston.

Sheu, L. J. (2011a). Thermal instability in a porous medium layer saturated with a visco-elastic nanofluid. *Transport in Porous Media, 88*(3), 461–477. doi:10.1007/s11242-011-9749-2

Sheu, L. J. (2011b). Linear stability of convection in a in a visco elastic nanofluid layer. *World Acad. Sc. Engg. Tech., 58*, 289–295.

Sheu, L. J., Tam, L. M., Chen, J. H., Chen, H. K., Lin, K. T., & Kang, Y. (2008). Chaotic convection of viscoelastic fluid in porous medium. *Chaos, Solitons, and Fractals, 37*(1), 113–124. doi:10.1016/j.chaos.2006.07.050

Shima, P. D., Philip, J., & Raj, B. (2009). Magnetically controllable nanofluid with tunable thermal conductivity and viscosity. *Applied Physics Letters, 95*(13), 133112. doi:10.1063/1.3238551

Shivakumara, I. S., & Suma, S. P. (2000). Effect of through flow and internal heat generation on the onset of convection in a fluid layer. *Acta Mechanica, 140*(3-4), 207–217. doi:10.1007/BF01182511

Shivakumara, I. S., & Dhananjaya, M. (2014). Penetrative Brinkman convection in an anisotropic porous layer saturated by a nanofluid. *Ain Shams Eng. Journal, 6*(2), 703–713. doi:10.1016/j.asej.2014.12.005

Shivakumara, I. S., Dhananjaya, M., & Ng, C. O. (2015). Thermal convective instability in an Oldroyd-B nanofluid saturated porous layer. *International Journal of Heat and Mass Transfer, 84*, 167–177. doi:10.1016/j.ijheatmasstransfer.2015.01.010

Shivakumara, I. S., Malashetty, M. S., & Chavaraddi, K. B. (2006). Onset of convection in a viscoelastic-fluid-saturated sparsely packed porous layer using a thermal nonequilibrium model. *Canadian Journal of Physics, 84*(11), 973–990. doi:10.1139/p06-085

Singh, M., & Mehta, C. B. (2013). Hall effect on Bénard convection of compressible viscoelastic fluid through porous medium.Journal of Fluids.

Singh, R., & Lillard, J. W. Jr. (2009). Nanoparticle- based targeted drug delivery. *Experimental and Molecular Pathology, 86*(3), 215–223. doi:10.1016/j.yexmp.2008.12.004 PMID:19186176

Sokolov, M., & Tanner, R. I. (1972). Convective stability of a general viscoelastic fluid heated from below. *Physics of Fluids, 15*(4), 534–539. doi:10.1063/1.1693945

Srivastava, L. P. (1971). Unsteady flow of Rivlin -Ericksen fluid with uniform distribution of dust particles through channels of different cross sections in the presence of time dependent pressure gradient. *Istanbul Teknil Universities Bulteni, 194*, 19–26.

Storesletten, L. (1998). Effects of anisotropy on convective flow through porous media. In Transport Phenomena in Porous Media, (pp. 261-283). Pergamon Press. doi:10.1016/B978-008042843-7/50011-8

Storesletten, L. (2004). Effects of anisotropy on convection in horizontal and inclined porous layer. In Emerging Technologies and Techniques Porous Media, (pp. 285-306). Kulwer Academic Publishers. doi:10.1007/978-94-007-0971-3_19

Straughan, B. (1991). *The energy method, stability and nonlinear convection*. New York: Springer-Verlag.

Straughan, B. (2008). *Stability and wave motion in porous media*. New York: Springer-Verlag.

Sulochana, C., Kumar, M. K. K., & Sandeep, N. (2015). Radiation and chemical reaction effects on MHD thermosolutal nanofluid flow over a vertical plate in porous medium. *Chemical and Process Engineering Research, 34*, 28–37.

Sunil & Singh, P. (2000). Thermal instability of a porous medium with relaxation and inertia in the presence of Hall effects. *Archive of Applied Mechanics, 70*(8–9), 649–658.

Sunil, Sharma, Y. D., Bharti, P. K., & Sharma, R. C. (2005). Thermosolutal instability of compressible Rivlin–Ericksen fluid with Hall currents. *Int. J. Appl. Mech. Eng., 10*(2), 329–343.

Taylor, R., Coulombe, S., Otanicar, T., Phelan, P., Gunawan, A., Lv, W., & Tyagi, H. et al. (2013). Small particles, big impacts: A review of the diverse applications of nanofluids. *Journal of Applied Physics, 113*(1), 011301. doi:10.1063/1.4754271

Thakur, R. C., & Rana, G. C. (2013). Effect of magnetic field on thermal instability of Oldroydian viscoelastic rotating fluid in porous medium. Int. *J. of Applied Mechanics and Engineering, 18*(2), 555–569.

Thomson, J. (1882). On changing tessellated structure in certain liquid.Proc. Phil. Soc. Glasgow, 13, 464-468.

Toms, B. A., & Strawbridge, D. J. (1953). Elastic and viscous properties of dilute solutions of polymethyl methacrylate in organic liquids. *Transactions of the Faraday Society*, *49*, 1225–1234. doi:10.1039/tf9534901225

Tong, T. W., & Subramanian, E. (1985). A boundary-layer analysis for natural-convection in porous enclosure: Use of the Brinkman-extended Darcy model. *International Journal of Heat and Mass Transfer*, *28*(3), 563–571. doi:10.1016/0017-9310(85)90179-6

Tsaia, C. Y., Chen, H. T., Ding, P. P., Chan, B., Luhd, T. Y., & Chen, P. H. (2004). Effect of structural character of gold nanoparticles in nanofluid on heat pipe thermal performance. *Materials Letters*, *58*(9), 1461–1471. doi:10.1016/j.matlet.2003.10.009

Turner, J. S. (1973). *Buoyancy effects in Fluids*. Cambridge University Press.

Turner, J. S. (1974). Double-diffusive phenomena. *Annual Review of Fluid Mechanics*, *6*(1), 37–54. doi:10.1146/annurev.fl.06.010174.000345

Tzeng, S. C., Lin, C. W. & Huang, K. D. (2005). Heat transfer enhancement of nanofluids in rotary blade coupling of four wheel drive vechicles. *Acta Mechanica, 179*(2), 11-23.

Tzou, D. Y. (2008a). Instability of nanofluids in natural convection. *ASME J. Heat Transf.*, *130*(7), 072401. doi:10.1115/1.2908427

Tzou, D. Y. (2008b). Thermal instability of nanofluids in natural convection. Int. *J. Heat Mass Transf.*, *51*(11-12), 2967–2979. doi:10.1016/j.ijheatmasstransfer.2007.09.014

Umavathi, J. C. (2015). Rayleigh–Bénard convection subject to time dependent wall temperature in a porous medium layer saturated by a nanofluid. *Meccanica, 50*(4), 981–994. doi:10.1007/s11012-014-0076-x

Umavathi, J. C., & Kumar, J. P. (2016). Onset of convection in a porous medium layer saturated with an Oldroyd nanofluid. *Journal of Heat Transfer*, *139*(1), 012401. doi:10.1115/1.4033698

Umavathi, J. C., & Mohite, M. B. (2014). Double-diffusive convective transport in a nanofluid-saturated porous layer with cross diffusion and variation of viscosity and conductivity. *Heat Transfer-Asian Research*, *43*(7), 628–652. doi:10.1002/htj.21102

Umavathi, J. C., & Mohite, M. B. (2016). Convective transport in a porous medium layer saturated with a Maxwell nanofluid. *Journal of King Saud University-Engineering Sciences*, *28*(1), 56–68. doi:10.1016/j.jksues.2014.01.002

Umavathi, J. C., Yadav, D., & Mohite, M. B. (2015). Linear and nonlinear stability analyses of double-diffusive convection in a porous medium layer saturated in a Maxwell nanofluid with variable viscosity and conductivity. *Elixir. Mech. Eng.*, *79*, 30407–30426.

Vadasz, P. (1997). *Flow in rotating porous media. Fluid Transport in porous Media*. Computational Mechanics Publication.

Vadasz, P. (1998a). Free convection in rotating porous media. *Transport Phenomena in Porous Media*, *1*, 285–312. doi:10.1016/B978-008042843-7/50012-X

Vadasz, P. (1998b). Coriolis effect on gravity-driven convection in a rotating porous layer heated from below. *Journal of Fluid Mechanics*, *376*, 351–375. doi:10.1017/S0022112098002961

Vadasz, P. (2000). Flow and thermal convection in rotating porous media. In K. Vafai (Ed.), *Handbook of porous media* (pp. 395–440). New York: Marcel Dekker, Inc. doi:10.1201/9780824741501.ch9

Vadasz, P. (2006). Heat conduction in nanofluid suspensions. *ASME J. Heat Transf.*, *128*(5), 465–477. doi:10.1115/1.2175149

Vafai, K. (2000). *Handbook of Porous Media*. New York: Marcel Dekker.

Vafai, K., & Hadim, H. A. (2000). *Handbook of Porous Media. M*. New York: Decker.

Vafai, K., & Tien, C. L. (1981). Boundary and inertia effects on flow and heat transfer in porous media. *International Journal of Heat and Mass Transfer*, *24*(2), 195–203. doi:10.1016/0017-9310(81)90027-2

Verma, S., & Tiwari, A. (2015). Progress of nanofluid application in solar collectors: A review. *Energy Conversion and Management*, *100*, 324–346.

Veronis, G. (1968). Large amplitude Bénard convection in rotating fluid. *Journal of Fluid Mechanics*, *31*, 113–139.

Vest, C. M., & Arpaci, V. (1969). Overstability of visco-elastic fluid layer heated from below. *Fluid Mech.*, *36*(03), 613–623. doi:10.1017/S0022112069001881

Wakif, A., Boulahia, Z., Zaydan, M., Yadil, N., & Sehaqui, R. (2016). The power series method to solve a magneto-convection problem in a Darcy-Brinkman porous medium saturated by an electrically conducting nanofluid layer. *International Journal of Innovation and Applied Studies*, *14*(4), 1048–1065.

Walters, K. (1960). The motion of an elastic-viscous liquid contained between coaxial cylinders. *The Quarterly Journal of Mechanics and Applied Mathematics*, *13*(4), 444–455. doi:10.1093/qjmam/13.4.444

Walters, K. (1962). Non-Newtonian effects in some elastic-viscous liquids whose behavior at small rates of shear is characterized by general linear equation of state. *The Quarterly Journal of Mechanics and Applied Mathematics*, *15*(1), 63–74. doi:10.1093/qjmam/15.1.63

Wang, X. Q., & Mujumdar, A. S. (2007). Heat transfer characteristics of nanofluids: A review. *International Journal of Thermal Sciences*, *46*(1), 1–19. doi:10.1016/j.ijthermalsci.2006.06.010

Wang, X., Xu, X., & Choi, S. (1999). Thermal conductivity of nanoparticle-fluid mixture. *Journal Thermphys Heat Transfer*, *13*, 474–480. doi:10.2514/2.6486

Wang, X. Q., & Majumdar, A. S. (2006). Heat transfer characteristics of nanofluids: A review. *International Journal of Thermal Sciences*, *46*(1), 1–19. doi:10.1016/j.ijthermalsci.2006.06.010

Wang, S., & Tan, W. C. (2008). Stability analysis of double-diffusive convection of Maxwell fluid in a porous medium heated from below. *Physics Letters. [Part A]*, *372*(17), 3046–3056. doi:10.1016/j.physleta.2008.01.024

Wang, S. W., & Tan, W. C. (2011). Stability analysis of Soret-driven double-diffusive convection of Maxwell fluid in a porous medium. *International Journal of Heat and Fluid Flow*, *32*(1), 88–94. doi:10.1016/j.ijheatfluidflow.2010.10.005

Wei, X., & Wang, L. (2010). Synthesis and thermal conductivity of microfluidic copper nanofluids. *Particuology*, *8*(3), 262–271. doi:10.1016/j.partic.2010.03.001

Williams, W. C., Bang, I. C., Forrest, E., Hu, L. W., & Buongiorno, J. (2006). Preparation and characterization of various nanofluids. *NSTI-Nanotech.*, *2*, 408–411.

Wong, V. K., & Leon, O. (2010). Applications of nanofluids: Current and future. *Advances in Mechanical Engineering*, *519659*, 1–11.

Wooding, R. A. (1960). Rayleigh instability of a thermal boundary layer in flow through a porous medium. *Journal of Fluid Mechanics*, *9*(02), 183–192. doi:10.1017/S0022112060001031

Xie, H., Wang, J., Xi, T., & Liu, Y. (2001). Study of thermal conductivity of SiC nanofluids. *J. Chi Ceramic Soc.*, *29*(4), 361–364.

Xie, H., Wang, J., Xi, T., & Liu, Y. (2002). Thermal conductivity of suspension containing nanosized SiC particles. *International Journal of Thermophysics*, 23571–23580.

Xuan, Y., & Li, Y. (2000). Heat transfer enhancement of nanofluids. *International Journal of Heat and Fluid Flow*, *21*(1), 58–64. doi:10.1016/S0142-727X(99)00067-3

Xuan, Y., & Li, Q. (2003). Investigation of Convective Heat transfer and flow features of nanofluids. *ASME Journal of Heat Transfer*, *125*(1), 151–155. doi:10.1115/1.1532008

Yadav, D. (2014). *Hydrodynamic and Hydromagnetic Instability in Nanofluids*. Lambert Academic Publishing.

Yadav, D., & Kim, M. C. (2015a). The onset of transient Soret-driven buoyancy convection in a nanoparticles suspension with particle concentration-dependent viscosity in a porous medium. *Journal of Porous Media*, *18*(4), 369–378. doi:10.1615/JPorMedia.v18.i4.10

Yadav, D., & Kim, M. C. (2015b). Linear and non-linear analyses of Soret-driven buoyancy convection in a vertically orientated Hele-Shaw cell with nanoparticles suspension. *Computers & Fluids*, *117*, 139–148. doi:10.1016/j.compfluid.2015.05.008

Yadav, D., & Lee, J. (2015a). The effect of local thermal non-equilibrium on the onset of Brinkman convection in a nanofluid saturated rotating porous layer. *Journal of Nanofluids*, *4*(3), 335–342. doi:10.1166/jon.2015.1159

Yadav, D., & Lee, J. (2015b). The onset of MHD nanofluid convection with Hall current effect. *Eur. Phys. J. Plus*, *130*(8), 162–184. doi:10.1140/epjp/i2015-15162-9

Yadav, D., Agrawal, G. S., & Bhargava, R. (2012a). The onset of convection in a binary nanofluid saturated porous layer. Int. J. *Theoretical and Applied Multiscale Mechanics*, *2*(3), 198–224. doi:10.1504/IJTAMM.2012.049931

Yadav, D., Bhargava, R., & Agrawal, G. S. (2011). Thermal instability of rotating nanofluid layer. *International Journal of Engineering Science*, *49*(11), 1171–1184. doi:10.1016/j.ijengsci.2011.07.002

Yadav, D., Bhargava, R., & Agrawal, G. S. (2012b). Boundary and internal heat source effects on the onset of Darcy- Brinkman convection in a porous layer saturated by nanofluid. *International Journal of Thermal Sciences*, *60*, 244–254. doi:10.1016/j.ijthermalsci.2012.05.011

Yadav, D., Bhargava, R., & Agrawal, G. S. (2013a). The onset of double diffusive nanofluid convection in a layer of a saturated porous medium with thermal conductivity and viscosity variation. *Journal of Porous Media*, *16*(2), 105–121. doi:10.1615/JPorMedia.v16.i2.30

Yadav, D., Bhargava, R., & Agrawal, G. S. (2013b). Numerical solution of a thermal instability problem in a rotating nanofluid layer. Int. *J Heat Mass Transf.*, *63*, 313–322. doi:10.1016/j.ijheatmasstransfer.2013.04.003

Yadav, D., Bhargava, R., & Agrawal, G. S. (2013c). Thermal instability in a nanofluid layer with vertical magnetic field. *Journal of Engineering Mathematics*, *80*(1), 147–164. doi:10.1007/s10665-012-9598-1

Yadav, D., Bhargava, R., Agrawal, G. S., Hwang, G. S., Lee, J., & Kim, M. C. (2014a). Magneto-convection in a rotating layer of nanofluid. Asia-Pac. *Chemical Engineering Journal*, *9*(5), 663–677.

Yadav, D., Bhargava, R., Agrawal, G. S., Yadav, N., Lee, J., & Kim, M. C. (2014b). Thermal instability in a rotating porous layer saturated by a non-Newtonian nanofluid with thermal conductivity and viscosity variation. *Microfluidics Nanofluidics*, *16*(1-2), 425–440. doi:10.1007/s10404-013-1234-5

Yadav, D., Lee, J., & Cho, H. H. (2015). Brinkman convection induced by purely internal heating in a rotating porous medium layer saturated by a nanofluid. *Powder Technology*, *286*, 592–601. doi:10.1016/j.powtec.2015.08.048

Yadav, D., Lee, D., Cho, H. H., & Lee, J. (2016a). The onset of double-diffusive nanofluid convection in a rotating porous medium layer with thermal conductivity and viscosity variation: A revised model. *Journal of Porous Media*, *19*(1), 1–16. doi:10.1615/JPorMedia.v19.i1.30

Yadav, D., Mohamed, R. A., Lee, J. & Cho, H.H. (2016b). Thermal convection in a Kuvshiniski viscoelastic nanofluid saturated porous layer. *Ain Shams Engineering Journal*.

Yoon, D. Y., Kim, M. C., & Choi, C. K. (2004). The onset of oscillatory convection in a horizontal porous layer saturated with viscoelastic Liquid. *Transport in Porous Media*, *55*(3), 275–284. doi:10.1023/B:TIPM.0000013328.69773.a1

Yu, W., & Xie, H. (2012). A review on nanofluids: Preparation, stability mechanisms, and applications. *Journal of Nanomaterials*, *2012*, 1–17.

Yu, W., France, D. M., Routbort, J. L., & Choi, S. U. S. (2008). Review and comparison of nanofluid thermal conductivity and heat transfer enhancements. *Heat Transfer Engineering*, *29*(5), 432–460. doi:10.1080/01457630701850851

Zhang, Z., Fu, C., & Tan, W. (2008). Linear and nonlinear stability analyses of thermal convection for Oldroyd-B fluids in porous media heated from below. *Physics of Fluids*, *20*(8), 084103-1, 084103–084112. doi:10.1063/1.2972154

Zhang, L. Xia, J., Zhao, Q., Liu, L. & Zhang, Z. (2010). *Functional graphene oxide as a nanocarrier for controlled loading and targeted delivery of mixed anticancer drugs*. Academic Press.

Zhang, L., Liu, J., He, G., Ye, Z., Fang, X., & Zhang, Z. (2014). Radiative properties of ionic liquid-based nanofluids for medium-to-high-temperature direct absorption solar collectors. *Solar Energy Materials and Solar Cells*, *130*, 521–528. doi:10.1016/j.solmat.2014.07.040

Zhu, H. T., & Yin, Y. S. (2004). A novel one-step chemical method preparation of copper nanofluids. *Journal of Colloid and Interface Science*, *227*(1), 100–130. doi:10.1016/j.jcis.2004.04.026 PMID:15276044

Zierep, J., & Qertel, J. (1982). *Convective transport and instability phenomena*. Karlsruhe: Braun.

Chapter 1
Thermal Convection in a Horizontal Layer of Maxwellian Visco-Elastic Nanofluid

INTRODUCTION

The fundamental stress-strain-velocity relations of the classical hydrodynamics are not applicable to vast number of highly viscous fluids. The materials such as paints, plastics, polymers, gel and more exotic one such as magma, saturated soils and Earth's lithosphere behaves as visco-elastic fluids. Visco-elastic fluids have range of unlikely behaviors. There are many visco-elastic fluids and Maxwell visco-elastic fluid is one of them. The investigation of such fluids is desirable because of their applications in chemical engineering and material industries. The problem of convective instability of visco-elastic fluid heated from below was first studied by Green (1968). The stability of a horizontal layer of Maxwellian visco-elastic fluid heated from below was studied by Vest and Arpaci (1969). Thermal instability of a Maxwell visco-elastic fluid in the presence of rotation was investigated Bhatia et al. (1972) and found that rotation has a destabilizing influence in contrast to its stabilizing effect on a viscous Newtonian fluid.

DOI: 10.4018/978-1-68318-006-7.ch001

The term 'nanofluid' refers to a fluid containing a suspension of nanoscale particles. This type of fluid is a mixture of a regular fluid, with a very small amount of suspended metallic or metallic oxide nanoparticles or nanotubes, which was first coined by (Choi, 1995). Nanoparticles materials may be taken as oxide ceramics (Al_2O_3, CuO), metal carbides (SiC), nitrides (AlN, SiN) or metals (Al, Cu) etc and base fluids are taken as water, ethylene or tri-ethylene-glycols and other coolants, oil and other lubricants, bio-fluids, polymer solutions, other common fluids. Typical dimension of the nanoparticles is in the range of a few to about 100 nm. Due to the enhanced properties of nanofluids as such as thermal transfer fluids for instance, these fluids can be used in automotive industry, in medical sciences, in power plant cooling systems as well as in computers. On the basis of Buongiorno's model Buongiorno (2006), the stability of problems of the onset of convection in nanofluid layer has been studied by Nield and Kuznetsov (2010a, 2010b, 2011a), Kuznetsov and Nield, (2010a, 2010b, 2010c), Chand and Rana (2012a, 2012b, 2012c, 2014a) and Rana et al. (2014). The choice of the boundary conditions imposed in all these studies on nanoparticles fraction is somewhat arbitrary, it could be argued that zero-flux for nanoparticles volume fraction is more realistic. Nield and Kuznetsov (2014), Chand et al. (2015a), Chand and Rana (2014d, 2015d), studied the thermal instability of nanofluid by taking normal component of the nanoparticle flux zero at boundary which is more physically realistic.

Due to importance of Maxwell visco-elastic nanofluids, in this chapter an attempt has been made to study the linear stability analysis of a horizontal layer of Maxwellian visco-elastic nanofluids heated from below for more realistic boundary conditions (by assuming nanoparticle flux to be zero rather than prescribing the nanoparticle volume fraction on the impermeable boundaries). The value of the temperature can be imposed on the boundaries, but the nanoparticle fraction adjusts so that the nanoparticles flux is zero on the boundaries. Stability is discussed analytically as well as numerically.

MATHEMATICAL FORMULATIONS OF THE PROBLEM

The Physical Problem

In this chapter we shall investigate the thermal convection in a horizontal layer of Maxwellian visco-elastic nanofluid. The physical configuration of the problem to be considered as:

Figure 1. Physical configuration of the problem

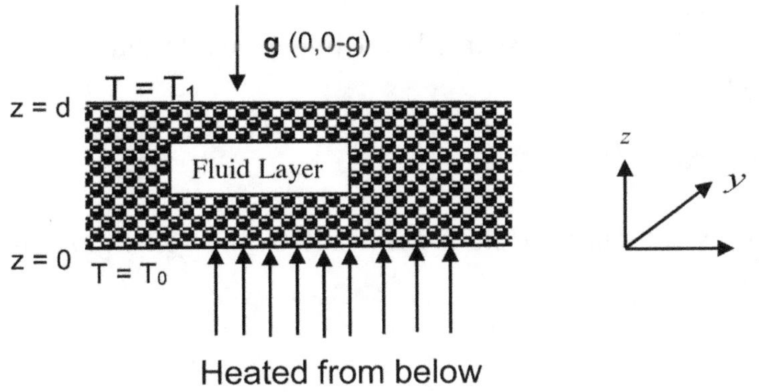

An infinite horizontal layer of Maxwellian visco-elastic nanofluid of thickness 'd' bounded by horizontal boundaries z = 0 and z = d. Fluid layer is acted upon by a gravity force **g**(0,0,-g) and is heated from below in such a way that horizontal boundaries z = 0 and z = d respectively maintained at a uniform temperature T_0 and T_1 ($T_0 > T_1$) as shown is Figure 1. The normal component of the nanoparticles flux has to vanish at an impermeable boundaries and the reference scale for temperature and nanoparticles fraction is taken to be T_1 and φ_0 respectively.

Assumptions

The mathematical equations describing the physical model are based upon the following assumptions:

1. Thermophysical properties of fluid expect for density in the buoyancy force (Boussinesq Hypothesis) are constant,
2. The fluid phase and nanoparticles are in thermal equilibrium state and thus, the heat flow has been described using one equation model,
3. Dilute mixture,
4. Nanoparticles are spherical,
5. Nanoparticles are non-magnetic,
6. No chemical reactions take place in fluid layer,
7. Negligible viscous dissipation,
8. Radiative heat transfer between the sides of wall is negligible when compared with other modes of the heat transfer,

9. Nanofluid is incompressible, electrically conducting, Newtonian and laminar flow,
10. Each boundary wall is assumed to be impermeable and perfectly thermal conducting.

Governing Equations

The Maxwell visco-elastic fluid is described by the constitutive relations as

$$T_{ij} = -p\delta_{ij} + \tau_{ij},$$

$$\left(1 + \lambda \frac{d}{dt}\right)\tau_{ij} = 2\mu e_{ij}, \tag{1.1}$$

$$e_{ij} = \frac{1}{2}\left(\frac{\partial u_i}{\partial x_j} + \frac{\partial u_j}{\partial x_i}\right),$$

where $T_{ij}, \tau_{ij}, e_{ij}, \delta_{ij}, p, v_i, x_i$ *and* μ denote respectively the stress tensor, shear stress tensor, rate of strain tensor, Kronecker delta, scalar pressure, velocity of fluid, position vector and viscosity of the fluid. $\frac{d}{dt}$ is the convection derivative and λ is the relaxation time. If $\lambda = 0$, then fluid is known as Newtonian viscous fluid.

According to the works of Chandrasekhar (1961), the basic hydrodynamic equations that govern the above described physical configuration are as follows

1. **Equation of Continuity:** Since this equation represents conservation of mass so the electromagnetic fields have no effect on it. Hence equation of continuity for incompressible fluid, is given by

$$\nabla \cdot \mathbf{v} = 0, \tag{1.2}$$

this implies that the density of a particle remains unchanged as we follow it with its motion.

2. **Equation of Motion:** The equation of motion is derived from the principle of conservation of linear momentum. Thus the equations of motion for a viscous, incompressible fluid is given by

$$\rho\frac{d\mathbf{v}}{dt} = -\nabla p + \rho\mathbf{X} + div\boldsymbol{\tau}\,, \tag{1.3}$$

where **X** is the external force acting on fluid and ρ is the density of fluid. Since external forces are of non-electromagnetic origin (gravity) only, then the equation of motion can be written as

$$\rho\frac{\mathrm{d}\mathbf{v}}{\mathrm{dt}} = -\nabla \mathrm{p} + \rho\mathbf{g} + div\boldsymbol{\tau}\,. \tag{1.4}$$

3. **Equation of Energy:** The equation of energy is derived from the law of conservation of angular momentum, and hence equation of energy for viscous compressible fluid is given by

$$\rho\frac{\partial}{\partial t}\left(c_v T\right) = \frac{\partial}{\partial x_j}\left(k_m\frac{\partial T}{\partial x_j}\right), \tag{1.5}$$

where c_v is the specific heat at constant volume, T is the temperature and k_m is the thermal conductivity. The viscous dissipation term, being small in magnitude, has not been included in equation (1.5).

4. **Equation of State:** The equation of state for viscous compressible fluid is given by

$$\rho = \rho_0\left(1+\alpha\left(T_0 - T\right)\right), \tag{1.6}$$

where α is the coefficient of the thermal expansion.

Using the constitutive relations (1.1) for Maxwell visco-elastic fluid, the equation of continuity and equation of motion for Maxwellian visco-elastic fluid are written as

$$\nabla\cdot\ \mathbf{v} = 0\,, \tag{1.7}$$

$$\rho\left(1+\lambda\frac{\partial}{\partial t}\right)\frac{d\mathbf{v}}{dt} = \left(1+\lambda\frac{\partial}{\partial t}\right)\left(-\nabla p + \rho g\right) + \mu\nabla^2\mathbf{v}\,. \tag{1.8}$$

According to the works of Buongiorno (2006), the density ρ is written as

$$\rho = \varphi\rho_{\mathrm{p}} + \left(1-\varphi\right)\rho_{\mathrm{f}}, \tag{1.9}$$

where φ is the volume fraction of the nanoparticles, ρ_p density of nanoparticles and ρ_f density of basefluid.

Taking the density of the nanofluid as that of the base fluid, the buoyancy term is approximated by

$$\rho\mathbf{g} \cong \left(\varphi\rho_{\mathrm{p}} + \left(1-\varphi\right)\left\{\rho\left(1-\alpha\left(\mathrm{T}-\mathrm{T}_0\right)\right)\right\}\right)\mathbf{g}, \tag{1.10}$$

where α is the coefficient of thermal expansion.

Using equation (1.10), the equation of motion for Maxwellian visco-elastic nanofluid is given as

$$\rho\left(1+\lambda\frac{\partial}{\partial t}\right)\frac{d\mathbf{v}}{dt} = \left(1+\lambda\frac{\partial}{\partial t}\right)\left(-\nabla p + \left(\varphi\rho_{\mathrm{p}} + \left(1-\varphi\right)\left\{\rho_{\mathrm{f}}\left(1-\alpha\left(\mathrm{T}-\mathrm{T}_0\right)\right)\right\}\right)\mathbf{g}\right) + \mu\nabla^2\mathbf{v}. \tag{1.11}$$

Due to Brownian motion and thermophoresis mass flux $\mathbf{j}_p$ of the nanoparticles in base fluid is given by

$$\mathbf{j}_p = -\rho_{\mathrm{p}}\mathrm{D}_{\mathrm{B}}\nabla\varphi - \rho_{\mathrm{p}}\left(\frac{\mathrm{D}_{\mathrm{T}}}{\mathrm{T}}\right)\nabla\mathrm{T}, \tag{1.12}$$

where $\mathrm{D}_{\mathrm{B}} = \dfrac{\mathrm{k}_{\mathrm{B}}\mathrm{T}}{3\pi\,\mu \mathrm{d}_{\mathrm{p}}}$ is the Brownian diffusion coefficient and $\mathrm{D}_{\mathrm{T}} = \left(\dfrac{\mu}{\acute{\mathrm{A}}}\right)\left(\dfrac{0.26\mathrm{k}}{2\mathrm{k}+\mathrm{k}_{\mathrm{p}}}\right)\varphi$ is the thermoporetic diffusion coefficient of the nanoparticles.

Equation of continuity for the nanoparticles is

$$\frac{\partial\varphi}{\partial \mathrm{t}} + \mathbf{v}\cdot\nabla\varphi = -\frac{1}{\rho_{\mathrm{p}}}\nabla\cdot\mathbf{j}_{\mathrm{p}} = \mathrm{D}_{\mathrm{B}}\nabla^2\varphi + \frac{\mathrm{D}_{\mathrm{T}}}{\mathrm{T}_1}\nabla^2\mathrm{T}. \tag{1.13}$$

Equation of energy for Maxwellian visco-elastic nanofluid is given by

$$\rho c \frac{\partial T}{\partial t} + \mathbf{v} \cdot \nabla T = k_m \nabla^2 T + (\rho c)_p \left(D_B \nabla \varphi \cdot \nabla T + \frac{D_T}{T_1} \nabla T \cdot \nabla T \right), \tag{1.14}$$

where ρc is heat capacity of fluid, $(\rho c)_p$ is heat capacity of nanoparticles, T_1 is the temperature of the fluid layer at z = d and k_m is thermal conductivity.

We assume that the temperature is constant and nanoparticles flux is zero on the boundaries. Thus boundary conditions Chandrasekhar (1961), Nield and Kuznetsov (2014) are

$$w = 0,\ T = T_0,\ \ D_B \frac{\partial \phi}{\partial z} + \frac{D_T}{T_1} \frac{\partial T}{\partial z} = 0 \ \text{ at } z = 0$$

and

$$w = 0,\ T = T_1,\ \ D_B \frac{\partial \phi}{\partial z} + \frac{D_T}{T_1} \frac{\partial T}{\partial z} = 0 \ \text{ at } \ z = d. \tag{1.15}$$

Introducing non-dimensional variables as

$$(x',y',z',) = \left(\frac{x,y,z}{d} \right),\ v'(u',v',w',) = V \left(\frac{u,v,w}{o} \right) d,\ t' = \frac{t^o}{d^2},\ p' = \frac{pd^2}{\mu \kappa},$$

$$\varphi' = \frac{(\varphi - \varphi_0)}{\varphi_0},\ T' = \frac{(T - T_1)}{(T_0 - T_1)},\ \text{where } \kappa = \frac{k_m}{\rho c} \text{ is thermal diffusivity of}$$

the fluid. (1.16)

Equations (1.7), (1.11), (1.13) and (1.14) in non-dimensional form can be written as

$$\nabla' \cdot \mathbf{v}' = 0, \tag{1.17}$$

$$\left(1 + F \frac{\partial}{\partial t'} \right) \frac{1}{\Pr} \frac{\partial \mathbf{v}'}{\partial t'} = \left(1 + F \frac{\partial}{\partial t'} \right) \left(-\nabla' p' \text{-} Rm \hat{e}_z + Ra T' \hat{e}_z - Rn \varphi' \hat{e}_z \right) + \nabla'^2 \mathbf{v}', \tag{1.18}$$

$$\frac{\partial \varphi'}{\partial t'} + \mathbf{v}' \cdot \nabla' \varphi' = \frac{1}{\mathrm{Le}} \nabla'^2 \varphi' + \frac{\mathrm{N_A}}{\mathrm{Le}} \nabla'^2 \mathrm{T}', \tag{1.19}$$

$$\frac{\partial \mathrm{T}'}{\partial t'} + \mathbf{v}' \cdot \nabla' \mathrm{T}' = \nabla'^2 \mathrm{T}' + \frac{\mathrm{N_B}}{\mathrm{Le}} \nabla' \varphi' \cdot \nabla' \mathrm{T}' + \frac{\mathrm{N_A N_B}}{\mathrm{Le}} \nabla' \mathrm{T}' \cdot \nabla' \mathrm{T}'. \tag{1.20}$$

Here the non-dimensional parameters are given as

$\mathrm{Pr} = \frac{\mu}{\rho \kappa}$ is the Prandtl number,

$Le = \frac{\kappa}{\mathrm{D_B}}$ is the Lewis number,

$F = \frac{\kappa \lambda}{d^2}$ is the stress relaxation parameter,

$\mathrm{Ra} = \frac{\rho g \alpha d^3 \left(\mathrm{T_0} - \mathrm{T_1}\right)}{\mu \kappa}$ is the Rayleigh number,

$\mathrm{Rm} = \frac{\left(\rho_p \varphi_0 + \rho\left(1\text{-}\varphi_0\right)\right) g d^3}{\mu \kappa}$ is the density Rayleigh number,

$\mathrm{Rn} = \frac{\left(\rho_p - \rho\right) \varphi_0 g d^3}{\mu \kappa}$ is the nanoparticles Rayleigh number,

$\mathrm{N_A} = \frac{\mathrm{D_T}\left(\mathrm{T_0}\text{-}\mathrm{T_1}\right)}{\mathrm{D_B T_1} \varphi_0}$ is the modified diffusivity ratio,

$\mathrm{N_B} = \frac{\left(\rho c\right)_p \varphi_0}{\left(\rho c\right)_f}$ is the modified particle-density increment.

In spirit of Oberbeck-Boussinesq approximation, equation (1.18) has been linearized by the neglect of a term proportional to the product of φ_0 and T. This approximation is valid in the case of small temperature gradients in a dilute suspension of nanoparticles.

The dimensionless boundary conditions are

$$\mathrm{w}' = 0, \quad \mathrm{T}' = 1, \quad \frac{\partial \phi'}{\partial z'} + N_A \frac{\partial T'}{\partial z'} = 0 \ \text{ at } \ z' = 0$$

and

$$\mathrm{w}'=0,\ \ \mathrm{T}'=0,\ \ \ \frac{\partial \phi'}{\partial \mathrm{z}'}+N_A\frac{\partial T'}{\partial \mathrm{z}'}=0\ \text{ at }\ \mathrm{z}'=1. \tag{1.21}$$

THE BASIC STATE AND ITS SOLUTIONS

The basic state was assumed to be quiescent and is given by

$$\mathrm{v}_i'\ \left(\mathrm{u}',\mathrm{v}',\mathrm{w}'\right)=0,$$

$$\mathrm{p}'=\mathrm{p}_b(\mathrm{z}),$$

$$\mathrm{T}'=\mathrm{T}_b(\mathrm{z}), \tag{1.22}$$

$$\varphi'=\varphi_b(\mathrm{z})\ \text{ and}$$

$$\rho=\rho_0\left(1+\alpha\left(\mathrm{T}\text{-}\mathrm{T}_0\right)\right).$$

Equations (1.17) – (1.20) reduce to

$$0=-\frac{d\mathrm{p}_b}{d\mathrm{z}'}\text{-Rm}+\mathrm{Ra}\mathrm{T}_b-\mathrm{Rn}\varphi_b, \tag{1.23}$$

$$\frac{d^2\mathrm{T}_b}{d\mathrm{z}'^2}+\frac{\mathrm{N_B}}{\mathrm{Le}}\frac{d\varphi_b}{d\mathrm{z}'}\frac{d\mathrm{T}_b}{d\mathrm{z}'}+\frac{\mathrm{N_A N_B}}{\mathrm{Le}}\left(\frac{d\mathrm{T}_b}{d\mathrm{z}'}\right)^2=0, \tag{1.24}$$

$$\frac{d^2\varphi_b}{d\mathrm{z}'^2}+\mathrm{N_A}\frac{d^2\mathrm{T}_b}{d\mathrm{z}'^2}=0, \tag{1.25}$$

Using boundary conditions (1.21), equation (1.25) gives

$$\varphi_b=-\mathrm{N_A}\mathrm{T}_b+\left(1-\mathrm{N_A}\right)\mathrm{z}'+\mathrm{N_A}. \tag{1.26}$$

On substituting the value of the φ_b from equation (1.26) in equation (1.24), we get

$$\frac{d^2T_b}{dz'}+\frac{\left(1\text{-}N_A\right)N_B}{Le}\frac{dT_b}{dz'}=0. \tag{1.27}$$

On integrating equation (1.27) with respect to z' and using boundary conditions (1.21), we get

$$T_b=\frac{1-e^{-\left(1-N_A\right)N_B\left(1-z'\right)/Le}}{1-e^{-\left(1-N_A\right)N_B/Le}}. \tag{1.28}$$

According to (Buongiorno, 2006), for most nanofluid investigated so far Le is large, is of order 10^2 - 10^3, while N_A is no greater than about 10. Then, the exponents in equation (1.28) are small. By expanding the exponential function into the power series and retaining up to the first order is negligible and thus a good approximation for the solution is given by

$$T_b=1-z'$$

and

$$\varphi_b=\phi_0+N_Az'$$

PERTURBATION SOLUTIONS

Let the initial basic state described by (1.22) is slightly perturbed so that perturbed state is given by

$$\left(u',v',w'\right)=0+\left(u'',v'',w''\right),$$

$$T'=T_b+T'', \tag{1.29}$$

$$\varphi'=\varphi_b+\varphi'',$$

$$p'=p_b+p'',$$

where $T_b = 1 - z'$, $\varphi_b = \phi_0 + N_A z'$ and (u'', v'', w''), T'', φ'' and p'' respectively the perturbations in initial velocity, temperature, volume fraction of the nanoparticles and pressure.

By substituting (1.29) in equations (1.16) – (1.20) and linearize by neglecting the product of the prime quantities, we obtained following equations

$$\nabla \cdot \mathbf{v} = 0, \tag{1.30}$$

$$\left(1 + F\frac{\partial}{\partial t}\right)\frac{1}{\Pr}\frac{\partial \mathbf{v}}{\partial t} = \left(1 + F\frac{\partial}{\partial t}\right)\left(-\nabla p + \mathrm{Ra}T\hat{e}_z - \mathrm{Rn}\varphi\hat{e}_z\right) + \nabla^2 \mathbf{v}, \tag{1.31}$$

$$\frac{\partial \varphi}{\partial t} + wN_A = \frac{1}{\mathrm{Le}}\nabla^2\varphi + \frac{N_A}{\mathrm{Le}}\nabla^2 T, \tag{1.32}$$

$$\frac{\partial T}{\partial t} - w = \nabla^2 T + \frac{N_B}{\mathrm{Le}}\left(N_A\frac{\partial T}{\partial z} - \frac{\partial \phi}{\partial z}\right) - \frac{2N_A N_B}{\mathrm{Le}}\frac{\partial T}{\partial z}. \tag{1.33}$$

Boundary conditions are

$$w = 0,\ T = 0,\quad \frac{\partial \phi}{\partial z} + N_A\frac{\partial T}{\partial z} = 0 \ \text{ at } z = 0\ ,1. \tag{1.34}$$

[Dashes (") have been suppressed for convenience]

Eliminating pressure term p from equation (1.31) by operating with curl twice on it, we have

$$\left(1 + F\frac{\partial}{\partial t}\right)\left(\frac{1}{\Pr}\frac{\partial}{\partial t}\left(\nabla^2 w\right) - \mathrm{Ra}\nabla_H^2 T + \mathrm{Rn}\nabla_H^2\varphi\right) - \nabla^4 w = 0, \tag{1.35}$$

where ∇_H^2, is two-dimensional Laplacian operator.

NORMAL MODE ANALYSIS

We shall now analyze an arbitrary perturbation into a complete set of normal modes and then examine the stability of each of those modes individually.

For the system of equations (1.35), (1.32)-(1.33) the analysis can be made in terms of two dimensional periodic wave numbers. Thus, assuming that the perturbed quantities are of the form

$$\left[w,T,\varphi\right]=\left[W(z),\Theta(z),\Phi(z)\right]\exp\left(ik_x x+ik_y y+nt\right), \tag{1.36}$$

where k_x, k_y are wave numbers in x and y direction and n is growth rate of disturbances.

Using equation (1.36), equations (1.35), (1.32) and (1.33) become

$$\left(D^2-a^2\right)\left(D^2-a^2-\frac{n\left(1+nF\right)}{Pr}\right)W-\left(1+nF\right)\left(a^2Ra\Theta-a^2Rn\Phi\right)=0, \tag{1.37}$$

$$N_A W-\frac{N_A}{Le}\left(D^2-a^2\right)\Theta-\left(\frac{1}{Le}\left(D^2-a^2\right)-n\right)\Phi=0, \tag{1.38}$$

$$W+\left(D^2-a^2-n-\frac{N_A N_B}{Le}D\right)\Theta-\frac{N_B}{Le}D\Phi=0, \tag{1.39}$$

where $D\equiv\frac{d}{dz}$ and $a=\sqrt{k_x^2+k_y^2}$ is the dimensionless resultant wave number.

The boundary conditions of the problem in view of normal mode analysis are

$$W=0,\ D^2W=0,\Theta=0,\ D\Phi+N_A D\Theta=0\ \text{ at }\quad z=0,1. \tag{1.40}$$

METHOD OF SOLUTION

The Galerkin weighted residuals method is used to obtain an approximate solution to the system of equations (1.37) – (1.39) with the boundary conditions (1.40). In this method, the test functions are the same as the base (trial) functions. Accordingly W, Θ and Φ are taken as

$$\mathrm{W} = \sum_{p=1}^{N} A_p \mathrm{W}_p,\ \Theta = \sum_{p=1}^{N} B_p \Theta_p, \Phi = \sum_{p=1}^{N} C_p \Phi_p, \tag{1.41}$$

where $\mathrm{W}_p = \Theta_p = \sin p\pi z, \Phi_p = -N_A \sin p\pi z$, A_p, B_p and C_p are unknown coefficients, p = 1, 2, 3,..., N and the base functions W_p, Θ_p, and Φ_p satisfying the boundary conditions (1.40). Using expression for W, Θ and Φ in equations (1.37) – (1.39) and multiplying the first equation by W_p the second equation by Θ_p and third equation by Φ_p and then integrating in the limits from zero to unity, we obtain a set of 3N linear homogeneous equations with 3N unknown A_p, B_p and C_p; p = 1, 2, 3,..., N. For existing of nontrivial solution, the vanishing of the determinant of coefficients produces the characteristics equation of the system in term of Rayleigh number Ra.

LINEAR STABILITY ANALYSIS

For the present formulation, we have considered the which system of equations (1.37) - (1.39) together with the boundary conditions (1.40) constitute a linear eigenvalue problem with variable coefficient for the growth rate of disturbance n of the system. Substituting equation (1.41) into the system of equations (1.37) -(1.39) and multiplying the first equation by W_p the second equation by Θ_p and third equation by Φ_p and then integrating in the limits from zero to unity and performing some integration by parts, one obtains the following matrix equation

$$\begin{bmatrix} \dfrac{\left(\pi^2+a^2\right)^2}{\left(1+nF\right)} + \dfrac{n}{\Pr}\left(\pi^2+a^2\right) & -a^2\mathrm{Ra} & -a^2\mathrm{N_A Rn} \\ 1 & -\left(\pi^2+a^2+n\right) & 0 \\ 1 & \dfrac{1}{\mathrm{Le}}\left(\pi^2+a^2\right) & -\left(\dfrac{1}{\mathrm{Le}}\left(\pi^2+a^2\right)+\mathrm{n}\right) \end{bmatrix} \begin{bmatrix} W_0 \\ \Theta_0 \\ \Phi_0 \end{bmatrix} = \begin{bmatrix} 0 \\ 0 \\ 0 \end{bmatrix}. \tag{1.42}$$

The non-trivial solution of the above matrix requires that

$$\mathrm{Ra} = \frac{\left(\pi^2 + a^2\right)}{\mathrm{a}^2\left(1 + nF\right)}\left(\left(\pi^2 + a^2\right) + \frac{\mathrm{n}\left(1 + nF\right)}{\mathrm{Pr}}\right)\left(\pi^2 + a^2 + n\right) - \frac{\left(\pi^2 + a^2\right) + Le\left(\pi^2 + a^2 + n\right)}{\left(\pi^2 + a^2\right) + \mathrm{nLe}}\mathrm{N_A Rn}. \tag{1.43}$$

The growth rate n is in general a complex quantity such that $\mathrm{n} = \omega_r + i\omega$, the system with $\omega_r < 0$ is always stable, while for $\omega_r > 0$ it will become unstable. For neutral stability, the real part of n is zero. Hence, we now write $\mathrm{n} = i\omega$, (where ω is real and is a dimensionless frequency) in equation (1.43), we have

$$\mathrm{Ra} = \Delta_1 + i\omega\Delta_2, \tag{1.44}$$

where

$$\Delta_1 = \frac{\left(\pi^2 + a^2\right)}{\mathrm{a}^2}\left(\frac{\left(\pi^2 + a^2\right)^2 + \omega^2 F\left(\pi^2 + a^2\right)}{1 + \omega^2 F^2} - \frac{\omega^2}{\mathrm{Pr}}\right) - \frac{\left(\pi^2 + a^2\right)^2\left(Le + 1\right) + Le\left(\left(\pi^2 + a^2\right)\right) + \omega^2 Le^2}{\left(\pi^2 + a^2\right)^2 + \omega^2 Le^2}\mathrm{N_A Rn} \tag{1.45}$$

and

$$\Delta_2 = \frac{\left(\pi^2 + a^2\right)^2}{\mathrm{a}^2}\left(\frac{1 - F\left(\pi^2 + a^2\right)}{1 + \omega^2 F^2} + \frac{1}{\mathrm{Pr}}\right) + \frac{Le^2\left(\pi^2 + a^2\right)}{\left(\pi^2 + a^2\right)^2 + \omega^2 Le^2}\mathrm{N_A Rn}. \tag{1.46}$$

Since Ra is a physical quantity, so it must be real. Hence, it follows from the equation (1.44) that either $\omega = 0$ (exchange of stability, steady state) or $\Delta_2 = 0$ ($\omega \neq 0$ overstability or oscillatory onset).

STATIONARY CONVECTION

For the case of stationary (non-oscillatory) convection $\mathrm{n} = \omega = 0$, thus equation (1.43) reduces to

$$\left(\mathrm{Ra}\right)_s = \frac{\left(\pi^2 + a^2\right)^3}{\mathrm{a}^2} - \left(1 + Le\right)\mathrm{N_A Rn}. \tag{1.47}$$

It is observed that stationary Rayleigh number is function of the Lewis number, the modified diffusivity ratio and the nanoparticles Rayleigh but independent of visco- elastic parameter, Prandtl number and modified particle-density increment. Thus Maxwell visco-elastic nanofluid behaves like an ordinary Newtonian nanofluid and instability is purely a phenomenon due to buoyancy coupled with the conservation of nanoparticles.

To find the critical value of $\left(\mathrm{Ra}\right)_s$, equation (1.47) is differentiated with respect to ‘a^2’ and then equated to zero. The minimum of first term of right-hand side of equation (1.47) is attained at $\mathrm{a_c} = \frac{\pi}{\sqrt{2}}$ and minimum value found to $\frac{27\pi^4}{4}$ so the corresponding critical Rayleigh number given by

$$\left(\mathrm{Ra}\right)_c = \frac{27\pi^4}{4} - \left(1 + Le\right)\mathrm{N_A Rn}. \tag{1.48}$$

In the absence of nanoparticles ($\mathrm{Rn} = \mathrm{Le} = \mathrm{N_A} = 0$), one recovers the well-known results that the critical Rayleigh number is equal to $\left(\mathrm{Ra}\right)_c = \frac{27\pi^4}{4}$.

This is good agreement of the result obtained by Nield and Kuznetsov (2010a).

Thus presence of the nanoparticles lowers the value of the critical Rayleigh number by usually by substantial amount. Also parameter $\mathrm{N_B}$ does not appear in the equation (1.48); thus instability is purely phenomena due to buoyancy coupled with conservation of nanoparticles. Thus average contribution of nanoparticles flux in the thermal energy equation is zero with one-term Galerkin approximation.

OSCILLATORY CONVECTION

For oscillatory convection $\Delta_2 = 0$ and $\omega \neq 0$, thus equation (1.46) gives a dispersion relation of the form

$$\mathrm{a_2}\left(\omega^2\right)^2 + \mathrm{a_1}\left(\omega^2\right) + \mathrm{a_0} = 0; \tag{1.49}$$

where

$$\mathrm{a}_2 = \frac{Le^2 F^2}{\mathrm{Pr}},$$

$$\mathrm{a}_1 = \frac{a^2 Le^2 F^2}{\pi^2 + a^2} N_A Rn + Le^2\left(1 - F\left(\pi^2 + a^2\right)\right) + F\left(\pi^2 + a^2\right) + Le^2,$$

$$\mathrm{a}_0 = \frac{a^2 Le^2}{\pi^2 + a^2} N_A Rn + \left(\pi^2 + a^2\right)\left(1 - F\left(\pi^2 + a^2\right)\right) + \frac{\left(\pi^2 + a^2\right)}{\mathrm{Pr}}.$$

Then equation (1.44) with $\Delta_2 = 0$ gives oscillatory Rayleigh number at the margin of stability as

$$\left(\mathrm{Ra}\right)_{\mathrm{osc}} = \frac{\left(\pi^2 + a^2\right)^2}{\mathrm{a}^2}\left(\frac{\left(\pi^2 + a^2\right) + \omega^2 F}{1 + \omega^2 F^2} - \frac{\omega^2}{\mathrm{Pr}}\right) - \frac{\left(\pi^2 + a^2\right)^2\left(Le + 1\right) + Le\left(\pi^2 + a^2\right) + \omega^2 Le^2}{\left(\pi^2 + a^2\right)^2 + \omega^2 Le^2} \mathrm{N_A Rn} \quad (1.50)$$

For the oscillatory convection to occur, ω^2 must be positive. If there are no positive roots of ω^2 in equation (1.49), then oscillatory convection is not possible. If there are positive roots of ω^2, the critical Rayleigh number for oscillatory convection can be obtained numerically minimizing equation (1.50) with respect to wave number, after substituting various values of physical parameters for ω^2 of equation (1.49) to determine the various effect of different parameter on the onset of oscillatory convection.

RESULTS AND DISCUSSION

To study the effect of Lewis number Le, modified diffusivity ratio N_A and nanoparticles Rayleigh number Rn on stationary convection, we examine the behavior of $\frac{\partial(\mathrm{Ra})_s}{\partial Le}$, $\frac{\partial(\mathrm{Ra})_s}{\partial N_A}$ and $\frac{\partial(\mathrm{Ra})_s}{\partial Rn}$ analytically.

From equation (1.47), we have

(i) $\dfrac{\left(\partial \mathrm{R}a\right)_s}{\partial Le} < 0,$

(ii) $\dfrac{\left(\partial \mathrm{R}a\right)_s}{\partial N_A} < 0,$

(iii) $\dfrac{\left(\partial \mathrm{R}a\right)_s}{\partial Rn} < 0.$

From these inequalities it is observed that Lewis number Le, modified diffusivity ratio N_A and nanoparticles Rayleigh number Rn have destabilizing effect on the stationary convection.

Expression for stationary Rayleigh number, which characterizes the stability of the system, is found to be depend upon Lewis number, modified diffusivity ratio and nanoparticles Rayleigh number. The computations are carried out for different values of parameters considered in the range $10^2 \le Ra \le 10^5$ (thermal Rayleigh number), $1 \le N_A \le 10$ (modified diffusivity ratio), $10^2 \le Le \le 10^4$ (Lewis number), $10^{-1} \le Rn \le 10^1$ (nanoparticles Rayleigh number).

The variations of the stationary thermal Rayleigh number with wave number have been plotted graphically as shown in Figures 2 - 4.

Figure 2 shows the variation of thermal Rayleigh number for different value of Lewis number Le and for the fixed value of other parameters. It is found that stationary Rayliegh number decreases as the values of Lewis number increases, indicating that Lewis number destabilizes the stationary convection. It is due to the fact that thermophoresis at a higher value of thermophoretic diffusivity is more supportable to the disturbance in nanofluids, while both thermophoresis and Brownian motion are driving forces in favor of the motion of nanoparticles. This is good agrement of the result obtained by Chand and Rana (2014d).

Figure 3 shows the variation of stationary Rayleigh number for different value of the modified diffusivity ratio N_A and fixed value of other parameters and it is found that Rayleigh number decreases with an increase in the value of the modified diffusivity ratio N_A, which means that the modified diffusivity ratio N_A destabilizes on the fluid layer. This may lead to an increase in volumetric fraction, which shows that Brownian motion of the nanoparticles will also increase, which may cause destabilizing effect. This is good agrement of the result obtained by Chand and Rana (2014d).

Figure 4 shows the variation of stationary Rayleigh number with wave number for different value of the nanoparticles Rayleigh number Rn and

Figure 2. Variation of the stationary Rayleigh number with wave number for different value of Lewis number

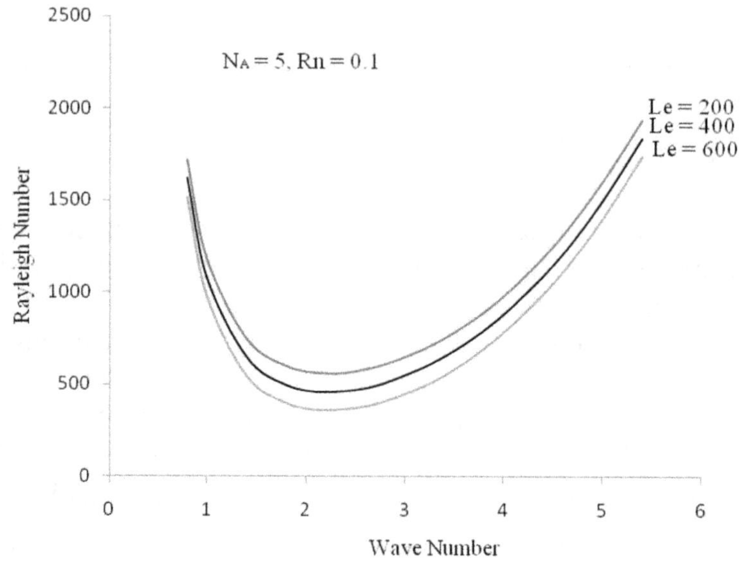

Figure 3. Variation of the stationary Rayleigh number with wave number for different value of modified diffusivity ratio

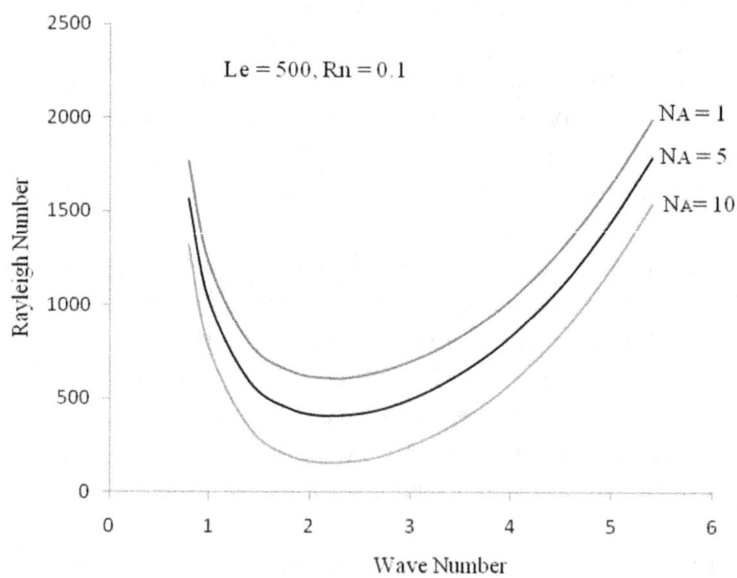

Figure 4. Variation of the stationary Rayleigh number with wave number for different value of concentration Rayleigh number

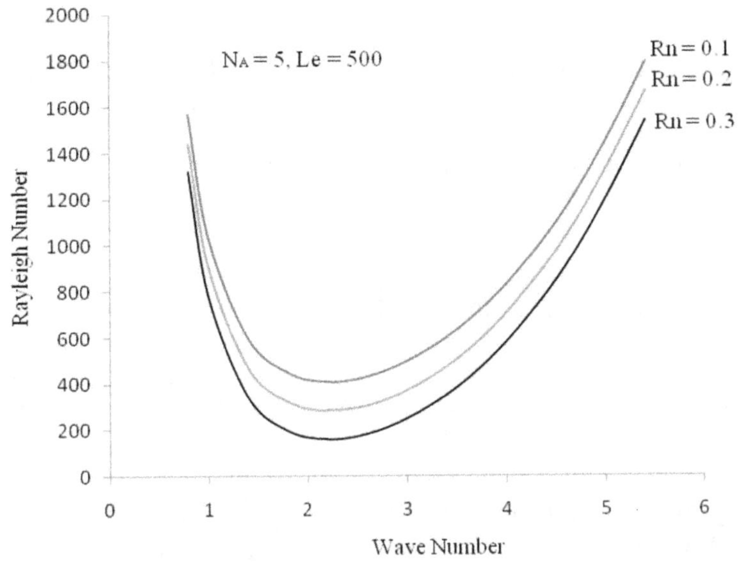

fixed value of other parameters and it is found that thermal Rayleigh number decreases with an increase in the value of the nanoparticles Rayleigh number Rn, which means that the nanoparticles Rayleigh number Rn has destabilizing effect on fluid layer. The destabilizing effect because the heavier nanoparticles moving through the base fluid makes more strong disturbances as compared with the lighter nanoparticles. This is the good agreement of the result obtained by Chand and Rana (2014d).

PRINCIPLE OF EXCHANGE OF STABILITY

Multiplying equation (1.37) by W* (the complex conjugate of W) and integrating over the range of z, i.e. $0 \leq z \leq 1$, we have

$$\int_0^1 W^*\left(D^2-a^2\right)^2 W dz-\frac{n\left(1+nF\right)}{\Pr}\int_0^1 W^*\left(D^2-a^2\right)\mathrm{W}dz\text{-}a^2\mathrm{Ra}\left(1+nF\right)\int_0^1 W^*\Theta dz$$
$$+a^2\mathrm{Rn}\left(1+nF\right)\int_0^1 W^*\Phi dz=0. \tag{1.51}$$

Now integrating equation (1.51) a suitable number of times and using the boundary conditions (1.40), we get

$$\int_0^1 \left(\left|D^2W\right|^2 + a^4\left|W\right|^2 + 2a^2\left|DW\right|^2\right)dz + \frac{n\left(1+nF\right)}{\Pr}\int_0^1 \left(\left|DW\right|^2 + a^2\left|W\right|^2\right)dz$$
$$-a^2\mathrm{Ra}\left(1+nF\right)\int_0^1 W^*\Theta\, dz + a^2\mathrm{Rn}\left(1+nF\right)\int_0^1 W^*\Phi\, dz = 0. \tag{1.52}$$

Taking the complex conjugate of the equation (1.39), we have

$$n^*\Theta^* - W^* = \left(D^2 - a^2\right)\Theta^* + \frac{N_B}{Le}\left(N_A D\Theta^* - D\Phi^*\right) - \frac{2N_A N_B}{Le} D\Theta^*. \tag{1.53}$$

Multiplying the equation (1.53) by Θ and integrating the over range of z, a suitable number of times and using boundary conditions (1.40), we get

$$n^*\int_0^1 \left|\Theta\right|^2 dz - \int_0^1 W^*\Theta\, dz = -\frac{1}{Le}\int_0^1 \left(\left|D\Theta\right|^2 + a^2\left|\Theta\right|^2\right)dz +$$
$$\frac{N_B}{Le}\int_0^1 \left(N_A\Theta D\Theta^* - \Theta D\Phi^*\right)dz - \frac{2N_A N_B}{Le}\int_0^1 \Theta D\Theta^* dz. \tag{1.54}$$

Taking the complex conjugate of the equation (1.38), we have

$$n^*\Phi^* + N_A W^* = \frac{1}{Le}\left(D^2 - a^2\right)\Phi^* + \frac{N_A}{Le}\left(D^2 - a^2\right)\Theta^*. \tag{1.55}$$

Multiplying equation (1.55) by Φ and integrating the range of z, a suitable number of times and using boundary conditions (1.40), we get

$$n^*\int_0^1 \left|\Phi\right|^2 dz + N_A\int_0^1 W^*\Phi\, dz = -\frac{1}{Le}\int_0^1 \left(\left|D\Phi\right|^2 + a^2\left|\Phi\right|^2\right)dz$$
$$-\frac{N_A}{Le}\int_0^1 \left(D\Phi D\Theta^* + a^2\Phi\Theta^*\right)dz. \tag{1.56}$$

From equations (1.54) and (1.56), we have

$$\begin{aligned}\int_0^1 W^*\Theta dz &= n^*\int_0^1 |\Theta|^2 dz + \frac{1}{Le}\int_0^1 \left(|D\Theta|^2 + a^2|\Theta|^2\right)dz \\ &+\frac{N_A N_B}{Le}\int_0^1 \Theta D\Theta^* dz + \frac{N_B}{Le}\int_0^1 \Theta D\Phi^* dz\end{aligned}, \tag{1.57}$$

$$\begin{aligned}N_A\int_0^1 W^*\Phi dz &= -n^*\int_0^1 |\Phi|^2 dz - \frac{1}{Le}\int_0^1 \left(|D\Phi|^2 + a^2|\Phi|^2\right)dz \\ &-\frac{N_A}{Le}\int_0^1 \left(D\Phi D\Theta^* + a^2\Phi\Theta^*\right)dz\end{aligned}. \tag{1.58}$$

Using equations (1.57) and (1.58), the equation (1.52) reduces to the following form

$$\begin{aligned}&\int_0^1 \left(|D^2W|^2 + a^4|W|^2 + 2a^2|DW|^2\right)dz + \frac{n(1+nF)}{\Pr}\int_0^1 \left(|DW|^2 + a^2|W|^2\right)dz \\ &-a^2\mathrm{Ra}(1+nF)\left(n^*\int_0^1 |\Theta|^2 dz + \frac{1}{Le}\int_0^1 \left(|D\Theta|^2 + a^2|\Theta|^2\right)dz + \frac{N_A N_B}{Le}\int_0^1 \Theta D\Theta^* dz + \frac{N_B}{Le}\int_0^1 \Theta D\Phi^* dz\right) \\ &-a^2\mathrm{Rn}(1+nF)\left(\frac{n^*}{N_A}\int_0^1 |\Phi|^2 dz + \frac{1}{N_A Le}\int_0^1 \left(|D\Phi|^2 + a^2|\Phi|^2\right)dz + \frac{1}{Le}\int_0^1 \left(D\Phi D\Theta^* + a^2\Phi\Theta^*\right)dz\right) = 0.\end{aligned} \tag{1.59}$$

Now putting $n = n_r + in_i$, in equation (1.59) and equating the imaginary parts of resulting equation, we get

$$\begin{aligned}&\frac{n_i}{\Pr}\int_0^1 \left(|DW|^2 + a^2|W|^2\right)dz + n_i a^2\mathrm{Ra}\left(\int_0^1 |\Theta|^2 dz\right) - \frac{a^2\mathrm{Ra}N_A N_B}{Le}\operatorname{Im}g\left(\int_0^1 \Theta D\Theta^* dz\right) \\ &-\frac{N_B a^2\mathrm{Ra}}{Le}\operatorname{Im}g\left(\int_0^1 \Theta D\Phi^* dz\right) + a^2 n_i\frac{\mathrm{Rn}}{N_A}\left(\int_0^1 |\Phi|^2 dz\right) - \frac{a^2 Rn}{Le}\operatorname{Im}g\left(\int_0^1 \left(D\Phi D\Theta^* + a^2\Phi\Theta^*\right)dz\right) = 0\end{aligned}. \tag{1.60}$$

But $\operatorname{Im}g\left(\int_0^1 \Theta D\Theta^* dz\right) \le \left|\int_0^1 \Theta D\Theta^* dz\right|,$

$$\le \int_0^1 |\Theta||D\Theta^*| dz,$$

$$\leq \int_0^1 |\Theta||D\Theta| dz,$$

Thus $\text{Img}\left(\int_0^1 \Theta D\Theta^* \, dz \right) \leq \frac{1}{2}\left[\int_0^1 |\Theta|^2 dz + \int_0^1 |D\Theta|^2 dz \right].$ (1.61)

Similarly $\text{Img}\left(\int_0^1 \Theta D\Phi^* \, dz \right) \leq \frac{1}{2}\left[\int_0^1 |\Theta|^2 dz + \int_0^1 |D\Phi|^2 dz \right],$ (1.62)

and

Img

$$\left(\int_0^1 \left(D\Phi D\Theta^* + a^2 \Phi\Theta^* \right) dz \right) \leq \frac{1}{2}\left[\int_0^1 |D\Phi|^2 dz + \int_0^1 |D\Theta|^2 dz + a^2 \int_0^1 |\Phi|^2 dz + a^2 \int_0^1 |\Theta|^2 dz \right]. \quad (1.63)$$

Using inequalities (1.61) - (1.63) in equation (1.60), we get

$$\frac{n_i}{\Pr}\int_0^1 \left(|DW|^2 + a^2 |W|^2 \right) dz + \left(n_i a^2 \text{Ra-} \frac{a^2 \text{RaN}_A N_B}{2Le} - \frac{\text{N}_B a^2 Ra}{2Le} - \frac{a^4 Rn}{2Le} \right)\left(\int_0^1 |\Theta|^2 dz \right)$$
$$+ a^2 Rn \left(\frac{\text{n}_i}{N_A} - \frac{a^2}{2Le} \right)\left(\int_0^1 |\Phi|^2 dz \right)$$
$$-\left(\frac{a^2 \text{RaN}_A N_B}{2Le} + \frac{\text{N}_B a^2 \text{Ra}}{2Le} + \frac{a^2 Rn}{2Le} \right)\left(\int_0^1 D\Theta dz \right) - \left(\frac{N_B a^2 Ra}{2Le} + \frac{a^2 Rn}{2Le} \right)\left(\int_0^1 D\Phi dz \right) \leq 0. \quad (1.64)$$

Since W (0) = W (1) = 0, and using Rayleigh-Ritz inequality, we have

$$\int_0^1 |W|^2 dz \leq \frac{1}{\pi^2}\int_0^1 |DW|^2 dz . \quad (1.65)$$

Also $\Phi(0) = 0 = \Phi(1)$ and $\Theta(0) = 0 = \Theta(1)$, and using Rayleigh-Ritz inequality, we have

$$\int_0^1 |\Phi|^2 dz \le \frac{1}{\pi^2}\int_0^1 |D\Phi|^2 dz, \tag{1.66}$$

$$\int_0^1 |\Theta|^2 dz \le \frac{1}{\pi^2}\int_0^1 |D\Theta|^2 dz. \tag{1.67}$$

Using inequalities (1.66) - (1.67) in the inequality (1.64), we have

$$\frac{n_i}{\Pr}\left(\pi^2+a^2\right)\int_0^1 |W|^2 dz$$
$$+\left(\left(n_i a^2\mathrm{Ra}-\frac{a^2\mathrm{RaN_A}N_B}{2Le}-\frac{\mathrm{N_B}a^2Ra}{2Le}-\frac{a^4Rn}{2Le}\right)-\pi^2\left(\frac{a^2\mathrm{RaN_A}N_B}{2Le}+\frac{\mathrm{N_B}\,a^2\mathrm{Ra}}{2Le}+\frac{a^2Rn}{2Le}\right)\right)\left(\int_0^1 |\Theta|^2 dz\right)$$
$$+\left(a^2Rn\left(\frac{\mathrm{n_i}}{N_A}-\frac{a^2}{2Le}\right)-\pi^2\left(\frac{N_Ba^2Ra}{2Le}+\frac{a^2Rn}{2Le}\right)\right)\left(\int_0^1 |\Phi|^2 dz\right)\le 0.$$

(1.68)

For the validity of "PES" put $n_i = 0$, (whatever may be the value of n_r i.e. positive, negative or zero) in (1.68), we get

$$\left(\left(-\frac{a^2\mathrm{RaN_A}N_B}{2Le}-\frac{\mathrm{N_B}a^2Ra}{2Le}-\frac{a^4Rn}{2Le}\right)-\pi^2\left(\frac{a^2\mathrm{RaN_A}N_B}{2Le}+\frac{\mathrm{N_B}\,a^2\mathrm{Ra}}{2Le}+\frac{a^2Rn}{2Le}\right)\right)\left(\int_0^1 |\Theta|^2 dz\right)$$
$$+\left(-\frac{a^4Rn}{2Le}-\pi^2\left(\frac{N_Ba^2Ra}{2Le}+\frac{a^2Rn}{2Le}\right)\right)\left(\int_0^1 |\Phi|^2 dz\right)\le 0.$$

(1.69)

The inequality (1.69) holds and hence "PES" is valid for the problem.

CASE OF OVERSTABILITY

Here we examine the possibility of as to whether instability may occur as overstability. Since we wish to determine the Rayleigh number for the onset of instability via a state of pure oscillation, it suffices to find the conditions for which equation (1.49) will admit the solution with real values of ω. All the coefficients a_0, a_1, a_2 in equation (1.49) are real.

Now the product of the roots of equation (1.49) $=\left(\frac{a_0}{a_2}\right)$ is positive.

a_2 is always positive and a_0 is negative if

$$\frac{a^2 Le^2}{\left(\pi^2+a^2\right)^2} N_A Rn + 1 + \frac{1}{\Pr} < F\left(\pi^2+a^2\right). \tag{1.70}$$

Thus inequality $\frac{a^2 Le^2}{\left(\pi^2+a^2\right)^2} N_A Rn + 1 + \frac{1}{\Pr} < F\left(\pi^2+a^2\right)$ is sufficient condition for the non-existence of overstability, the violation of which does not necessarily imply the occurrence of overstability.

CONCLUSION

Thermal instability in a horizontal layer of Maxwellian visco-elastic nanofluid is investigated. The flux of volume fraction of nanoparticles is taken to be zero on the isothermal boundaries and the eigen value problem is solved using the Galerkin residual method. The results have been presented both analytically and graphically.

The main conclusions derived from the present chapter are as follows

1. The instability purely phenomenon due to buoyancy coupled with the conservation of nanoparticle and is independent of the contribution of Brownian motion and thermophoresis.
2. For stationary convection Maxwellian visco-elastic nanofluid behaves like an ordinary Newtonian nanofluid.
3. Lewis number, modified diffusivity ratio and nanoparticles Rayleigh number destabilize the stationary convection.
4. Principle of Exchange of Stabilities is valid for the problem.
5. Sufficient condition for the non-existence of overstability is

$$\frac{a^2 Le^2}{\left(\pi^2+a^2\right)^2} N_A Rn + 1 + \frac{1}{\Pr} < F\left(\pi^2+a^2\right).$$

REFERENCES

Bhatia, P. K., & Steiner, J. M. (1972). Convective instability in a rotating viscoelastic fluid. *Zeitschrift für Angewandte Mathematik und Mechanik, 52*(6), 321–330. doi:10.1002/zamm.19720520601

Buongiorno, J. (2006). Convective transport in nanofluids. *ASME Journal of Heat Transfer, 128*(3), 240–250. doi:10.1115/1.2150834

Chand, R., & Rana, G. C. (2012a). Oscillating convection of nanofluid in porous medium. *Transport in Porous Media, 95*(2), 269–284. doi:10.1007/s11242-012-0042-9

Chand, R., & Rana, G. C. (2012b). On the onset of thermal convection in rotating nanofluid layer saturating a Darcy-Brinkman porous medium. *International Journal of Heat and Mass Transfer, 55*(21-22), 5417–5424. doi:10.1016/j.ijheatmasstransfer.2012.04.043

Chand, R., & Rana, G. C. (2012c). Thermal instability of Rivlin-Ericksen elastico-viscous nanofluid saturated by a porous medium. *Journal of Fluids Engineering, 134*(12), 121203. doi:10.1115/1.4007901

Chand, R., & Rana, G. C. (2014a). Hall effect on the thermal instability in a horizontal layer of nanofluid. *Journal of Nanofluids, 3*(3), 247–253. doi:10.1166/jon.2014.1100

Chand, R., & Rana, G. C. (2014d). Thermal instability in a Brinkman porous medium saturated by nanofluid with no nanoparticle flux on boundaries. *Special Topics & Reviews in Porous Media: An International Journal, 5*(4), 277–286. doi:10.1615/SpecialTopicsRevPorousMedia.v5.i4.10

Chand, R., & Rana, G. C. (2015d). Thermal instability in a horizontal layer of Walter's (Model B') visco-elastic nanofluid-a more realistic approach. *Applications and Applied Mathematics: An International Journal, 10*(2), 1027–1042.

Chandrasekhar, S. (1961). *Hydrodynamic and Hydromagnetic Stability*. New York: Dover Publication.

Choi, S. (1995). Enhancing thermal conductivity of fluids with nanoparticles. In D. A. Siginer & H. P. Wang (Eds.), *Developments and applications of non-Newtonian flows* (Vol. 231, pp. 99–105). MD: ASME FED.

Green, T. (1968). Oscillating convection in an elasticoviscous liquid. *Physics of Fluids, 11*(7), 1410–1412. doi:10.1063/1.1692123

Kuznetsov, A. V., & Nield, D. A. (2010a). Effect of local thermal non-equilibrium on the onset of convection in a porous medium layer saturated by a nanofluid. *Transport in Porous Media, 83*(2), 425–436. doi:10.1007/s11242-009-9452-8

Kuznetsov, A. V., & Nield, D. A. (2010b). Thermal instability in a porous medium layer saturated by a nanofluid: Brinkman Model. *Transport in Porous Media, 81*(3), 409–422. doi:10.1007/s11242-009-9413-2

Kuznetsov, A. V., & Nield, D. A. (2010c). The onset of double-diffusive nanofluid convection in a layer of a saturated porous medium. *Transport in Porous Media, 85*(3), 941–951. doi:10.1007/s11242-010-9600-1

Nield, D. A., & Kuznetsov, A. V. (2010a). The onset of convection in a horizontal nanofluid layer of finite depth. *European Journal of Mechanics - B/ Fluids, 29*(3), 217–223. doi:10.1016/j.euromechflu.2010.02.003

Nield, D. A., & Kuznetsov, A. V. (2010b). The onset of convection in a layer of cellular porous material: Effect of temperature-dependent conductivity arising from radiative transfer. *Journal of Heat Transfer, 132*(7), 074503. doi:10.1115/1.4001125

Nield, D. A., & Kuznetsov, A. V. (2011a). The onset of double-diffusive convection in a nanofluid layer. *International Journal of Heat and Fluid Flow, 32*(4), 771–776. doi:10.1016/j.ijheatfluidflow.2011.03.010

Nield, D. A., & Kuznetsov, A. V. (2014). Thermal instability in a porous medium layer saturated by a nanofluid: A revised model, Int. *J. Heat and Mass Transf., 68*, 211–214. doi:10.1016/j.ijheatmasstransfer.2013.09.026

Rana, G. C., Thakur, R. C., & Kumar, S. (2012). Thermosolutal convection in compressible Walters (model B) fluid permeated with suspended particles in a Brinkman porous medium. *The Journal of Computational Multiphase Flows, 4*(2), 211–224. doi:10.1260/1757-482X.4.2.211

Vest, C. M., & Arpaci, V. (1969). Overstability of visco-elastic fluid layer heated from below. *Fluid Mech., 36*(03), 613–623. doi:10.1017/S0022112069001881

Chapter 2
Effect of Rotation on the Onset of Thermal Convection in a Layer of Maxwellian Visco-Elastic Nanofluid

INTRODUCTION

The effect of rotation on the thermal instability is important in certain chemical engineering and biochemical engineering. Thermal instability in rotating non-Newtonian fluid has considerable interest due to its wide range of applications in engineering, including rotating machineries such as nuclear reactors, petroleum industry, biochemical and geophysical problems. To attain the improved performance of such applications the use of nanofluids with higher thermal conductivities can be considered as a working medium. Thermal instability in rotating fluids about a vertical axis combines the element of thermal buoyancy and rotation induced Coriolis and centrifugal forces. Due to the Coriolis force on the thermal instability problem another parameter namely Taylor number is introduced in this problem. Taylor number is a non-dimensional number which is a measure of rotation rate. It is apparent that thermal instability in rotating Maxwellian nanofluids will play an important role in many physical phenomenon concerning with geophysics, astrophysics and oceanography.

DOI: 10.4018/978-1-68318-006-7.ch002

Thermal instability problem for a regular fluid with rotation or without rotation was first discussed by Chandrasekhar (1961). Thermal instability problem in rotating micro polar fluids have been studied by Qin & Kaloni (1992) and found that rotation has a stabilizing effect for low values of Taylor number. Thermal instability problems for nanofluid with rotation were studied by Yadav et al. (2011, 2013b), Chand and Rana (2012b), Chand (2013a), Rana et al. (2014c), Chand et al. (2015d), Rana and Chand (2015b), Rana and Agarwal (2015).

In the present chapter we have extend our study to find the effect of rotation on the thermal instability of Maxwellian visco-elastic nanofluid layer for more realistic boundary conditions. Stability is discussed analytically as well as numerically using Galerkin-type weighted residuals method. It has been observed that rotation has stabilizing effect on the fluid layer.

MATHEMATICAL FORMULATIONS OF THE PROBLEM

The Physical Problem

In this chapter we shall investigate the effect of rotation on the onset of thermal convection in a horizontal layer Maxwellian visco-elastic nanofluid. The physical configuration of the problem to be considered as:

An infinite horizontal layer of Maxwell visco-elastic nanofluid of thickness 'd' bounded by horizontal boundaries $z = 0$ and $z = d$. A Cartesian coordinate system (x, y, z) is chosen with the origin at the bottom of the fluid layer and the z- axis normal to the fluid layer. Fluid layer is rotating uniform about z-axis with angular velocity $\Omega(0, 0, \Omega)$ and is acted upon by gravity force $\mathbf{g}(0, 0, -g)$. Fluid layer is heated from below in such a way that horizontal boundaries $z = 0$ and $z = d$ respectively maintained at a uniform temperature T_0 and T_1 $(T_0 > T_1)$. The normal component of the nanoparticles flux has to vanish at an impermeable boundary and the temperature T is taken to be T_0 at $z = 0$ and T_1 at $z = d$, $(T_0 > T_1)$ as shown in Figure 1. The reference scale for temperature and nanoparticles fraction is taken to be T_1 and φ_0 respectively.

The following analysis is confined to a narrow and very long fluid layer. Since the significance of the centrifugal acceleration depends on the offset distance from the centre of rotation therefore for the layer which is adjacent to the rotation axis (i.e. $x = 0$, $y = 0$), the impact of the centrifugal acceleration to be zero. Due to the fact that here a narrow and very long fluid layer is considered, centrifugal effects can be neglected in the momentum equation.

Figure 1. Physical configuration of the problem

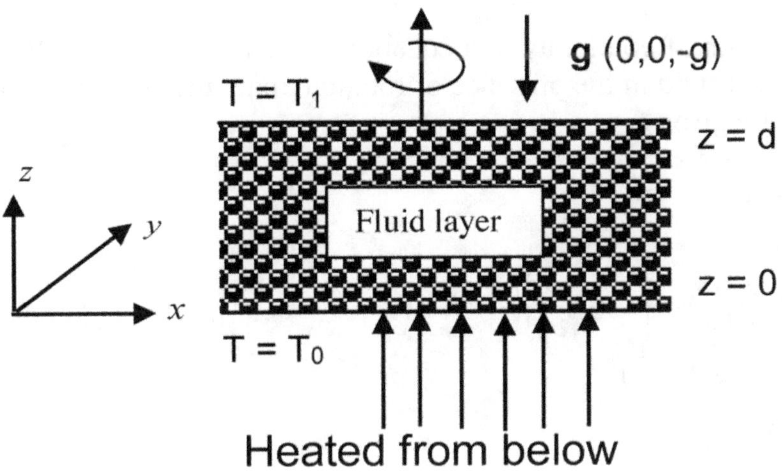

Assumptions

The mathematical equations describing the physical model are based upon the following assumptions:

1. Thermophysical properties of fluid expect for density in the buoyancy force (Boussinesq Hypothesis) are constant,
2. The fluid phase and nanoparticles are in thermal equilibrium state and thus, the heat flow has been described using one equation model,
3. Dilute mixture,
4. Nanoparticles are spherical,
5. Nanoparticles are non-magnetic,
6. No chemical reactions take place in fluid layer,
7. Negligible viscous dissipation,
8. Radiative heat transfer between the sides of wall is negligible when compared with other modes of the heat transfer,
9. Nanofluid is incompressible, electrically conducting, Newtonian and laminar flow,
10. Each boundary wall is assumed to be impermeable and perfectly thermal conducting,
11. Angular velocity of fluid layer is assumed to be constant.

Governing Equations

The equation of continuity and equation of motion for Maxwellian visco-elastic nanofluid in the presence of rotation under the Boussinesq approximation are given as

$$\nabla \cdot \mathbf{v} = 0, \tag{2.1}$$

$$\rho\left(1+\lambda\frac{\partial}{\partial t}\right)\frac{d\mathbf{v}}{dt} = \left(1+\lambda\frac{\partial}{\partial t}\right)\left(-\nabla p + \left(\varphi\rho_{\mathrm{p}} + (1-\varphi)\left\{\rho_{\mathrm{f}}\left(1-\alpha\left(\mathrm{T}-\mathrm{T}_0\right)\right)\right\}\right)\mathbf{g}\right), \\ +\mu\nabla^2\mathbf{v} + 2\rho\left(\mathbf{v}\times\Omega\right) \tag{2.2}$$

where $\frac{\mathrm{d}}{\mathrm{dt}} = \frac{\partial}{\partial \mathrm{t}} + (\mathbf{v}\cdot\nabla)$ represents the convection derivative; **v** is the velocity of fluid, p is the pressure, ρ_0 is the density of nanofluid at lower layer, ρ_p is the density of nanoparticles, λ is the relaxation time, φ is the volume fraction of the nanoparticles, T is the temperature, α is coefficient of the thermal expansion, **g** is acceleration due to gravity, **Ω** is angular velocity of fluid and μ is the viscosity.

Equation of energy for Maxwellian visco-elastic nanofluid is given by

$$\rho c\frac{\partial \mathrm{T}}{\partial \mathrm{t}} + \mathbf{v}\cdot\nabla\mathrm{T} = \mathrm{k}_{\mathrm{m}}\nabla^2\mathrm{T} + (\rho c)_{\mathrm{p}}\left(\mathrm{D}_{\mathrm{B}}\nabla\varphi\cdot\nabla\mathrm{T} + \frac{\mathrm{D}_{\mathrm{T}}}{\mathrm{T}_1}\nabla\mathrm{T}\cdot\nabla\mathrm{T}\right), \tag{2.3}$$

where ρc is heat capacity of fluid, $(\rho c)_p$ is heat capacity of nanoparticles, T_1 is the temperature of the fluid layer at z = d, and k_m is thermal conductivity.

Equation of continuity for the nanoparticles is given by

$$\frac{\partial\varphi}{\partial \mathrm{t}} + \mathbf{v}\cdot\nabla\varphi = \mathrm{D}_{\mathrm{B}}\nabla^2\varphi + \frac{\mathrm{D}_{\mathrm{T}}}{\mathrm{T}_1}\nabla^2\mathrm{T}, \tag{2.4}$$

where D_B is the Brownian diffusion coefficient, given by Einstein-Stokes equation and D_T is the thermoporetic diffusion coefficient of the nanoparticles.

We assume that the temperature is constant and nanoparticles flux is zero on the boundaries. Thus boundary conditions Chandrasekhar (1961), Nield and Kuznetsov (2014) are

$$w = 0,\ T = T_0,\quad D_B \frac{\partial \phi}{\partial z} + \frac{D_T}{T_1}\frac{\partial T}{\partial z} = 0 \ \text{ at } z = 0 \text{ and}$$

$$w = 0,\ T = T_1,\quad D_B \frac{\partial \phi}{\partial z} + \frac{D_T}{T_1}\frac{\partial T}{\partial z} = 0 \quad \text{at} \quad z = d. \tag{2.5}$$

Introducing non-dimensional variables as

$$(x',y',z') = \left(\frac{x,y,z}{d}\right),\ v'(u',v',w',) = V\left(\frac{u,v,w}{\kappa}\right)d,\ t' = \frac{t^{o}}{d^2},\ p' = \frac{pd^2}{\mu\kappa},$$

$$\varphi' = \frac{(\varphi - \varphi_0)}{\varphi_0},\ T' = \frac{(T - T_1)}{(T_0 - T_1)},$$

where $\kappa = \frac{k_m}{\rho c}$ is the thermal diffusivity of the fluid.

Equations (2.1) - (2.5) in non-dimensional form can be written as

$$\nabla' \cdot \mathbf{v}' = 0, \tag{2.6}$$

$$\left(1 + F\frac{\partial}{\partial t'}\right)\frac{1}{\Pr}\frac{\partial \mathbf{v}'}{\partial t'} = \left(1 + F\frac{\partial}{\partial t'}\right)\left(-\nabla' p' \text{-} Rm\hat{e}_z + RaT'\hat{e}_z - Rn\varphi'\hat{e}_z\right) + \nabla'^2\mathbf{v}' + \sqrt{Ta}\left(v'\hat{e}_x \text{-} u'\hat{e}_y\right), \tag{2.7}$$

$$\frac{\partial \varphi'}{\partial t'} + \mathbf{v}' \cdot \nabla'\varphi' = \frac{1}{Le}\nabla'^2\varphi' + \frac{N_A}{Le}\nabla'^2 T', \tag{2.8}$$

$$\frac{\partial T'}{\partial t'} + \mathbf{v}' \cdot \nabla' T' = \nabla'^2 T' + \frac{N_B}{Le}\nabla'\varphi' \cdot \nabla' T' + \frac{N_A N_B}{Le}\nabla' T' \cdot \nabla' T'. \tag{2.9}$$

Here the non-dimensional parameters are as follows

$\Pr = \frac{\mu}{\rho\kappa}$ is the Prandtl number,

$Le = \dfrac{\kappa}{D_B}$ is the Lewis number,

$F = \dfrac{\kappa\lambda}{d^2}$ is the stress relaxation parameter,

$\mathrm{Ra} = \dfrac{\rho g \alpha d^3 \left(T_0 - T_1\right)}{\mu\kappa}$ is the Rayleigh number,

$\mathrm{Rm} = \dfrac{\left(\rho_p \varphi_0 + \rho\left(1\text{-}\varphi_0\right)\right) g d^3}{\mu\kappa}$ is the density Rayleigh number,

$\mathrm{Rn} = \dfrac{\left(\rho_p - \rho\right)\varphi_0 g d^3}{\mu\kappa}$ is the nanoparticles Rayleigh number,

$\mathrm{Ta} = \left(\dfrac{2\Omega d^2}{\nu}\right)^2$ is the Taylor number,

$N_A = \dfrac{D_T\left(T_0\text{-}T_1\right)}{D_B T_1 \varphi_0}$ is the modified diffusivity ratio,

$N_B = \dfrac{\left(\rho c\right)_p \varphi_0}{\left(\rho c\right)_f}$ is the modified particle-density increment.

In spirit of Oberbeck-Boussinesq approximation, equation (2.7) has been linearized by the neglect of a term proportional to the product of φ_0 and T. This approximation is valid in the case of small temperature gradients in a dilute suspension of nanoparticles.

The dimensionless boundary conditions are

$$w' = 0, \quad T' = 1, \quad \frac{\partial \phi'}{\partial z'} + N_A \frac{\partial T'}{\partial z'} = 0 \ \text{at} \ \ z' = 0 \ \text{and}$$

$$w' = 0, \ T' = 0, \ \frac{\partial \phi'}{\partial z'} + N_A \frac{\partial T'}{\partial z'} = 0 \ \text{at} \ z' = 1. \tag{2.10}$$

THE BASIC STATE AND ITS SOLUTIONS

The basic state was assumed to be quiescent and is given by

$$v_i'\ \left(u',v',w'\right) = 0,$$

$$p' = p_b(z),$$

$$T' = T_b(z), \tag{2.11}$$

$$\varphi' = \varphi_b(z) \ \text{ and}$$

$$\rho = \rho_0\left(1+\alpha\left(T\text{-}T_0\right)\right).$$

Equations (2.6) – (2.9) reduce to

$$0 = -\frac{dp_b}{dz'}\text{-Rm} + \text{Ra}T_b - \text{Rn}\varphi_b, \tag{2.12}$$

$$\frac{d^2T_b}{dz'^2} + \frac{N_B}{Le}\frac{d\varphi_b}{dz'}\frac{dT_b}{dz'} + \frac{N_A N_B}{Le}\left(\frac{dT_b}{dz'}\right)^2 = 0, \tag{2.13}$$

$$\frac{d^2\varphi_b}{dz'^2} + N_A\frac{d^2T_b}{dz'^2} = 0, \tag{2.14}$$

Using boundary conditions in (2.10), equation (2.14) gives

$$\varphi_b = -N_A T_b + \left(1-N_A\right)z' + N_A. \tag{2.15}$$

On substituting the value of the φ_b from equation (2.15) in equation (2.13), we get

$$\frac{d^2T_b}{dz'} + \frac{\left(1\text{-}N_A\right)N_B}{Le}\frac{dT_b}{dz'} = 0. \tag{2.16}$$

On integrating equation (2.16) with respect to z' and using boundary conditions (2.10), we get

$$T_b = \frac{1 - e^{-(1-N_A)N_B(1-z)/Le}}{1 - e^{-(1-N_A)N_B/Le}}. \tag{2.17}$$

According to Buongiorno (2006), for most nanofluid investigated so far Lewis number Le is large, is of order 10^2-10^3, while N_A is no greater than about 10. Then, the exponents in equation (2.17) are small. By expanding the exponential function into the power series and retaining up to the first order is negligible. Thus a good approximation for the basic solution is given by

$$T_b = 1 - z'$$

and

$$\varphi_b = \phi_0 + N_A z'$$

PERTURBATION SOLUTIONS

Let the initial basic state described by (2.11) is slightly perturbed so that perturbed state is given by

$$(u', v', w') = 0 + (u'', v'', w''),$$

$$T' = T_b + T'', \tag{2.18}$$

$$\varphi' = \varphi_b + \varphi'',$$

$$p' = p_b + p'',$$

where $T_b = 1 - z'$, $\varphi_b = \phi_0 + N_A z'$ and (u'', v'', w''), T'', φ'' and p'' respectively the perturbations in the initial velocity, temperature, volume fraction of the nanoparticles and pressure.

By substituting (2.18) in equations (2.6) – (2.9) and linearize by neglecting the product of the prime quantities, we obtained following equations

$$\nabla \cdot \mathbf{v} = 0, \tag{2.19}$$

$$\left(1+F\frac{\partial}{\partial t}\right)\frac{1}{\Pr}\frac{\partial \mathbf{v}}{\partial t} = \left(1+F\frac{\partial}{\partial t}\right)\left(-\nabla p + \mathrm{Ra}T\hat{e}_z - \mathrm{Rn}\varphi\hat{e}_z\right) + \nabla^2\mathbf{v} + \sqrt{\mathrm{Ta}}\left(v\hat{e}_x\text{-}u\hat{e}_y\right), \tag{2.20}$$

$$\frac{\partial \varphi}{\partial t} + wN_A = \frac{1}{\mathrm{Le}}\nabla^2\varphi + \frac{N_A}{\mathrm{Le}}\nabla^2 T, \tag{2.21}$$

$$\frac{\partial T}{\partial t} - w = \nabla^2 T + \frac{N_B}{\mathrm{Le}}\left(N_A\frac{\partial T}{\partial z} - \frac{\partial \phi}{\partial z}\right) - \frac{2N_A N_B}{\mathrm{Le}}\frac{\partial T}{\partial z}. \tag{2.22}$$

Boundary conditions are

$$w = 0,\ T = 0,\quad \frac{\partial \phi}{\partial z} + N_A\frac{\partial T}{\partial z} = 0 \quad \text{at } z = 0\ ,1. \tag{2.23}$$

[Dashes (") have been suppressed for convenience]
Eliminating pressure term p from equation (2.20), we have

$$\left(1+F\frac{\partial}{\partial t}\right)\frac{1}{\Pr}\frac{\partial}{\partial t}\nabla^2 w - \nabla^4 w - \left(1+F\frac{\partial}{\partial t}\right)\left(\mathrm{Ra}\nabla_H^2 T\text{-}\mathrm{Rn}\nabla_H^2\varphi\right) + \sqrt{\mathrm{Ta}}\frac{\partial \xi}{\partial z} = 0, \tag{2.24}$$

where $\nabla_H^2 = \frac{\partial^2}{\partial x^2} + \frac{\partial^2}{\partial y^2}$ is the two-dimensional Laplacian operator and $\xi = \frac{\partial v}{\partial x} - \frac{\partial u}{\partial y}$ is the vorticity.

Also from equation (2.20), we have

$$\left(1+F\frac{\partial}{\partial t}\right)\left(\nabla^2 - \frac{1}{\Pr}\frac{\partial}{\partial t}\right)\xi = \text{-}\sqrt{Ta}\frac{\partial w}{\partial z}. \tag{2.25}$$

Now eliminating ξ from equations (2.24) and (2.25), we have

$$\begin{pmatrix} \left(1+F\dfrac{\partial}{\partial t}\right)^2\left(\nabla^2-\dfrac{1}{\Pr}\dfrac{\partial}{\partial t}\right)\dfrac{1}{\Pr}\dfrac{\partial}{\partial t}\nabla^2 \mathrm{w}-\left(1+F\dfrac{\partial}{\partial t}\right)\left(\nabla^2-\dfrac{1}{\Pr}\dfrac{\partial}{\partial t}\right)\nabla^4 \mathrm{w} \\ -\mathrm{Ra}\nabla_H^2\left(1+F\dfrac{\partial}{\partial t}\right)^2\left(\nabla^2-\dfrac{1}{\Pr}\dfrac{\partial}{\partial t}\right)\mathrm{T}+\mathrm{Rn}\nabla_H^2\left(1+F\dfrac{\partial}{\partial t}\right)^2\left(\nabla^2-\dfrac{1}{\Pr}\dfrac{\partial}{\partial t}\right)\varphi \end{pmatrix}-\mathrm{Ta}\frac{\partial^2 \mathrm{w}}{\partial z^2}=0. \tag{2.26}$$

NORMAL MODE ANALYSIS

We shall now analyze an arbitrary perturbation into a complete set of normal modes and then examine the stability of each of those modes individually. For the system of equations (2.19), (2.21), (2.22) and (2.26) the analysis can be made in terms of two dimensional wave numbers. Thus, assuming that the perturbed quantities are of the form

$$\left[\mathrm{w},\mathrm{T},\varphi\right]=\left[\mathrm{W}(z),\Theta(z),\Phi(z)\right]\exp\left(ik_x x+ik_y y+nt\right), \tag{2.27}$$

where k_x, k_y are wave numbers in x and y direction and n is growth rate of disturbances.

Using equation (2.27), equations (2.26), (2.21) and (2.22) become

$$\begin{aligned} &\left(\left(D^2-a^2\right)\left(1+nF\right)\left(D^2-a^2-\frac{n}{\Pr}\right)\left(D^2-a^2-\frac{n\left(1+nF\right)}{\Pr}\right)+TaD^2\right)W \\ &-\left(1+nF\right)^2\left(D^2-a^2-\frac{n}{\Pr}\right)\left(a^2\mathrm{Ra}\Theta-a^2\mathrm{Rn}\Phi\right)=0, \end{aligned} \tag{2.28}$$

$$N_A W-\frac{N_A}{Le}\left(D^2-a^2\right)\Theta-\left(\frac{1}{Le}\left(D^2-a^2\right)-n\right)\Phi=0, \tag{2.29}$$

$$W+\left(D^2-a^2-n-\frac{N_A N_B}{\mathrm{Le}}D\right)\Theta\text{-}\frac{N_B}{\mathrm{Le}}D\Phi=0, \tag{2.30}$$

where $D\equiv\dfrac{d}{dz}$ and $a=\sqrt{k_x^2+k_y^2}$ is the dimensionless resultant wave number.

The boundary conditions of the problem in view of normal mode analysis are

$$W = 0,\ D^2W = 0,\tilde{}\ = 0,\ D\Phi + N_A D\Theta = 0 \quad \text{at} \quad z = 0,1. \tag{2.31}$$

METHOD OF SOLUTION

The Galerkin weighted residuals method is used to obtain an approximate solution to the system of equations (2.28) – (2.30) with the corresponding boundary conditions (2.31). In this method, the test functions are the same as the base (trial) functions. Accordingly W, Θ and Φ are taken as

$$W = \sum_{p=1}^{N} A_p W_p,\ \Theta = \sum_{p=1}^{N} B_p \Theta_p, \Phi = \sum_{p=1}^{N} C_p \Phi_p, \tag{2.32}$$

where $W_p = \Theta_p = \sin p\pi z, \Phi_p = -N_A \sin p\pi z$, A_p, B_p and C_p are unknown coefficients, p = 1, 2, 3,..., N and the base functions W_p, Θ_p and Φ_p satisfying the boundary conditions (2.31). Using expression for W, Θ and Φ in equations (2.28) – (2.30) and multiplying the first equation by W_p the second equation by Θ_p and third equation by Φ_p and then integrating in the limits from zero to unity, we obtain a set of 3N linear homogeneous equations with 3N unknown A_p, B_p and C_p; p = 1, 2, 3,..., N. For existing of nontrivial solution, the vanishing of the determinant of coefficients produces the characteristics equation of the system in term of Rayleigh number Ra.

LINEAR STABILITY ANALYSIS

For the present formulation, we have considered the which system of equations (2.28) – (2.30) together with the boundary conditions (2.31) constitute a linear eigenvalue problem with variable coefficient for the growth rate of disturbance n of the system. Substituting equation (2.32) into the system of equations (2.28) – (2.30) and multiplying the first equation by W_p the second equation by Θ_p and third equation by Φ_p and then integrating in the limits from zero to unity and performing some integration by parts, one obtains the following matrix equation

$$\begin{bmatrix} \frac{(\pi^2+a^2)}{(1+nF)}\left(\pi^2+a^2+\frac{n(1+nF)}{\mathrm{Pr}}\right)+\frac{\pi^2 Ta}{(1+nF)^2\left(\pi^2+a^2+\frac{n}{\mathrm{Pr}}\right)} & -\mathrm{a}^2\mathrm{Ra} & -\mathrm{a}^2\mathrm{N_A Rn} \\ 1 & -(\pi^2+a^2+n) & 0 \\ 1 & \frac{1}{\mathrm{Le}}(\pi^2+a^2) & -\left(\frac{1}{\mathrm{Le}}(\pi^2+a^2)+\mathrm{n}\right) \end{bmatrix} \begin{bmatrix} W_0 \\ \Theta_0 \\ \Phi_0 \end{bmatrix} = \begin{bmatrix} 0 \\ 0 \\ 0 \end{bmatrix}. \tag{2.33}$$

The non-trivial solution of the above matrix requires that

$$\mathrm{Ra} = \frac{1}{\mathrm{a}^2}\left(\frac{(\pi^2+a^2)}{(1+nF)}\left(\pi^2+a^2+\frac{n(1+nF)}{\mathrm{Pr}}\right)+\frac{\pi^2 Ta}{(1+nF)^2\left(\pi^2+a^2+\frac{n}{\mathrm{Pr}}\right)}\right)(\pi^2+a^2+n) - \frac{(\pi^2+a^2)+Le(\pi^2+a^2+n)}{(\pi^2+a^2)+\mathrm{nLe}}\mathrm{N_A Rn}. \tag{2.34}$$

The growth rate n is in general a complex quantity such that $n = \omega_r + i\omega$, the system with $\omega_r < 0$ is always stable, while for $\omega_r > 0$ it will become unstable. For neutral stability, the real part of n is zero. Hence, we now write $n = i\omega$, (where ω is real and is a dimensionless frequency) in equation (2.34), we have

$$\mathrm{Ra} = \Delta_1 + i\omega\Delta_2, \tag{2.35}$$

where

$$\Delta_1 = \frac{(\pi^2+a^2)}{\mathrm{a}^2}\left(\frac{(\pi^2+a^2)^2+\omega^2 F(\pi^2+a^2)}{1+\omega^2F^2}-\frac{\omega^2}{\mathrm{Pr}}\right) + \frac{\pi^2 Ta\left((\pi^2+a^2)\left((1-\omega^2F^2)(\pi^2+a^2)-\frac{2\omega^2 F}{\mathrm{Pr}}\right)+\omega^2\left(2F(\pi^2+a^2)+\frac{(1-\omega^2F^2)}{\mathrm{Pr}}\right)\right)}{\left((1-\omega^2F^2)(\pi^2+a^2)-\frac{2\omega^2F}{\mathrm{Pr}}\right)^2+\omega^2\left(2F(\pi^2+a^2)+\frac{(1-\omega^2F^2)}{\mathrm{Pr}}\right)^2} - \frac{(\pi^2+a^2)^2(Le+1)+Le((\pi^2+a^2))+\omega^2Le^2}{(\pi^2+a^2)^2+\omega^2Le^2}\mathrm{N_A Rn} \tag{2.36}$$

and

$$\Delta_2 = \frac{\left(\pi^2 + a^2\right)^2}{a^2}\left(\frac{1 - F\left(\pi^2 + a^2\right)}{1 + \omega^2 F^2} + \frac{1}{Pr}\right)$$

$$+ \frac{\pi^2 Ta\left(\left(\left(1 - \omega^2 F^2\right)\left(\pi^2 + a^2\right) - \frac{2\omega^2 F}{Pr}\right) + \left(\pi^2 + a^2\right)\left(2F\left(\pi^2 + a^2\right) + \frac{\left(1 - \omega^2 F^2\right)}{Pr}\right)\right)}{\left(\left(1 - \omega^2 F^2\right)\left(\pi^2 + a^2\right) - \frac{2\omega^2 F}{Pr}\right)^2 + \omega^2\left(2F\left(\pi^2 + a^2\right) + \frac{\left(1 - \omega^2 F^2\right)}{Pr}\right)^2}$$

$$- \frac{Le^2\left(\pi^2 + a^2\right)}{\left(\pi^2 + a^2\right)^2 + \omega^2 Le^2} N_A Rn. \tag{2.37}$$

Since Ra is a physical quantity, so it must be real. Hence, it follows from the equation (2.35) that either $\omega = 0$ (exchange of stability, steady state) or $\Delta_2 = 0$ ($\omega \neq 0$ overstability or oscillatory onset).

STATIONARY CONVECTION

For the case of stationary (non- oscillatory) convection, $n = \omega = 0$, thus equation (2.34) reduces to

$$\left(Ra\right)_s = \frac{\left(\pi^2 + a^2\right)^3 + \pi^2 Ta}{a^2} - \left(1 + Le\right) N_A Rn. \tag{2.38}$$

It is observed that stationary Rayleigh number Ra is function of the Taylor number Ta (rotation), Lewis number Le, the modified diffusivity ratio N_A and the nanoparticles Rayleigh Rn but independent of visco- elastic parameter F, Prandtl number Pr and modified particle- density increment N_B. Thus Maxwellian visco-elastic nanofluid behaves like an ordinary Newtonian nanofluid and instability is purely a phenomenon due to buoyancy coupled with the conservation of nanoparticles.

In the absence of rotation (Ta = 0) equation (2.38) reduces to

$$\left(\mathrm{Ra}\right)_s = \frac{\left(\pi^2 + a^2\right)^3}{\mathrm{a}^2} - \left(1 + Le\right)\mathrm{N}_\mathrm{A}\mathrm{Rn}. \tag{2.39}$$

This is the good agreement of the result (1.47) obtained in Chapter 1.

To find the critical value of $\left(\mathrm{Ra}\right)_\mathrm{s}$, equation (2.38) is differentiated with respect to 'a^2' and then equated to zero. The minimum of first term of right-hand side of equation (2.38) is attained at $\mathrm{a}_\mathrm{c} = \frac{\pi}{\sqrt{2}}$ and minimum value found to $\frac{27\pi^4}{4}$ so the corresponding critical Rayleigh number given by

$$\left(\mathrm{Ra}\right)_c = \frac{27\pi^4 + \pi^2 Ta}{4} - \left(1 + Le\right)\mathrm{N}_\mathrm{A}\mathrm{Rn}\,. \tag{2.40}$$

In the absence of rotation and nanoparticles (Ta = Rn = Le = N_A = 0), one recovers the well-known results that the critical Rayleigh-number is equal to $\left(\mathrm{Ra}\right)_c = \frac{27\pi^4}{4}$.

This is good agreement of the result obtained by Chandrasekhar (1961).

Thus presence of the nanoparticles lowers the value of the critical Rayleigh number by usually by substantial amount. Also parameter N_B does not appear in the equation (2.40); thus instability is purely phenomena due to buoyancy coupled with conservation of nanoparticles. Thus average contribution of nanoparticles flux in the thermal energy equation is zero with one-term Galerkin approximation.

RESULTS AND DISCUSSION

To study the effect of Rotation (Ta), Lewis number Le, modified diffusivity ratio N_A and nanoparticles Rayleigh number Rn on stationary convection, we examine the behavior of $\frac{\partial\left(\mathrm{Ra}\right)_s}{\partial Ta}, \frac{\partial\left(\mathrm{Ra}\right)_s}{\partial Le}, \frac{\partial\left(\mathrm{Ra}\right)_s}{\partial N_A}$ and $\frac{\partial\left(\mathrm{Ra}\right)_s}{\partial Rn}$ analytically.

From equation (2.38), we have

1. $\dfrac{\left(\partial \mathrm{R}a\right)_s}{\partial Ta} > 0,$
2. $\dfrac{\left(\partial \mathrm{R}a\right)_s}{\partial Le} < 0,$
3. $\dfrac{\left(\partial \mathrm{R}a\right)_s}{\partial N_A} < 0,$
4. $\dfrac{\left(\partial \mathrm{R}a\right)_s}{\partial Rn} < 0.$

These inequalities imply that Taylor number Ta has stabilizing effect while Lewis number Le, modified diffusivity ratio N_A and nanoparticles Rayleigh number Rn have destabilizing effect on the stationary convection.

Expression for stationary Rayleigh number, which characterizes the stability of the system, is found to be depend upon Taylor number (rotation), Lewis number, modified diffusivity ratio and nanoparticles Rayleigh number. The computations are carried out for different values of parameters considered in the range $10^2 \leq Ra \leq 10^5$ (Rayleigh number), $10 \leq Ta \leq 10^4$ (Taylor number), $1 \leq N_A \leq 10$ (modified diffusivity ratio), $10^2 \leq Le \leq 10^4$ (Lewis number), $10^{-1} \leq Rn \leq 10^1$ (nanoparticles Rayleigh number).

The variations of the stationary thermal Rayleigh number with wave number have been plotted graphically as shown in Figures 2 - 5.

Figure 2 shows only the variation of stationary Rayleigh number with wave number for different value of the Taylor number with fixed value of other parameters and it is found that the Rayleigh number increases with an increase in the value of the Taylor number, which imply that rotation delay the onset of stationary convection. It is due to the fact Coriolis force due to rotation drags the perturbed transverse motion and kinetic energy is dissipated by the viscosity, which intern increases the critical electric strength required for the onset of convection. Rotation acts so as to suppress the vertical motion, and hence thermal convection, by restricting the motion to the horizontal plane. The corresponding critical wave number a_c is plotted in Fig. 2 and indicates that an increase in the value of Taylor number Ta tends to increase a_c. Thus its effect is to reduce the size of convection cells. This is good agrement of the result obtained by Chand and Rana (2012b), Chand (2013a), Yadav et al. (2011).

Figure 3 shows the variation of thermal Rayleigh number for different value of Lewis number Le and for the fixed value of other parameters. It is found that stationary Rayliegh number decreases as the value of Lewis

Figure 2. Variation of the stationary Rayleigh number with wave number for different value of Taylor number

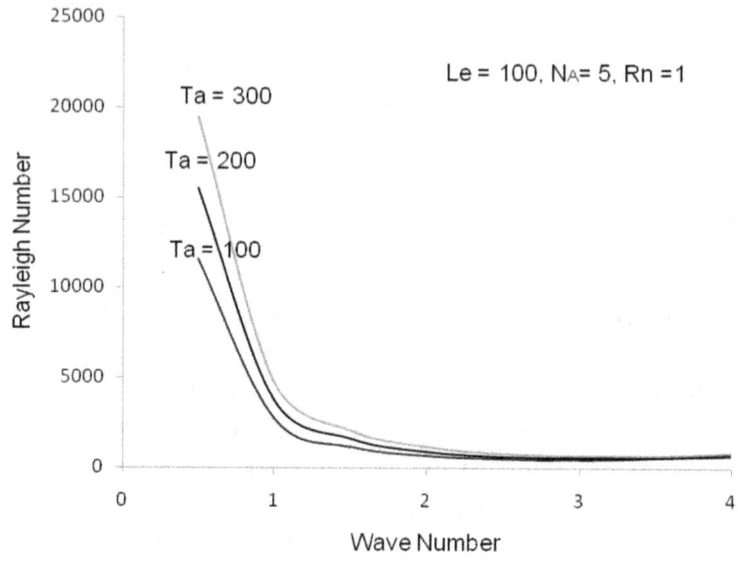

Figure 3. Variation of the stationary Rayleigh number with wave number for different value of Lewis number

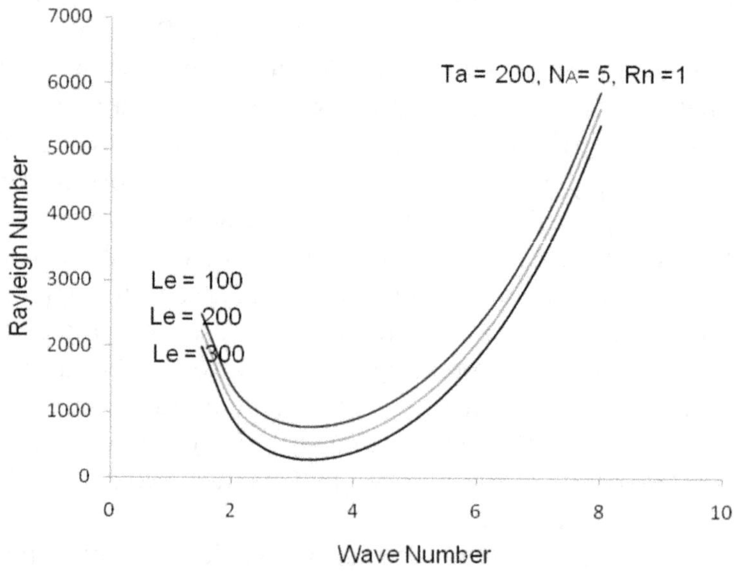

Figure 4. Variation of the stationary Rayleigh number with wave number for different value of modified diffusivity ratio

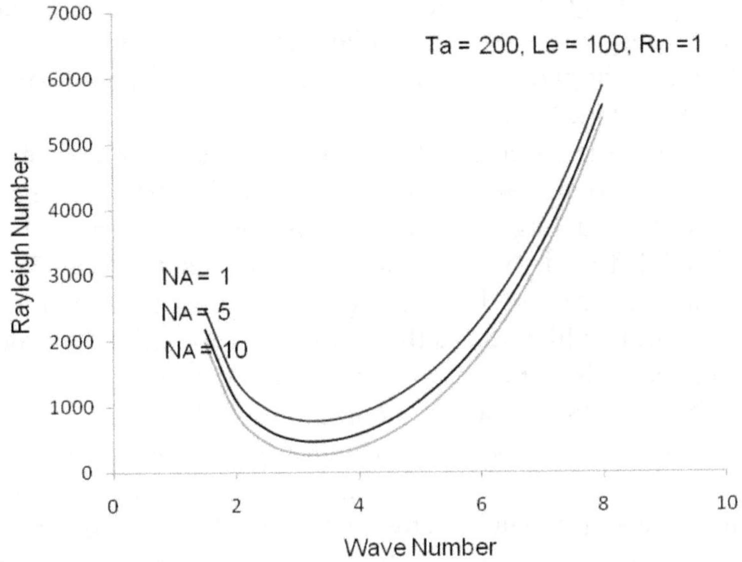

Figure 5. Variation of the stationary Rayleigh number with wave number for different value of nanoparticles Rayleigh number

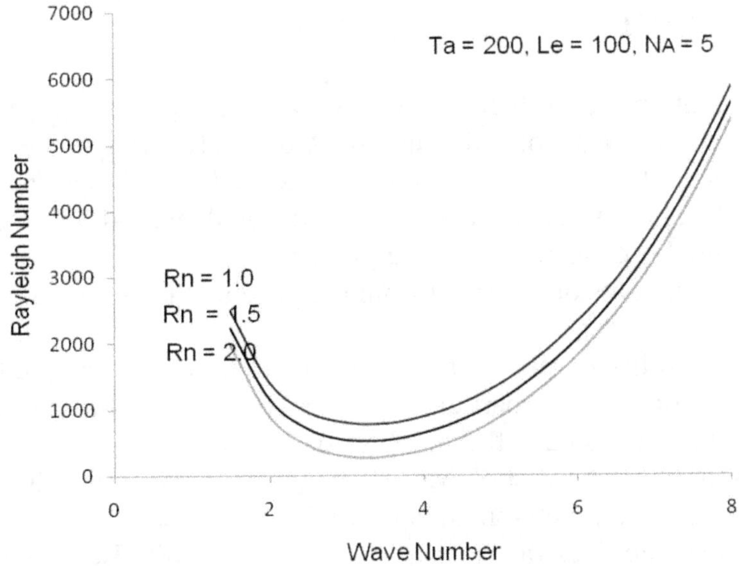

number increases, indicating that Lewis number destabilizes the stationary convection. It is due to the fact that thermophoresis at a higher value of thermophoretic diffusivity is more supportable to the disturbance in nanofluids, while both thermophoresis and Brownian motion are driving forces in favor of the motion of nanoparticles. This is good agrement of the result obtained by Chand and Rana (2014d).

Figure 4 shows the variation of stationary Rayleigh number for different value of the modified diffusivity ratio N_A and fixed value of other parameters and it is found that Rayleigh number decreases with an increase in the value of the modified diffusivity ratio N_A, which means that the modified diffusivity ratio N_A destabilizes on the fluid layer. This may lead to an increase in volumetric fraction, which shows that Brownian motion of the nanoparticles will also increase, which may cause destabilizing effect. This is good agrement of the result obtained by Chand and Rana (2014d).

Figure 5 shows the variation of stationary Rayleigh number with wave number for different value of the nanoparticles Rayleigh number Rn and fixed value of other parameters and it is found that thermal Rayleigh number decreases with an increase in value of the nanoparticles Rayleigh number Rn, which means that the nanoparticles Rayleigh number Rn has destabilizing effect on fluid layer. It has destabilizing effect because the heavier nanoparticles moving through the base fluid makes more strong disturbances as compared with the lighter nanoparticles.

CONCLUSION

Effect of rotation on the thermal convection in a horizontal layer of Maxwellian visco-elastic nanofluid is investigated. The flux of volume fraction of nanoparticles is taken to be zero on the isothermal boundaries and the eigen value problem is solved using the Galerkin residual method. The results have been presented both analytically and graphically.

The main conclusions derived from the present chapter are as follows:

1. The instability purely phenomenon due to buoyancy coupled with the conservation of nanoparticle and is independent of the contribution of Brownian motion and thermophoresis.
2. For stationary convection Maxwellian visco-elastic nanofluid behaves like an ordinary Newtonian nanofluid.
3. Rotation stabilizes the stationary convection while Lewis number Le, modified diffusivity ratio N_A and nanoparticles Rayleigh number Rn destabilizes the stationary convection.

REFERENCES

Chand, R. (2013a). Thermal instability of rotating nanofluid. *Int. J. Appl. Math and Mech.*, *9*(3), 70–90.

Chand, R., & Rana, G. C. (2012b). On the onset of thermal convection in rotating nanofluid layer saturating a Darcy-Brinkman porous medium. *International Journal of Heat and Mass Transfer*, *55*(21-22), 5417–5424. doi:10.1016/j.ijheatmasstransfer.2012.04.043

Chand, R., & Rana, G. C. (2014d). Thermal instability in a Brinkman porous medium saturated by nanofluid with no nanoparticle flux on boundaries. *Special Topics & Reviews in Porous Media: An International Journal*, *5*(4), 277–286. doi:10.1615/SpecialTopicsRevPorousMedia.v5.i4.10

Chand, R., Rana, G. C., & Kango, S. K. (2015d). Effect of variable gravity on thermal instability of rotating nanofluid in porous medium. *FME Transactions*, *43*(1), 62–69. doi:10.5937/fmet1501062c

Chandrasekhar, S. (1961). *Hydrodynamic and Hydromagnetic Stability*. New York: Dover Publication.

Qin, Y., & Kaloni, P. N. (1992). A thermal instability problem in a rotating micropolar fluid. *International Journal of Engineering Science*, *30*(9), 1117–1126. doi:10.1016/0020-7225(92)90061-K

Rana, G. C., & Chand, R. (2015b). Onset of thermal convection in a rotating nanofluid layer saturating a Darcy-Brinkman porous medium: A more realistic model. *Journal of Porous Media*, *18*(6), 629–635. doi:10.1615/JPorMedia.v18.i6.60

Rana, P., & Agarwal, S. (2015). Convection in a binary nanofluid saturated rotating porous layer. *Journal of Nanofluids*, *4*(1), 59–65. doi:10.1166/jon.2015.1123

Yadav, D., Bhargava, R., & Agrawal, G. S. (2011). Thermal instability of rotating nanofluid layer. *International Journal of Engineering Science*, *49*(11), 1171–1184. doi:10.1016/j.ijengsci.2011.07.002

Yadav, D., Bhargava, R., & Agrawal, G. S. (2013b). Numerical solution of a thermal instability problem in a rotating nanofluid layer. Int. *J Heat Mass Transf.*, *63*, 313–322. doi:10.1016/j.ijheatmasstransfer.2013.04.003

Chapter 3
Thermal Convection in a Layer of Maxwellian Visco-Elastic Nanofluid in the Presence of Vertical Magnetic Field

INTRODUCTION

The effect of magnetic field on the thermal instability in nanofluids has its relevance and importance in chemical engineering, biochemical engineering, industry and many physical phenomenon concerning with geophysics and astrophysics. These applications include design of chemical processing equipment, formation and dispersion of fog, distributions of temperature and moisture over agricultural fields and groves of fruit trees and damage of crops due to freezing and pollution of the environment etc. Magneto convection for the classical Rayleigh Bénard problem for a fluid layer combines the element of thermal buoyancy and magnetic field induced Lorentz force. Due to this Lorentz force term on the Rayleigh Bénard convection another non dimensional parameter called Chandrasekhar number is introduced in the problem. Investigation of the effects of magnetic field on the onset of convection was started several decades ago. Chandrasekhar (1961) studied in

DOI: 10.4018/978-1-68318-006-7.ch003

detail the thermal convection in a hydromagnetics. Patil and Rudraiah (1973), Alchaar et al. (1995) considered the problem of thermosolutal convection in the presence of magnetic field for different boundary conditions. Magnetic field plays an important role in the Rayleigh-Bénard convection in a layer of nanofluid and finds applications in biomedical engineering such as MRI, plethora of engineering, power plant cooling systems as well as in computers. Magnetic field plays an important role in engineering and industrial applications. Various problems on effect of magnetic field on nanofluids have been considered by researchers in the past e.g. Chand (2013b), Mahajan and Arora (2013), Yadav et al. (2013c), Gupta et al. (2013, 2015) and found that magnetic field has stabilizing effect on the fluid layer.

In the present chapter we investigated the effect of magnetic field on thermal instability of Maxwellian nanofluid layer for more realistic boundary conditions. It has been observed magnetic field has stabilizing effect on the fluid layer.

MATHEMATICAL FORMULATIONS OF THE PROBLEM

The Physical Problem

In this chapter we shall investigate the effect of magnetic field on the onset of thermal convection in a horizontal layer Maxwellian visco-elastic nanofluid. The physical configuration of the problem to be considered as

Consider an infinite horizontal layer of Maxwellian visco-elastic nanofluid of thickness 'd' bounded by horizontal boundaries z = 0 and z = d. A Cartesian coordinate system (x, y, z) is chosen with the origin at the bottom of the fluid layer and the z- axis normal to the fluid layer. Fluid layer is acted upon by a uniform vertical magnetic field **H**(0, 0, H) and is acted upon by a gravity force **g**(0, 0,-g). Fluid layer is heated from below in such a way that horizontal boundaries z = 0 and z = d respectively maintained at a uniform temperature T_0 and T_1 ($T_0 > T_1$). The normal component of the nanoparticles flux has to vanish at an impermeable boundary and the temperature T is taken to be T_0 at z = 0 and T_1 at z = d ($T_0 > T_1$) as shown in Figure 1. The reference scale for temperature and nanoparticles fraction is taken to be T_1 and φ_0 respectively.

Figure 1. Physical configuration of the problem

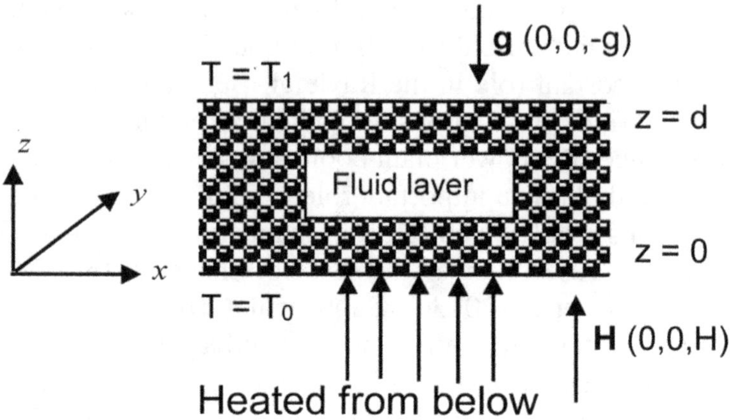

Assumptions

The mathematical equations describing the physical model are based upon the following assumptions:

1. Thermophysical properties of fluid expect for density in the buoyancy force (Boussinesq Hypothesis) are constant,
2. The fluid phase and nanoparticles are in thermal equilibrium state and thus, the heat flow has been described using one equation model,
3. Dilute mixture,
4. Nanoparticles are spherical,
5. Nanoparticles are non-magnetic,
6. No chemical reactions take place in fluid layer,
7. Negligible viscous dissipation,
8. Radiative heat transfer between the sides of wall is negligible when compared with other modes of the heat transfer,
9. Nanofluid is incompressible, electrically conducting, Newtonian and laminar flow,
10. Each boundary wall is assumed to be impermeable and perfectly thermal conducting.

GOVERNING EQUATIONS

The governing equations of for Maxwellian visco-elastic nanofluid in the presence of magnetic field under the Boussinesq approximation are given as

$$\nabla \cdot \mathbf{v} = 0, \tag{3.1}$$

$$\rho\left(1+\lambda\frac{\partial}{\partial t}\right)\frac{d\mathbf{v}}{dt} = \left(1+\lambda\frac{\partial}{\partial t}\right)\left(-\nabla p + \left(\varphi\rho_p + (1-\varphi)\left\{\rho_f\left(1-\alpha\left(T-T_0\right)\right)\right\}\right)\mathbf{g}\right)$$
$$+\mu\nabla^2\mathbf{v} + \frac{\mu_e}{4\pi}\left(\mathbf{H}\cdot\nabla\right)\mathbf{H} \quad , \tag{3.2}$$

where $\frac{d}{dt} = \frac{\partial}{\partial t} + \left(\mathbf{v}\cdot\nabla\right)$ is stands for convection derivative while **v**, p, ρ, μ, μ_e and α stands for fluid velocity, hydrostatic pressure, density of nanofluid, viscosity, magnetic permeability and the coefficient of thermal expansion respectively and λ is the relaxation time.

Equation of energy for Maxwell visco-elastic nanofluid is given by

$$\rho c\frac{\partial T}{\partial t} + \mathbf{v}\cdot\nabla T = k_m\nabla^2 T + \left(\rho c\right)_p\left(D_B\nabla\varphi\cdot\nabla T + \frac{D_T}{T_1}\nabla T\cdot\nabla T\right), \tag{3.3}$$

where ρc is heat capacity of fluid, $(\rho c)_p$ is heat capacity of nanoparticles, T_1 is the temperature of the fluid layer at z = d and k_m is the thermal conductivity. Equation of continuity for the nanoparticles is

$$\frac{\partial\varphi}{\partial t} + \mathbf{v}\cdot\nabla\varphi = D_B\nabla^2\varphi + \frac{D_T}{T_1}\nabla^2 T, \tag{3.4}$$

where D_B is the Brownian diffusion coefficient, given by Einstein-Stokes equation and D_T is the thermoporetic diffusion coefficient of the nanoparticles.

Maxwell equations are

$$\frac{d\mathbf{H}}{dt} = \left(\mathbf{H}\cdot\nabla\right)\mathbf{v} + \eta\nabla^2\mathbf{v}, \tag{3.5}$$

$$\nabla\cdot\mathbf{H} = 0, \tag{3.6}$$

where η is the electrical resitivity.

We assume that the temperature is constant and nanoparticles flux is zero on the boundaries. Thus boundary conditions are

$$w = 0,\ T = T_0,\ \ D_B \frac{\partial \phi}{\partial z} + \frac{D_T}{T_1}\frac{\partial T}{\partial z} = 0 \ \text{ at } z = 0 \text{ and}$$

$$w = 0,\ T = T_1,\ \ D_B \frac{\partial \phi}{\partial z} + \frac{D_T}{T_1}\frac{\partial T}{\partial z} = 0 \ \text{ at } \ z = d. \tag{3.7}$$

Introducing non-dimensional variables as

$$(x',y',z') = \left(\frac{x,y,z}{d}\right),\ \ v'(u',v',w') = V\left(\frac{u,v,w}{o}\right)d,\ \ t' = \frac{t^{o}}{d^2},\ \ p' = \frac{pd^2}{\mu\kappa},\ \ \varphi' = \frac{\left(\varphi - \varphi_0\right)}{\varphi_0},$$

$$T' = \frac{\left(T - T_1\right)}{\left(T_0 - T_1\right)},\ \ H' = \frac{H}{|H|},$$

where $\kappa = \frac{k_m}{\rho c}$ is thermal diffusivity of the fluid.

Equations (3.1) - (3.7) in non-dimensional form can be written as

$$\nabla' \cdot \mathbf{v}' = 0, \tag{3.8}$$

$$\left(1 + F\frac{\partial}{\partial t'}\right)\frac{1}{\Pr}\frac{\partial \mathbf{v}'}{\partial t'} =$$
$$\left(1 + F\frac{\partial}{\partial t'}\right)\left(-\nabla' p' \text{-} Rm\hat{e}_z + RaT'\hat{e}_z - Rn\varphi'\hat{e}_z\right) + \nabla'^2 \mathbf{v}' + \frac{\Pr}{\Pr_M} Q\left(\mathbf{H}' \cdot \nabla'\right)\mathbf{H}', \tag{3.9}$$

$$\frac{\partial \varphi'}{\partial t'} + \mathbf{v}' \cdot \nabla' \varphi' = \frac{1}{Le}\nabla'^2 \varphi' + \frac{N_A}{Le}\nabla'^2 T', \tag{3.10}$$

$$\frac{\partial T'}{\partial t'} + \mathbf{v}' \cdot \nabla' T' = \nabla'^2 T' + \frac{N_B}{Le}\nabla' \varphi' \cdot \nabla' T' + \frac{N_A N_B}{Le}\nabla' T' \cdot \nabla' T', \tag{3.11}$$

$$\frac{d\mathbf{H}'}{dt'} = \left(\mathbf{H}' \cdot \nabla'\right)\mathbf{v}' + \frac{\mathrm{Pr}}{\mathrm{Pr}_M}\nabla'^2\mathbf{v}', \tag{3.12}$$

$$\nabla' \cdot \mathbf{H}' = 0. \tag{3.13}$$

Here the non-dimensional parameters are given as follows

$\mathrm{Pr} = \frac{\mu}{\rho\kappa}$ is the Prandtl number,

$\mathrm{Pr}_M = \frac{\mu}{\rho\eta}$ is the magnetic Prandtl number,

$Le = \frac{\kappa}{\mathrm{D_B}}$ is the Lewis number,

$F = \frac{\kappa\lambda}{d^2}$ is the stress relaxation parameter,

$\mathrm{Ra} = \frac{\rho g\alpha d^3\left(\mathrm{T_0} - \mathrm{T_1}\right)}{\mu\kappa}$ is the Rayleigh number,

$\mathrm{Rm} = \frac{\left(\rho_p\varphi_0 + \rho\left(1\text{-}\varphi_0\right)\right)gd^3}{\mu\kappa}$ is the density Rayleigh number,

$\mathrm{Rn} = \frac{\left(\rho_p - \rho\right)\varphi_0 gd^3}{\mu\kappa}$ is the nanoparticles Rayleigh number,

$\mathrm{Q} = \frac{\mu_e H^2 d^2}{4\pi\rho\upsilon\eta}$ is the Chandrasekhar number,

$\mathrm{N_A} = \frac{\mathrm{D_T}\left(\mathrm{T_0}\text{-}\mathrm{T_1}\right)}{\mathrm{D_B}\mathrm{T_1}\varphi_0}$ is the modified diffusivity ratio,

$\mathrm{N_B} = \frac{\left(\rho c\right)_p\varphi_0}{\left(\rho c\right)_f}$ is the modified particle-density increment.

In spirit of Oberbeck-Boussinesq approximation, equation (3.9) has been linearized by the neglect of a term proportional to the product of φ_0 and T.

This approximation is valid in the case of small temperature gradients in a dilute suspension of nanoparticles.

The dimensionless boundary conditions are

$$w' = 0, \quad T' = 1, \quad \frac{\partial \phi'}{\partial z'} + N_A \frac{\partial T'}{\partial z'} = 0 \text{ at } \quad z' = 0 \text{ and}$$

$$w' = 0, \; T' = 0, \quad \frac{\partial \phi'}{\partial z'} + N_A \frac{\partial T'}{\partial z'} = 0 \text{ at } z' = 1. \tag{3.14}$$

THE BASIC STATE AND ITS SOLUTIONS

The basic state was assumed to be quiescent and is given by

$$v_i' \left(u', v', w'\right) = 0,$$

$$p' = p_b(z),$$

$$H' = He_z,$$

$$T' = T_b(z), \tag{3.15}$$

$$\varphi' = \varphi_b(z) \text{ and}$$

$$\rho = \rho_0\left(1 + \alpha\left(T\text{-}T_0\right)\right).$$

An approximate solution for basic state is given by

$$T_b = 1 - z' \tag{3.16}$$

and

$$\varphi_b = \phi_0 + N_A z' \tag{3.17}$$

PERTURBATION SOLUTIONS

Let the initial basic state described by (3.15) is slightly perturbed so that perturbed state is given by

$$(\mathrm{u}',\mathrm{v}',\mathrm{w}') = 0 + (\mathrm{u}'',\mathrm{v}'',\mathrm{w}''),$$

$$\mathrm{H}' = He_z + \mathrm{H}''(\mathrm{H}''\mathrm{x},\mathrm{H}''\mathrm{y},\mathrm{H}''\mathrm{z}),$$

$$\mathrm{T}' = \mathrm{T}_b + \mathrm{T}'', \tag{3.18}$$

$$\mathrm{p}' = \mathrm{p}_b + \mathrm{p}'',$$

$$\varphi' = \varphi_b + \varphi'',$$

where $\mathrm{T}_b = 1 - \mathrm{z}'$, $\varphi_b = \phi_0 + N_A \mathrm{z}'$ and $(\mathrm{u}'',\mathrm{v}'',\mathrm{w}'')$, $\mathrm{H}''(\mathrm{H}''\mathrm{x},\mathrm{H}''\mathrm{y},\mathrm{H}''\mathrm{z})$, T'', φ'' and p'' respectively the perturbations in the initial velocity, magnetic field, temperature, pressure and volume fraction of the nanoparticles.

By substituting (3.18) in equations (3.8) – (3.13) and linearize by neglecting the product of the prime quantities, we obtained following equations

$$\nabla \cdot \mathbf{v} = 0, \tag{3.19}$$

$$\left(1 + F\frac{\partial}{\partial t}\right)\frac{1}{\mathrm{Pr}}\frac{\partial \mathbf{v}}{\partial t} = \left(1 + F\frac{\partial}{\partial t}\right)\left(-\nabla \mathrm{p} + \mathrm{Ra}\mathrm{T}\hat{\mathrm{e}}_z - \mathrm{Rn}\varphi\hat{\mathrm{e}}_z\right) + \nabla^2\mathbf{v} + \frac{\mathrm{Pr}}{\mathrm{Pr}_M}\frac{Q\partial \mathbf{H}}{\partial z}, \tag{3.20}$$

$$\frac{\partial \varphi}{\partial \mathrm{t}} + wN_A = \frac{1}{\mathrm{Le}}\nabla^2\varphi + \frac{\mathrm{N_A}}{\mathrm{Le}}\nabla^2\mathrm{T}, \tag{3.21}$$

$$\frac{\partial \mathrm{T}}{\partial \mathrm{t}} - w = \nabla^2\mathrm{T} + \frac{\mathrm{N_B}}{\mathrm{Le}}\left(N_A\frac{\partial T}{\partial z} - \frac{\partial \phi}{\partial z}\right) - \frac{2\mathrm{N_A}\mathrm{N_B}}{\mathrm{Le}}\frac{\partial T}{\partial z}, \tag{3.22}$$

$$\frac{\partial \mathrm{H}}{\partial \mathrm{t}} = \frac{\partial w}{\partial z} + Q\nabla^2 H, \tag{3.23}$$

$$\nabla \cdot H = 0, \tag{3.24}$$

where $\nabla_H^2 = \frac{\partial^2}{\partial x^2} + \frac{\partial^2}{\partial y^2}$ is the two-dimensional Laplacian operator.

[Dashes (“) have been suppressed for convenience]

It will be noted that the parameter Rm is not involved in these and subsequent equations. It is just a measure of the basic static pressure gradient.

Now by eliminating pressure term p and magnetic field term H from equations (3.20) by making use of equations (3.19), (3.23) and (3.24), we have

$$\left(\left(\frac{\Pr}{\Pr_M}\nabla^2 - \frac{\partial}{\partial t}\right)\left(\nabla^2 - \left(1 + F\frac{\partial}{\partial t}\right)\frac{1}{\Pr}\frac{\partial}{\partial t}\right) - \frac{\Pr}{\Pr_M}QD^2\right)\nabla^2 w$$
$$+\left(1 + F\frac{\partial}{\partial t}\right)\left(\frac{\Pr}{\Pr_M}\nabla^2 - \frac{\partial}{\partial t}\right)\left(Ra\nabla_H^2 T - \mathrm{Rn}\nabla_H^2 \varphi\right) = 0. \tag{3.25}$$

Boundary conditions are

$$w = 0,\ T = 0,\quad \frac{\partial \phi}{\partial z} + N_A \frac{\partial T}{\partial z} = 0 \ \text{ at } z = 0\ ,1. \tag{3.26}$$

NORMAL MODE ANALYSIS

Analyzing the disturbances into the normal modes and assuming that the perturbed quantities are of the form

$$\left[w, T, \varphi\right] = \left[W(z), \Theta(z), \Phi(z)\right] \exp\left(ik_x x + ik_y y + nt\right), \tag{3.27}$$

where k_x, k_y are wave numbers in x and y direction and n is growth rate of disturbances.

Using equation (3.27), equations (3.29), (3.25) and (3.26) become

$$\left(\left(\frac{\Pr}{\Pr_M}\left(D^2 - a^2\right) - n\right)\left(D^2 - a^2 - \frac{\left(1 + nF\right)n}{\Pr}\right) - \frac{\Pr}{\Pr_M}QD^2\right)\left(D^2 - a^2\right)W$$
$$-\left(\frac{\Pr}{\Pr_M}\left(D^2 - a^2\right) - n\right)\left(1 + nF\right)\left(a^2 Ra\Theta - a^2\mathrm{Rn}\Phi\right) = 0, \tag{3.28}$$

$$N_A W - \frac{N_A}{Le}\left(D^2 - a^2\right)\Theta - \left(\frac{1}{Le}\left(D^2 - a^2\right) - n\right)\Phi = 0, \tag{3.29}$$

$$W + \left(D^2 - a^2 - n - \frac{N_A N_B}{Le} D\right)\Theta - \frac{N_B}{Le} D\Phi = 0, \tag{3.30}$$

where $D \equiv \frac{d}{dz}$ and $a = \sqrt{k_x^2 + k_y^2}$ is the dimensionless resultant wave number.

The boundary conditions of the problem in view of normal mode analysis are

$$W = 0,\ D^2W = 0, \Theta = 0,\ D\Phi + N_A D\Theta = 0 \ \text{ at } \ z = 0,1. \tag{3.31}$$

METHOD OF SOLUTION

The Galerkin weighted residuals method is used to obtain an approximate solution to the system of equations (3.28) – (3.30) with the corresponding boundary conditions (3.31). In this method, the test functions are the same as the base (trial) functions. Accordingly W, Θ and Φ are taken as

$$W = \sum_{p=1}^{N} A_p W_p,\ \Theta = \sum_{p=1}^{N} B_p \Theta_p, \Phi = \sum_{p=1}^{N} C_p \Phi_p, \tag{3.32}$$

where $W_p = \Theta_p = \sin p\pi z, \Phi_p = -N_A \sin p\pi z$, A_p, B_p and C_p are unknown coefficients, p = 1, 2, 3,..., N and the base functions W_p, Θ_p, and Φ_p satisfying the boundary conditions (3.31). Using expression for W, Θ and Φ in equations (3.28) – (3.30) and multiplying the first equation by W_p the second equation by Θ_p and third equation by Φ_p and then integrating in the limits from zero to unity, we obtain a set of 3N linear homogeneous equations with 3N unknown A_p, B_p and C_p; p = 1, 2, 3,..., N. For existing of nontrivial solution, the vanishing of the determinant of coefficients produces the characteristics equation of the system in term of Rayleigh number Ra.

LINEAR STABILITY ANALYSIS

For the present formulation, we have considered the which system of equations (3.28) – (3.30) together with the boundary conditions (3.31) constitute a linear eigenvalue problem with variable coefficient for the growth rate of disturbance n of the system. Substituting equation (3.32) into the system of equations (3.28) - (3.30) and multiplying the first equation by W_p the second equation by Θ_p and third equation by Φ_p and then integrating in the limits from zero to unity and performing some integration by parts, one obtains the following matrix equation

$$\begin{bmatrix} \frac{(\pi^2+a^2)}{(1+nF)}\left(\pi^2+a^2+\frac{n(1+nF)}{\mathrm{Pr}}\right)+\frac{\frac{\mathrm{Pr}}{\mathrm{Pr}_M}\pi^2 Q}{\frac{\mathrm{Pr}}{\mathrm{Pr}_M}(\pi^2+a^2)+n} & -a^2\mathrm{Ra} & -a^2 N_A \mathrm{Rn} \\ 1 & -(\pi^2+a^2+n) & 0 \\ 1 & \frac{1}{\mathrm{Le}}(\pi^2+a^2) & -\left(\frac{1}{\mathrm{Le}}(\pi^2+a^2)+n\right) \end{bmatrix} \begin{bmatrix} W_0 \\ \Theta_0 \\ \Phi_0 \end{bmatrix} = \begin{bmatrix} 0 \\ 0 \\ 0 \end{bmatrix}. \tag{3.33}$$

The non-trivial solution of the above matrix requires that

$$\mathrm{Ra} = \frac{1}{a^2}\left(\frac{(\pi^2+a^2)}{(1+nF)}\left(\pi^2+a^2+\frac{n(1+nF)}{\mathrm{Pr}}\right)+\frac{\frac{\mathrm{Pr}}{\mathrm{Pr}_M}\pi^2 Q\frac{(\pi^2+a^2)}{(1+nF)}}{\frac{\mathrm{Pr}}{\mathrm{Pr}_M}(\pi^2+a^2)+n}\right)(\pi^2+a^2+n) - \frac{(\pi^2+a^2)+Le(\pi^2+a^2+n)}{(\pi^2+a^2)+n\mathrm{Le}}N_A\mathrm{Rn}. \tag{3.34}$$

The growth rate n is in general a complex quantity such that $n=\omega_r+i\omega$, the system with $\omega_r<0$ is always stable, while for $\omega_r>0$ it will become unstable. For neutral stability, the real part of n is zero. Hence, we now write $n = i\omega$, (where ω is real and is a dimensionless frequency) in equation (3.34), we have

$$\mathrm{Ra} = \Delta_1 + i\omega\Delta_2, \tag{3.35}$$

where

$$\Delta_1 = \frac{\left(\pi^2 + a^2\right)}{\mathrm{a}^2}\left(\frac{\left(\pi^2 + a^2\right)^2 + \omega^2 F\left(\pi^2 + a^2\right)}{1 + \omega^2 F^2} - \frac{\omega^2}{\mathrm{Pr}}\right)$$
$$+ \frac{\pi^2 Q \frac{\mathrm{Pr}}{\mathrm{Pr}_M}\left(\left(\pi^2 + a^2\right)\left(\left(1 - \omega^2 F^2\right)\left(\pi^2 + a^2\right) - \frac{2\omega^2 F}{\mathrm{Pr}}\right) + \omega^2\left(2F\left(\pi^2 + a^2\right) + \frac{\left(1 - \omega^2 F^2\right)}{\mathrm{Pr}}\right)\right)}{\left(\left(1 - \omega^2 F^2\right)\left(\pi^2 + a^2\right) - \frac{2\omega^2 F}{\mathrm{Pr}}\right)^2 + \omega^2\left(2F\left(\pi^2 + a^2\right) + \frac{\left(1 - \omega^2 F^2\right)}{\mathrm{Pr}}\right)^2}$$
$$- \frac{\left(\pi^2 + a^2\right)^2 \left(Le + 1\right) + Le\left(\left(\pi^2 + a^2\right)\right) + \omega^2 Le^2}{\left(\pi^2 + a^2\right)^2 + \omega^2 Le^2}\mathrm{N_A Rn}$$

(3.36)

and

$$\Delta_2 = \frac{\left(\pi^2 + a^2\right)^2}{\mathrm{a}^2}\left(\frac{1 - F\left(\pi^2 + a^2\right)}{1 + \omega^2 F^2} + \frac{1}{\mathrm{Pr}}\right)$$
$$+ \frac{\pi^2 Q \frac{\mathrm{Pr}}{\mathrm{Pr}_M}\left(\left(\left(1 - \omega^2 F^2\right)\left(\pi^2 + a^2\right) - \frac{2\omega^2 F}{\mathrm{Pr}}\right) + \left(\pi^2 + a^2\right)\left(2F\left(\pi^2 + a^2\right) + \frac{\left(1 - \omega^2 F^2\right)}{\mathrm{Pr}}\right)\right)}{\left(\left(1 - \omega^2 F^2\right)\left(\pi^2 + a^2\right) - \frac{2\omega^2 F}{\mathrm{Pr}}\right)^2 + \omega^2\left(2F\left(\pi^2 + a^2\right) + \frac{\left(1 - \omega^2 F^2\right)}{\mathrm{Pr}}\right)^2}$$
$$- \frac{Le^2\left(\pi^2 + a^2\right)}{\left(\pi^2 + a^2\right)^2 + \omega^2 Le^2}\mathrm{N_A Rn}.$$

(3.37)

Since Ra is a physical quantity, so it must be real. Hence, it follows from the equation (3.35) that either $\omega = 0$ (exchange of stability, steady state) or $\Delta_2 = 0$ ($\omega \neq 0$ overstability or oscillatory onset).

STATIONARY CONVECTION

For the case of stationary convection $n = \omega = 0$, equation (3.34) reduces to

$$\left(\mathrm{Ra}\right)_s = \frac{\left(\pi^2 + a^2\right)^3 + \pi^2 Q}{\mathrm{a}^2} - \left(1 + Le\right)\mathrm{N_A Rn}. \tag{3.38}$$

It is observed that stationary Rayleigh number Ra is function of the Chandrasekhar number Q (Magnetic field), Lewis number Le, the modified diffusivity ratio N_A and the nanoparticles Rayleigh Rn but independent of visco-elastic parameter F, Prandtl number Pr and modified particle-density increment N_B. Thus Maxwellian visco-elastic nanofluid behaves like an ordinary Newtonian nanofluid and instability is purely a phenomenon due to buoyancy coupled with the conservation of nanoparticles.

In the absence of magnetic field (Q =0) equation (3.38) reduces to

$$\left(\mathrm{Ra}\right)_s = \frac{\left(\pi^2 + a^2\right)^3}{\mathrm{a}^2} - \left(1 + Le\right)\mathrm{N_A Rn}. \tag{3.39}$$

This is the good agreement of the result (1.47) obtained in Chapter 1.

To find the critical value of $\left(\mathrm{Ra}\right)_s$, equation (3.38) is differentiated with respect to 'a^2' and then equated to zero. The minimum of first term of right-hand side of equation (3.38) is attained at $\mathrm{a_c} = \frac{\pi}{\sqrt{2}}$ and minimum value found to $\frac{27\pi^4}{4}$ so the corresponding critical Rayleigh number given by

$$\left(\mathrm{Ra}\right)_c = \frac{27\pi^4 + \pi^2 Q}{4} - \left(1 + Le\right)\mathrm{N_A Rn}. \tag{3.40}$$

In the absence of magnetic field and nanoparticles (Q = Rn = Le = N_A = 0), one recovers the well-known results that the critical Rayleigh number is equal to $\left(\mathrm{Ra}\right)_c = \frac{27\pi^4}{4}$.

This is good agreement of the result obtained by Chandrasekhar (1961).

Thus presence of the nanoparticles lowers the value of the critical Rayleigh number by usually by substantial amount. Also parameter N_B does not appear in the equation (3.40); thus instability is purely phenomena due to buoyancy coupled with conservation of nanoparticles. Thus average contribution of nanoparticles flux in the thermal energy equation is zero.

RESULTS AND DISCUSSION

To study the effect of magnetic field (Chandrasekhar number Q), Lewis number Le, modified diffusivity ratio N_A and nanoparticles Rayleigh number Rn on stationary convection, we examine the behavior of $\frac{\partial(\mathrm{R}a)_s}{\partial Q}, \frac{\partial(\mathrm{R}a)_s}{\partial Le}, \frac{\partial(\mathrm{R}a)_s}{\partial N_A}$ and $\frac{\partial(\mathrm{R}a)_s}{\partial Rn}$ analytically.

From equation (3.38), we have

(i) $\frac{(\partial\mathrm{R}a)_s}{\partial Q} > 0,$

(ii) $\frac{(\partial\mathrm{R}a)_s}{\partial Le} < 0,$

(iii) $\frac{(\partial\mathrm{R}a)_s}{\partial N_A} < 0,$

(iv) $\frac{(\partial\mathrm{R}a)_s}{\partial Rn} < 0.$

These inequalities imply that magnetic field has stabilizing effect while Lewis number Le, modified diffusivity ratio N_A and nanoparticles Rayleigh number Rn destabilizing effect on the stationary convection.

Expression for stationary Rayleigh number, which characterizes the stability of the system, is found to be depend upon magnetic field, Lewis number, modified diffusivity ratio and nanoparticles Rayleigh number. The computations are carried out for different values of parameters considered in the range $10^2 \leq Ra \leq 10^5$ (thermal Rayleigh number), $10 \leq Q \leq 10^3$ (Chandrasekhar number), $1 \leq N_A \leq 10$ (modified diffusivity ratio), $10^2 \leq Le \leq 10^4$ (Lewis number), $10^{-1} \leq Rn \leq 10^1$ (nanoparticles Rayleigh number).

The variations of the stationary thermal Rayleigh number with wave number have been plotted graphically as shown in Figures 2 – 5

Figure 2 shows the variation of thermal Rayleigh number for different value of Chandrasekhar number and for the fixed value of other parameters. It is found that stationary Rayliegh number increases as the values of Chandrasekhar number increases, indicating that Chandrasekhar number stabilizes the stationary convection.

Figure 3 shows the variation of thermal Rayleigh number for different value of Lewis number Le and for the fixed value of other parameters. It is found that stationary Rayliegh number decreases as the values of Lewis

Figure 2. Variation of the stationary Rayleigh number with wave number for different value of Chandrasekhar number

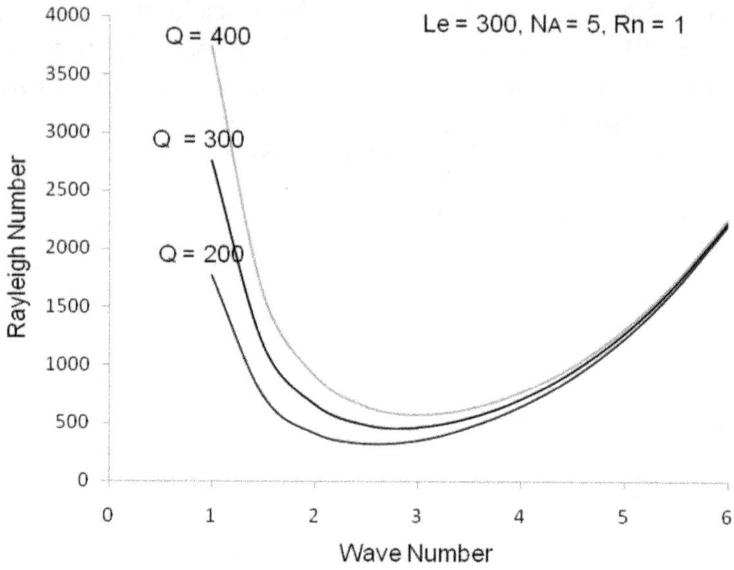

Figure 3. Variation of the stationary Rayleigh number with wave number for different value of Lewis number

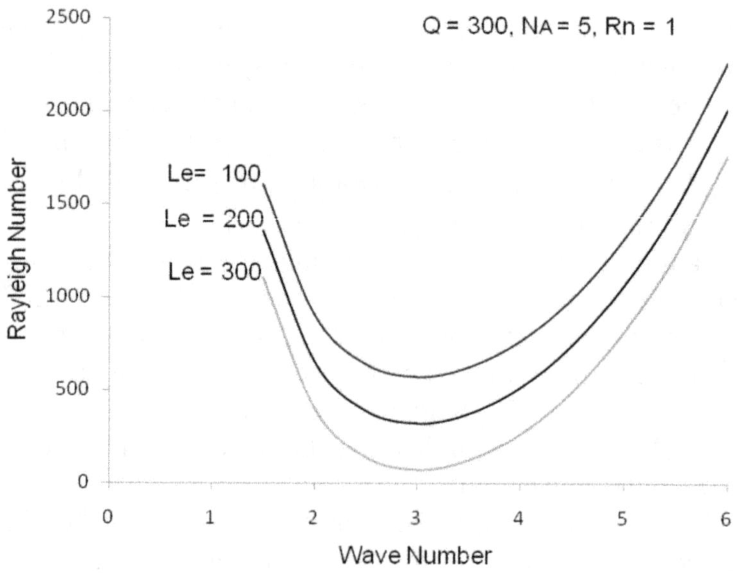

number increases, indicating that Lewis number destabilizes the stationary convection. It is due to the fact that thermophoresis at a higher value of thermophoretic diffusivity is more supportable to the disturbance in nanofluids, while both thermophoresis and Brownian motion are driving forces in favor of the motion of nanoparticles. This is good agrement of the result obtained by Chand and Rana (2014d, 2015a).

Figure 4 shows the variation of stationary Rayleigh number for different value of the modified diffusivity ratio N_A and fixed value of other parameters and it is found that Rayleigh number decreases with an increase in the value of the modified diffusivity ratio $N_{A,}$ which means that the modified diffusivity ratio N_A destabilizes on the fluid layer. This may lead to an increase in volumetric fraction, which shows that Brownian motion of the nanoparticles will also increase, which may cause destabilizing effect. This is good agrement of the result obtained by Chand and Rana (2014d, 2015a).

Figure 5 shows the variation of stationary Rayleigh number with wave number for different value of the nanoparticles Rayleigh number Rn and fixed value of other parameters and it is found that thermal Rayleigh number decreases with an increase in value of the nanoparticles Rayleigh number Rn, which means that the nanoparticles Rayleigh number Rn has destabilizing effect on fluid layer. It has destabilizing effect because the heavier nanoparticles

Figure 4. Variation of the stationary Rayleigh number with wave number for different value of modified diffusivity ratio

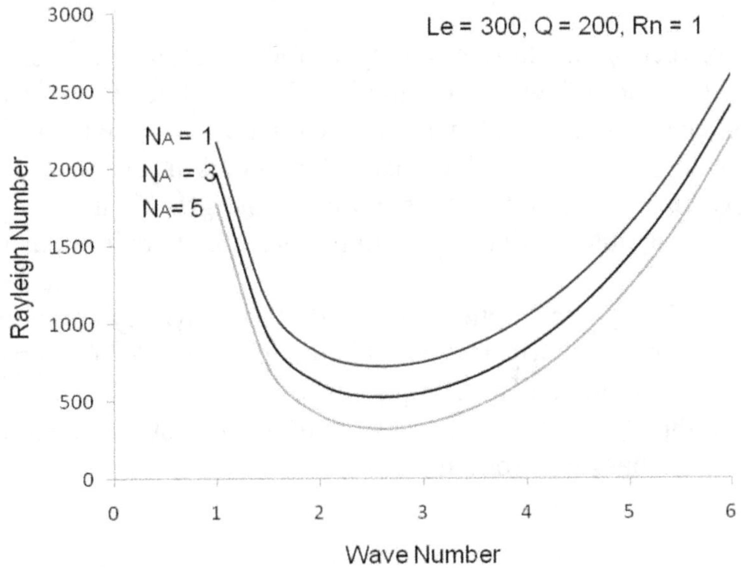

Figure 5. Variation of the stationary Rayleigh number with wave number for different value of nanoparticles Rayleigh number

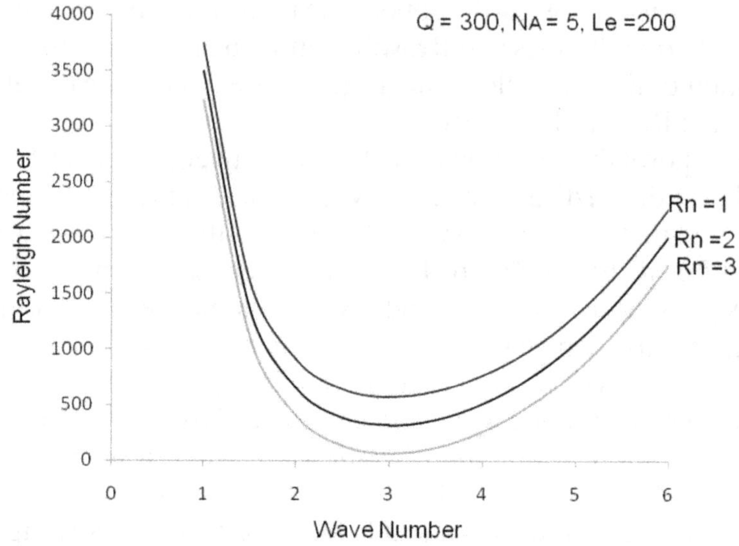

moving through the base fluid makes more strong disturbances as compared with the lighter nanoparticles. This is good agrement of the result obtained by Chand and Rana (2014d, 2015a).

CONCLUSION

Thermal convection in a horizontal layer of Maxwellian visco-elastic nanofluid in the presence of vertical magnetic field is studied. The flux of volume fraction of nanoparticles is taken to be zero on the isothermal boundaries and the eigen value problem is solved using the Galerkin residual method. The results have been presented both analytically and graphically.

The main conclusions derived from the present chapter are as follows

1. The instability purely phenomenon due to buoyancy coupled with the conservation of nanoparticle and is independent of the contribution of Brownian motion and thermophoresis.
2. For stationary convection Maxwellian visco-elastic nanofluid behaves like an ordinary Newtonian nanofluid.

3. Magnetic field stabilizes the stationary convection while Lewis number, modified diffusivity ratio and nanoparticles Rayleigh number Rn destabilizes the stationary convection.

REFERENCES

Alchaar, S., Vasseur, P., & Bilgen, E. (1995). Effect of a magnetic field on the onset of convection in a porous medium. *Heat and Mass Transfer*, *30*(4), 259–267. doi:10.1007/BF01602772

Chand, R. (2013b). On the onset of Rayleigh-Bénard convection in a layer of nanofluid in Hydromagnetics. *Int. J. of Nanoscience*, *12*(6), 1350038. doi:10.1142/S0219581X13500385

Chand, R., & Rana, G. C. (2014d). Thermal instability in a Brinkman porous medium saturated by nanofluid with no nanoparticle flux on boundaries. *Special Topics & Reviews in Porous Media: An International Journal*, *5*(4), 277–286. doi:10.1615/SpecialTopicsRevPorousMedia.v5.i4.10

Chand, R., & Rana, G. C. (2015a). Magneto convection in a layer of nanofluid in porous medium- a more realistic approach. *Journal of Nanofluids*, *4*(2), 196–202. doi:10.1166/jon.2015.1142

Chandrasekhar, S. (1961). *Hydrodynamic and Hydromagnetic Stability*. New York: Dover Publication.

Gupta, U., Ahuja, J., & Wanchoo, R. K. (2013). Magneto convection in a nanofluid layer. *International Journal of Heat and Mass Transfer*, *64*, 1163–1171. doi:10.1016/j.ijheatmasstransfer.2013.05.035

Gupta, U., Sharma, J., & Sharma, V. (2015). Instability of binary nanofluid with magnetic field. *Applied Mathematics and Mechanics*, *36*(6), 693–706. doi:10.1007/s10483-015-1941-6

Mahajan, A., & Arora, M. (2013). Convection in rotating magnetic nanofluids. *Applied Mathematics and Computation*, *219*(11), 6284–6296. doi:10.1016/j.amc.2012.12.012

Patil, R. P., & Rudraiah, N. (1973). Stability of Hydromagnetic thermo convective flow through porous medium. Trans. ASME. *Journal of Applied Mechanics*, *40*(4), 879–884. doi:10.1115/1.3423181

Yadav, D., Bhargava, R., & Agrawal, G. S. (2013c). Thermal instability in a nanofluid layer with vertical magnetic field. *Journal of Engineering Mathematics*, *80*(1), 147–164. doi:10.1007/s10665-012-9598-1

Chapter 4
Combined Effect of Rotation and Magnetic Field on the Thermal Convection in a Layer of Maxwellian Visco-Elastic Nanofluid

INTRODUCTION

The effects of magnetic field and rotation on the thermal instability in nanofluids have its relevance and importance in engineering and industry. Magneto convection for the classical Rayleigh Bénard problem for a fluid layer combines the element of thermal buoyancy and magnetic field induced Lorentz force. Due to this Lorentz force term on the Rayleigh-Bénard convection another non dimensional parameter called Chandrasekhar number

DOI: 10.4018/978-1-68318-006-7.ch004

is introduced in the problem. Thermal instability in rotating fluids about a vertical axis combines the element of thermal buoyancy and rotation induced Coriolis and centrifugal forces. Due to the Coriolis force on the thermal instability problem another parameter namely Taylor number is introduced in this problem. Taylor number is a non-dimensional number which is a measure of rotation rate.

The magnetic field in the presence of rotation is known as 'Magnetohydrodynamic' which has many applications in industry such as crystal growth, metal casting and liquid metal cooling blankets for fusion reactors. Thermal instability of a rotating nanofluid layer in the presence of magnetic fluid is studied by Mahajan and Arora (2013) and found that magnetic field and rotation stabilize the fluid layer. Some other aspects of magnetic nanofluid are studied by Parekh and Lee (2011) and Patel (2012).

In the present chapter an attempt has been made to study the combined effect of magnetic field and rotation on the thermal instability of Maxwellian visco-elastic nanofluid layer for more realistic boundary conditions. It has been observed that both magnetic field and rotation have stabilizing effect on the fluid layer.

MATHEMATICAL FORMULATIONS OF THE PROBLEM

The Physical Problem

In this chapter we shall investigate the combined effect of rotation and magnetic field on the onset of thermal convection in a horizontal layer Maxwellian visco-elastic nanofluid. The physical configuration of the problem to be considered as:

An infinite horizontal layer of Maxwellian visco-elastic nanofluid of thickness 'd' bounded by horizontal boundaries $z = 0$ and $z = d$. A Cartesian coordinate system (x, y, z) is chosen with the origin at the bottom of the fluid layer and the z- axis normal to the fluid layer. Fluid layer is rotating uniform about z-axis with angular velocity $\Omega(0, 0,\Omega)$ and is acted upon by a uniform vertical magnetic field $\mathbf{H}(0, 0, H)$. Fluid layer also acted upon by a gravity force $\mathbf{g}(0, 0,-g)$ and heated from below in such a way that horizontal boundaries $z = 0$ and $z = d$ respectively maintained at a uniform temperature T_0 and T_1 $(T_0 > T_1)$. The normal component of the nanoparticles flux has to vanish at an impermeable boundary and the temperature T is taken to be T_0 at $z = 0$ and T_1 at $z = d$, $(T_0 > T_1)$ as shown in Figure 1. The reference scale for temperature and nanoparticles fraction is taken to be T_1 and φ_0 respectively.

Figure 1. Physical configuration of the problem

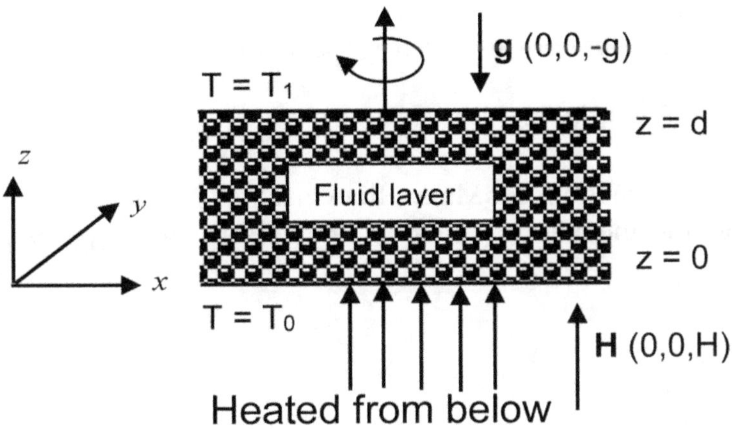

The following analysis is confined to a narrow and very long fluid layer. Since the significance of the centrifugal acceleration depends on the offset distance from the centre of rotation therefore for the layer which is adjacent to the rotation axis (i.e. x = 0, y = 0), the impact of the centrifugal acceleration to be zero. Due to the fact that here a narrow and very long fluid layer is considered, centrifugal effects can be neglected in the momentum equation.

Assumptions

The mathematical equations describing the physical model are based upon the following assumptions:

1. Thermophysical properties of fluid expect for density in the buoyancy force (Boussinesq Hypothesis) are constant,
2. The fluid phase and nanoparticles are in thermal equilibrium state and thus, the heat flow has been described using one equation model,
3. Dilute mixture,
4. Nanoparticles are spherical,
5. Nanoparticles are non-magnetic,
6. No chemical reactions take place in fluid layer,
7. Negligible viscous dissipation,
8. Radiative heat transfer between the sides of wall is negligible when compared with other modes of the heat transfer,
9. Nanofluid is incompressible, electrically conducting, Newtonian and laminar flow,

10. Each boundary wall is assumed to be impermeable and perfectly thermal conducting,
11. Angular velocity of fluid layer is assumed to be constant.

Governing Equations

The governing equations for Maxwellian visco-elastic nanofluid in the presence of rotation and magnetic field under the Boussinesq approximation are given as

$$\nabla \cdot \mathbf{v} = 0, \tag{4.1}$$

$$\rho\left(1+\lambda\frac{\partial}{\partial t}\right)\frac{d\mathbf{v}}{dt} = \left(1+\lambda\frac{\partial}{\partial t}\right)\left(-\nabla p + \left(\varphi\rho_p + (1-\varphi)\left\{\rho_f\left(1-\alpha\left(T-T_0\right)\right)\right\}\right)\mathbf{g}\right)$$

$$+\mu\nabla^2\mathbf{v} + 2\rho\left(\mathbf{v}\times\Omega\right) + \frac{\mu_e}{4\pi}\left(\mathbf{H}\cdot\nabla\right)\mathbf{H}, \tag{4.2}$$

where $\frac{d}{dt} = \frac{\partial}{\partial t} + (\mathbf{v}\cdot\nabla)$ is stands for convection derivative while q, p, ρ, μ, μ_e and α stands for fluid velocity, hydrostatic pressure, density of nanofluid, viscosity, magnetic permeability and the coefficient of thermal expansion respectively.

Equation of energy for Maxwellian visco-elastic nanofluid is given by

$$\rho c\frac{\partial T}{\partial t} + \mathbf{v}\cdot\nabla T = k_m\nabla^2 T + (\rho c)_p\left(D_B\nabla\varphi\cdot\nabla T + \frac{D_T}{T_1}\nabla T\cdot\nabla T\right), \tag{4.3}$$

where ρc is heat capacity of fluid, $(\rho c)_p$ is heat capacity of nanoparticles, T_1 is the temperature of the fluid layer at z = d and k_m is thermal conductivity.

Equation of continuity for the nanoparticles is given by

$$\frac{\partial\varphi}{\partial t} + \mathbf{v}\cdot\nabla Æ = D_B\nabla^2\varphi + \frac{D_T}{T_1}\nabla^2 T, \tag{4.4}$$

where D_B is the Brownian diffusion coefficient, given by Einstein-Stokes equation and D_T is the thermoporetic diffusion coefficient of the nanoparticles.

Maxwell equations are

$$\frac{d\mathbf{H}}{dt} = \left(\mathbf{H}\cdot\nabla\right)\mathbf{q} + \eta\nabla^2\mathbf{v}\,, \tag{4.5}$$

$$\nabla\cdot\mathbf{H} = 0, \tag{4.6}$$

where η is the electrical resitivity.

We assume that the temperature is constant and nanoparticles flux is zero on the boundaries. Thus boundary conditions are

$$\mathrm{w} = 0,\ \mathrm{T} = \mathrm{T}_0,\ \ \mathrm{D_B}\frac{\partial\phi}{\partial \mathrm{z}} + \frac{\mathrm{D_T}}{T_1}\frac{\partial T}{\partial \mathrm{z}} \quad = 0 \ \ \text{at } \mathrm{z} = 0$$

and

$$\mathrm{w} = 0,\ \mathrm{T} = \mathrm{T}_1,\ \ \mathrm{D_B}\frac{\partial\phi}{\partial \mathrm{z}} + \frac{\mathrm{D_T}}{T_1}\frac{\partial T}{\partial \mathrm{z}} \quad = 0 \ \ \text{at } \ \mathrm{z} = \mathrm{d}. \tag{4.7}$$

Introducing non-dimensional variables as

$$(\mathrm{x}',\mathrm{y}',\mathrm{z}',) = \left(\frac{\mathrm{x,y,z}}{\mathrm{d}}\right),\ \ \mathrm{v}'(\mathrm{u}',\mathrm{v}',\ \mathrm{w}',) = \upsilon\left(\frac{\mathrm{u,v,w}}{\kappa}\right)\mathrm{d},\ \ \mathrm{t}' = \frac{\mathrm{t}\kappa}{\mathrm{d}^2},\ \ \mathrm{p}' = \frac{\mathrm{pd}^2}{\mu\kappa},$$

$$\varphi' = \frac{\left(\varphi - \varphi_0\right)}{\varphi_0},\ \ \mathrm{T}' = \frac{\left(\mathrm{T} - \mathrm{T}_1\right)}{\left(\mathrm{T}_0 - \mathrm{T}_1\right)},\ \ \mathrm{H}' = \frac{H}{|H|}$$

where

$\kappa = \dfrac{\mathrm{k_m}}{\rho \mathrm{c}}$ is thermal diffusivity of the fluid.

Equations (4.1) - (4.7) in non-dimensional form can be written as

$$\nabla'\cdot\mathbf{v}' = 0, \tag{4.8}$$

$$\left(1 + F\frac{\partial}{\partial t'}\right)\frac{1}{\mathrm{Pr}}\frac{\partial\mathbf{v}'}{\partial \mathrm{t}'} = \left(1 + F\frac{\partial}{\partial t'}\right)\left(-\nabla'\mathrm{p}'\text{-}\mathrm{Rm}\hat{\mathrm{e}}_z + \mathrm{RaT}'\hat{\mathrm{e}}_z - \mathrm{Rn}\varphi'\hat{\mathrm{e}}_z\right) + \nabla'^2\mathbf{v}'$$
$$+ \frac{\mathrm{Pr}}{\mathrm{Pr}_M}Q\left(\mathbf{H}'.\nabla'\right)\mathbf{H}' + \sqrt{\mathrm{Ta}}\left(\mathrm{v}'\hat{\mathrm{e}}_x\text{-}\mathrm{u}'\hat{\mathrm{e}}_y\right), \tag{4.9}$$

$$\frac{\partial \varphi'}{\partial t'} + \mathbf{v}' \cdot \nabla' \varphi' = \frac{1}{\mathrm{Le}} \nabla'^2 \varphi' + \frac{\mathrm{N_A}}{\mathrm{Le}} \nabla'^2 \mathrm{T}', \tag{4.10}$$

$$\frac{\partial \mathrm{T}'}{\partial t'} + \mathbf{v}' \cdot \nabla' \mathrm{T}' = \nabla'^2 \mathrm{T}' + \frac{\mathrm{N_B}}{\mathrm{Le}} \nabla' \varphi' \cdot \nabla' \mathrm{T}' + \frac{\mathrm{N_A N_B}}{\mathrm{Le}} \nabla' \mathrm{T}' \cdot \nabla' \mathrm{T}', \tag{4.11}$$

$$\frac{d\mathbf{H}'}{dt'} = \left(\mathbf{H}' \cdot \nabla'\right) \mathbf{v}' + \frac{\mathrm{Pr}}{\mathrm{Pr}_M} \nabla'^2 \mathbf{v}', \tag{4.12}$$

$$\nabla' \cdot \mathbf{H}' = 0. \tag{4.13}$$

Here the non-dimensional parameters are given as follows

$\mathrm{Pr} = \frac{\mu}{\rho\kappa}$ is the Prandtl number,

$\mathrm{Pr}_M = \frac{\mu}{\rho\eta}$ is the magnetic Prandtl number,

$Le = \frac{\kappa}{\mathrm{D_B}}$ is the Lewis number,

$F = \frac{\kappa\lambda}{d^2}$ is the stress relaxation parameter,

$\mathrm{Ra} = \frac{\rho g \alpha d^3 \left(\mathrm{T_0} - \mathrm{T_1}\right)}{\mu\kappa}$ is the Rayleigh number,

$\mathrm{Rm} = \frac{\left(\rho_p \varphi_0 + \rho\left(1-\varphi_0\right)\right) g d^3}{\mu\kappa}$ is the density Rayleigh number,

$\mathrm{Rn} = \frac{\left(\rho_p - \rho\right) \varphi_0 g d^3}{\mu\kappa}$ is the nanoparticles Rayleigh number,

$\mathrm{Ta} = \left(\frac{2\Omega d^2}{\nu}\right)^2$ is the Taylor number,

$\mathrm{Q} = \frac{\mu_e H^2 d^2}{4\pi\rho\upsilon\eta}$ is the Chandrasekhar number,

$$N_A = \frac{D_T(T_0\text{-}T_1)}{D_B T_1 \varphi_0}$$ is the modified diffusivity ratio,

$$N_B = \frac{(\rho c)_p \varphi_0}{(\rho c)_f}$$ is the modified particle-density increment.

In spirit of Oberbeck-Boussinesq approximation, equation (4.9) has been linearized by the neglect of a term proportional to the product of φ_0 and T. This approximation is valid in the case of small temperature gradients in a dilute suspension of nanoparticles.

The dimensionless boundary conditions are

$$w' = 0,\quad T' = 1,\quad \frac{\partial \phi'}{\partial z'} + N_A \frac{\partial T'}{\partial z'} = 0 \text{ at } z' = 0$$

and

$$w' = 0,\ T' = 0,\ \frac{\partial \phi'}{\partial z'} + N_A \frac{\partial T'}{\partial z'} = 0 \text{ at } z' = 1. \tag{4.14}$$

THE BASIC STATE AND ITS SOLUTIONS

The basic state was assumed to be quiescent and is given by

$$v_i'\left(u', v', w'\right) = 0,$$

$$p' = p_b(z),$$

$$H' = He_z,$$

$$T' = T_b(z), \tag{4.15}$$

$$\varphi' = \varphi_b(z) \text{ and}$$

$$\rho = \rho_0\left(1 + \alpha\left(T\text{-}T_0\right)\right).$$

Equations (4.8) – (4.11) reduce to

$$0 = -\frac{dp_b}{dz'}\text{-Rm} + \text{Ra}T_b - \text{Rn}\varphi_b, \tag{4.16}$$

$$\frac{d^2T_b}{dz'^2} + \frac{N_B}{\text{Le}}\frac{d\varphi_b}{dz'}\frac{dT_b}{dz'} + \frac{N_A N_B}{\text{Le}}\left(\frac{dT_b}{dz'}\right)^2 = 0, \tag{4.17}$$

$$\frac{d^2\varphi_b}{dz'^2} + N_A\frac{d^2T_b}{dz'^2} = 0. \tag{4.18}$$

Using boundary conditions in (4.14), equation (4.18) gives

$$\varphi_b = -N_A T_b + (1 - N_A)z' + N_A. \tag{4.19}$$

On substituting the value of the φ_b from equation (4.19) in equation (4.17), we get

$$\frac{d^2T_b}{dz'} + \frac{(1\text{-}N_A)N_B}{\text{Le}}\frac{dT_b}{dz'} = 0. \tag{4.20}$$

On integrating equation (4.20) with respect to z' and using boundary conditions (4.14), we get

$$T_b = \frac{1 - e^{-(1-N_A)N_B(1-z')/Le}}{1 - e^{-(1-N_A)N_B/Le}}. \tag{4.21}$$

According to Buongiorno (2006), for most nanofluid investigated so far Le is large, is of order 10^2-10^3, while N_A is no greater than about 10. Then, the exponents in equation (4.21) are small. By expanding the exponential function into the power series and retaining up to the first order is negligible. Thus an approximate solution for the basic state is given by

$T_b = 1 - z'$ and

$\varphi_b = \phi_0 + N_A z'$

PERTURBATION SOLUTIONS

Let the initial basic state described by (4.15) is slightly perturbed so that perturbed state is given by

$$(u',v',w') = 0 + (u'',v'',w''),$$

$$H' = He_z + H''(H''x, H''y, H''z),$$

$$T' = T_b + T'', \tag{4.22}$$

$$p' = p_b + p'',$$

$$\varphi' = \varphi_b + \varphi'',$$

where $T_b = 1 - z'$, $\varphi_b = \phi_0 + N_A z$ and (u'',v'',w''), $H''(H''x,H''y,H''z)$, T'', φ'' and p'' respectively the perturbations in the initial velocity, magnetic field, temperature, pressure and volume fraction of the nanoparticles.

By substituting (4.22) in equations (4.8) - (4.13) and linearize by neglecting the product of the prime quantities, we obtained following equations

$$\nabla \cdot \mathbf{v} = 0, \tag{4.23}$$

$$\left(1 + F\frac{\partial}{\partial t}\right)\frac{1}{\Pr}\frac{\partial \mathbf{v}}{\partial t} = \left(1 + F\frac{\partial}{\partial t}\right)\left(-\nabla p + \mathrm{Ra}T\hat{e}_z - \mathrm{Rn}\varphi\hat{e}_z\right) + \nabla^2\mathbf{v}$$

$$+\frac{\Pr}{\Pr_M}Q\frac{\partial \mathbf{H}}{\partial z} + \sqrt{\mathrm{Ta}}\left(v\hat{e}_x \text{-} u\hat{e}_y\right), \tag{4.24}$$

$$\frac{\partial \varphi}{\partial t} + wN_A = \frac{1}{\mathrm{Le}}\nabla^2\varphi + \frac{N_A}{\mathrm{Le}}\nabla^2 T, \tag{4.25}$$

$$\frac{\partial T}{\partial t} - w = \nabla^2 T + \frac{N_B}{\mathrm{Le}}\left(N_A\frac{\partial T}{\partial z} - \frac{\partial \phi}{\partial z}\right) - \frac{2N_A N_B}{\mathrm{Le}}\frac{\partial T}{\partial z}, \tag{4.26}$$

$$\frac{\partial H}{\partial t} = \frac{\partial w}{\partial z} + Q\nabla^2 H, \tag{4.27}$$

$$\nabla \cdot H = 0. \tag{4.28}$$

Boundary conditions are

$$w = 0, T = 0, \quad \frac{\partial \phi}{\partial z} + N_A \frac{\partial T}{\partial z} = 0 \text{ at } z = 0,1. \tag{4.29}$$

[Dashes (") have been suppressed for convenience]

Eliminating pressure term p and magnetic field term H from equation (4.24) by making use of the equations (4.23), (4.27) and (4.28), we have

$$\begin{aligned}&\left(\begin{aligned}&\left((1+nF)\left(\nabla^2-\frac{n}{\Pr}\right)\left(\nabla^2-\frac{n(1+nF)}{\Pr}\right)\left(\frac{\Pr}{\Pr_M}\nabla^2-n\right)\nabla^2\right)\\&+TaD^2\left(\frac{\Pr}{\Pr_M}\nabla^2-n\right)-\left(\frac{\Pr}{\Pr_M}QD^2\right)\left(\nabla^2-\frac{n}{\Pr}\right)\nabla^2\end{aligned}\right)W\\&-(1+nF)^2\left(\nabla^2-\frac{n}{\Pr}\right)\left(\frac{\Pr}{\Pr_M}\nabla^2-n\right)\left(a^2\mathrm{Ra}\Theta-a^2\mathrm{Rn}\Phi\right)=0.\end{aligned} \tag{4.30}$$

NORMAL MODE ANALYSIS

Analyzing the disturbances into the normal modes and assuming that the perturbed quantities are of the form

$$[w, T, \varphi] = [W(z), \Theta(z), \Phi(z)] \exp\left(ik_x x + ik_y y\right), \tag{4.31}$$

where k_x, k_y are wave numbers in x and y direction.

Using equation (4.31), equations (4.30), (4.25) and (4.26) become

$$\begin{aligned}&\left(\begin{aligned}&\left((1+nF)\left(D^2-a^2-\frac{n}{\Pr}\right)\left(D^2-a^2-\frac{n(1+nF)}{\Pr}\right)\left(\frac{\Pr}{\Pr_M}(D^2-a^2)-n\right)(D^2-a^2)\right)\\&+TaD^2\left(\frac{\Pr}{\Pr_M}(D^2-a^2)-n\right)-\left(\frac{\Pr}{\Pr_M}QD^2\right)(D^2-a^2)\left(D^2-a^2-\frac{n}{\Pr}\right)\end{aligned}\right)W\\&-(1+nF)^2\left(D^2-a^2-\frac{n}{\Pr}\right)\left(\frac{\Pr}{\Pr_M}(D^2-a^2)-n\right)\left(a^2\mathrm{Ra}\Theta-a^2\mathrm{Rn}\Phi\right)=0,\end{aligned} \tag{4.32}$$

$$N_A W - \frac{N_A}{Le}\left(D^2 - a^2\right)\Theta - \left(\frac{1}{Le}\left(D^2 - a^2\right) - n\right)\Phi = 0, \tag{4.33}$$

$$W + \left(D^2 - a^2 - n - \frac{N_A N_B}{Le} D\right)\Theta - \frac{N_B}{Le} D\Phi = 0, \tag{4.34}$$

where $D \equiv \frac{d}{dz}$ and $a = \sqrt{k_x^2 + k_y^2}$ is the dimensionless resultant wave number.

The boundary conditions of the problem in view of normal mode analysis are

$$W = 0, D^2W = 0, \Theta = 0, D\Phi + N_A D\Theta = 0 \text{ at } z = 0,1. \tag{4.35}$$

METHOD OF SOLUTION

The Galerkin weighted residuals method is used to obtain an approximate solution to the system of equations (4.32) - (4.34) with boundary conditions (4.35). In this method, the test functions are the same as the base (trial) functions. Accordingly W, Θ and Φ are taken as

$$W = \sum_{p=1}^{N} A_p W_p, \Theta = \sum_{p=1}^{N} B_p \Theta_p, \Phi = \sum_{p=1}^{N} C_p \Phi_p, \tag{4.36}$$

where $W_p = \Theta_p = \sin p\pi z, \Phi_p = -N_A \sin p\pi z$, A_p, B_p and C_p are unknown coefficients, p = 1, 2, 3,..., N and the base functions W_p, Θ_p and Φ_p satisfying the boundary conditions (4.35). Using expression for W, Θ and Φ in equations (4.32) – (4.34) and multiplying the first equation by W_p the second equation by Θ_p and third equation by Φ_p and then integrating in the limits from zero to unity, we obtain a set of 3N linear homogeneous equations with 3N unknown A_p, B_p and C_p; p = 1, 2, 3,..., N. For existing of nontrivial solution, the vanishing of the determinant of coefficients produces the characteristics equation of the system in term of Rayleigh number Ra.

LINEAR STABILITY ANALYSIS

For the present formulation, we have considered the which system of equations (4.32) - (4.34) together with the boundary conditions (4.35) constitute a linear eigenvalue problem with variable coefficient for the growth rate of disturbance n of the system. Substituting equation (4.36) into the system of equations (4.32) - (4.34) and multiplying the first equation by W_p the second equation by Θ_p and third equation by Φ_p and then integrating in the limits from zero to unity and performing some integration by parts, one obtains the following matrix equation

$$\begin{bmatrix} (\pi^2+a^2)+\dfrac{\pi^2 Ta}{(\pi^2+a^2)}+\pi^2 Q & -a^2 Ra & -a^2 N_A Rn \\ 1 & -(\pi^2+a^2) & 0 \\ 1 & \dfrac{1}{Le}(\pi^2+a^2) & -\dfrac{1}{Le}(\pi^2+a^2) \end{bmatrix}\begin{bmatrix} W_0 \\ \Theta_0 \\ \Phi_0 \end{bmatrix}=\begin{bmatrix} 0 \\ 0 \\ 0 \end{bmatrix}. \tag{4.37}$$

The non-trivial solution of the above matrix requires that

$$Ra=\frac{1}{a^2}\left((\pi^2+a^2)^3+\pi^2 Ta+(\pi^2+a^2)\pi^2 Q\right)-(1+Le)N_A Rn. \tag{4.38}$$

It is observed that stationary Rayleigh number Ra is function of the Taylor number Ta (rotation), Chandrasekhar number Q (magnetic field), Lewis number Le, the modified diffusivity ratio N_A and the nanoparticles Rayleigh Rn but independent of visco- elastic parameter F, Prandtl number Pr, magnetic Prandtl number Pr_M and modified particle-density increment N_B. Thus Maxwellian visco-elastic nanofluid behaves like an ordinary Newtonian nanofluid and instability is purely a phenomenon due to buoyancy coupled with the conservation of nanoparticles.

In the absence of magnetic field (Q = 0) equation (4.38) reduces to

$$Ra=\frac{(\pi^2+a^2)^3+\pi^2 Ta}{a^2}-(1+Le)N_A Rn. \tag{4.39}$$

This is the good agreement of the result (2.38) obtained in Chapter 2.

In the absence of rotation (Ta = 0) equation (4.38) reduces to

$$Ra=\frac{(\pi^2+a^2)^3+(\pi^2+a^2)\pi^2 Q}{a^2}-(1+Le)N_A Rn. \tag{4.40}$$

This is the good agreement of the result (3.38) obtained in Chapter 3.

In the absence of both magnetic field and rotation (Q = Ta = 0) equation (4.38) reduces to

$$Ra = \frac{\left(\pi^2 + a^2\right)^3}{a^2} - \left(1 + Le\right) N_A Rn. \tag{4.41}$$

This is the good agreement of the result (1.47) obtained in Chapter 1.

To find the critical value of Ra, equation (4.38) is differentiated with respect to 'a^2' and then equated to zero. The minimum value of the Rayleigh number Ra occurs at the critical wave number $a = a_c$ where a_c satisfies the equation.

$$2\left(a_c^2\right)^3 + 3\pi^2\left(a_c^2\right)^2 - \left(\pi^2 Ta + \pi^4 Q + \pi^6\right) = 0\,. \tag{4.42}$$

It is important to note that the critical wave number a_c depends on the Taylor number Ta and Chandrasekhar number Q.

In the absence of Taylor number Ta and Chandrasekhar number Q (Ta = Q = 0), minimum value of the Rayleigh number $(Ra)_c$ attained at $a_c = \frac{\pi}{\sqrt{2}}$ and minimum value found to $\frac{27\pi^4}{4}$ so the corresponding critical Rayleigh number given by

$$\left(Ra\right)_c = \frac{27\pi^4}{4} - \left(1 + Le\right) N_A Rn\,. \tag{4.43}$$

In the absence of rotation, magnetic field and nanoparticles (Ta = Q = Rn = Le = N_A = 0), one recovers the well- known results that the critical Rayleigh number is equal to $\left(Ra\right)_c = \frac{27\pi^4}{4}$.

This is good agreement of the result obtained by Chandrasekhar (1961).

Thus presence of the nanoparticles lowers the value of the critical Rayleigh number by usually by substantial amount. Also parameter N_B does not appear in the equation (4.38); thus instability is purely phenomena due to buoyancy coupled with conservation of nanoparticles. Thus average contribution of nanoparticles flux in the thermal energy equation is zero.

RESULTS AND DISCUSSION

To study the effect of rotation (Taylor number Ta), magnetic field, (Chandrasekhar number Q), Lewis number Le, modified diffusivity ratio N_A and nanoparticles Rayleigh number Rn on stationary convection, we examine the behavior of $\frac{\partial(\mathrm{R}a)_s}{\partial Ta}, \frac{\partial(\mathrm{R}a)_s}{\partial Q}, \frac{\partial(\mathrm{R}a)_s}{\partial Le}, \frac{\partial(\mathrm{R}a)_s}{\partial N_A}$ and $\frac{\partial(\mathrm{R}a)_s}{\partial Rn}$ analytically.

From equation (2.38), we have

(i) $\frac{(\partial \mathrm{R}a)_s}{\partial Ta} > 0,$

(ii) $\frac{(\partial \mathrm{R}a)_s}{\partial Q} > 0,$

(iii) $\frac{(\partial \mathrm{R}a)_s}{\partial Le} < 0,$

(iv) $\frac{(\partial \mathrm{R}a)_s}{\partial N_A} < 0,$

(v) $\frac{(\partial \mathrm{R}a)_s}{\partial Rn} < 0.$

These inequalities imply that rotation and magnetic field stabilize the fluid layer while Lewis number Le, modified diffusivity ratio N_A and nanoparticles Rayleigh number Rn destabilizes the fluid layer.

Expression for stationary Rayleigh number, which characterizes the stability of the system, is found to be depend upon Taylor number (rotation), Chandrasekhar number (magnetic field), Lewis number, modified diffusivity ratio and nanoparticles Rayleigh number. The computations are carried out for different values of parameters considered in the range $10^2 \le Ra \le 10^5$ (thermal Rayleigh number), $10 \le Ta \le 10^4$ (Taylor number), $10 \le Q \le 10^4$ (Chandrasekhar number), $1 \le N_A \le 10$ (modified diffusivity ratio), $10^2 \le Le \le 10^4$ (Lewis number), $10^{-1} \le Rn \le 10^1$ (nanoparticles Rayleigh number).

The variations of the stationary thermal Rayleigh number with wave number have been plotted graphically as shown in Figures 2 – 6.

Figure 2 shows only the variation of stationary Rayleigh number with wave number for different value of the Taylor number with fixed value of other parameters and it is found that the Rayleigh number increases with an increase in the value of the Taylor number, which imply that rotation delay the onset of stationary convection. It is due to the fact Coriolis force due to

Figure 2. Variation of the Rayleigh number with wave number for different value of Taylor number

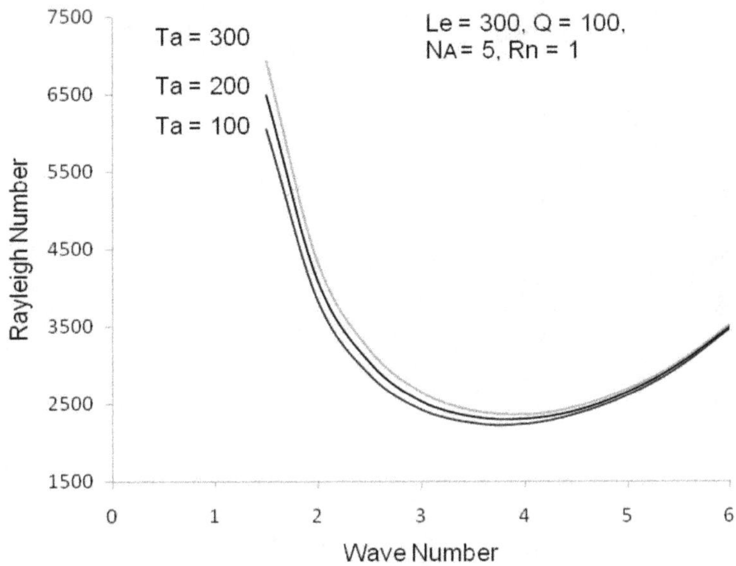

rotation drags the perturbed transverse motion and kinetic energy is dissipated by the viscosity, which intern increases the critical electric strength required for the onset of convection. This is good agrement of the result obtained by Chand and Rana (2012b).

Figure 3 shows the variation of thermal Rayleigh number for different value of Chandrasekhar number and for the fixed value of other parameters. It is found that stationary Rayliegh number increases as the values of Chandrasekhar number increases, indicating that magnetic field stabilizes the stationary convection. This is good agrement of the result obtained by Chand (2013b).

Figure 4 shows the variation of thermal Rayleigh number for different value of Lewis number Le and for the fixed value of other parameters. It is found that stationary Rayliegh number decreases as the values of Lewis number increases, indicating that Lewis number destabilizes the stationary convection. It is due to the fact that thermophoresis at a higher value of thermophoretic diffusivity is more supportable to the disturbance in nanofluids, while both thermophoresis and Brownian motion are driving forces in favor of the motion of nanoparticles. This is good agrement of the result obtained by Chand and Rana (2015a, 2014d).

Figure 3. Variation of the Rayleigh number with wave number for different value of Chandrasekhar number

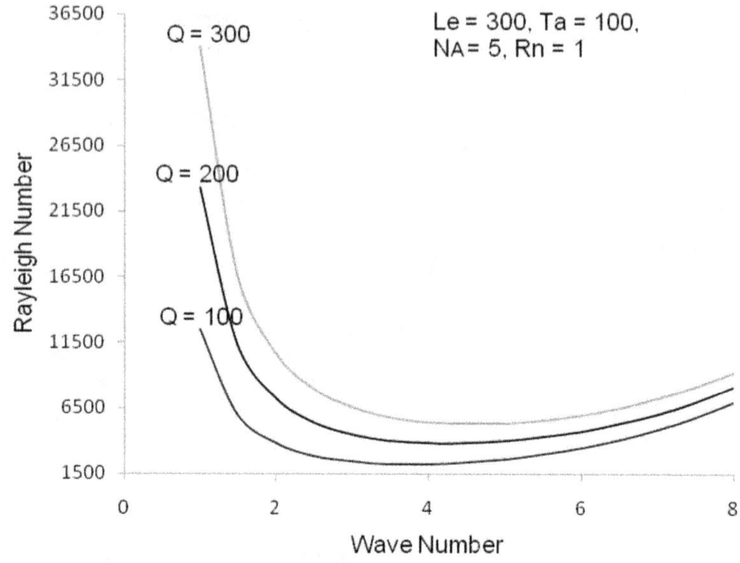

Figure 4. Variation of the Rayleigh number with wave number for different value of Lewis number

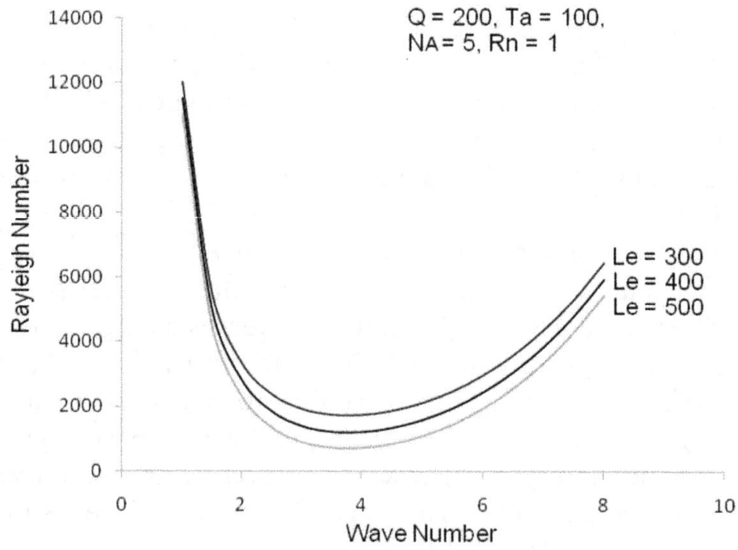

Figure 5 shows the variation of stationary Rayleigh number for different value of the modified diffusivity ratio N_A and fixed value of other parameters and it is found that Rayleigh number decreases with an increase in the value of the modified diffusivity ratio N_A, which means that the modified diffusivity ratio N_A destabilizes on the fluid layer. This may lead to an increase in volumetric fraction, which shows that Brownian motion of the nanoparticles will also increase, which may cause destabilizing effect. This is good agrement of the result obtained by Chand & Rana (2015a, 2014d).

Figure 6 shows the variation of stationary Rayleigh number with wave number for different value of the nanoparticles Rayleigh number Rn and fixed value of other parameters and it is found that thermal Rayleigh number decreases with an increase in value of the nanoparticles Rayleigh number Rn, which means that the nanoparticles Rayleigh number Rn has destabilizing effect on fluid layer. It has destabilizing effect because the heavier nanoparticles moving through the base fluid makes more strong disturbances as compared with the lighter nanoparticles.

Figure 5. Variation of the Rayleigh number with wave number for different value of modified diffusivity ratio

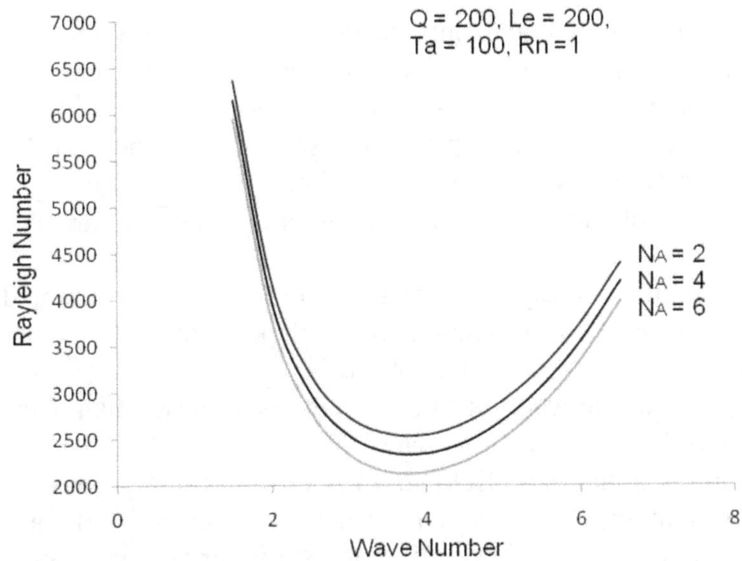

Figure 6. Variation of the Rayleigh number with wave number for different value of nanoparticles Rayleigh number

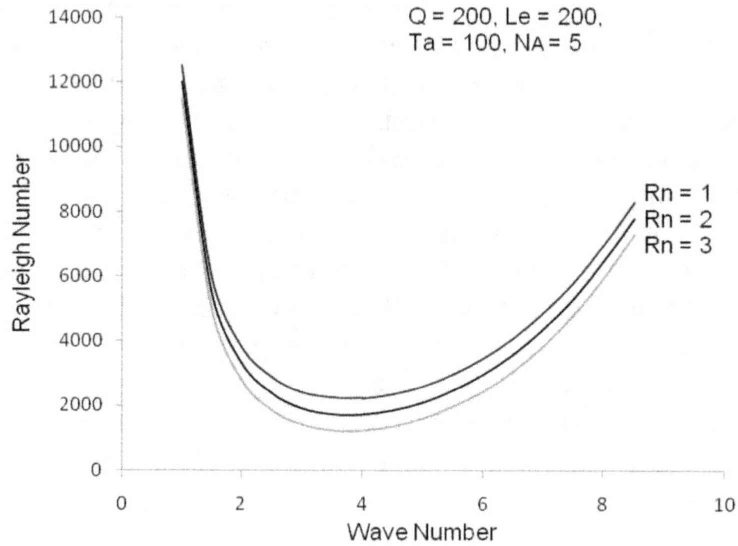

CONCLUSION

Thermal convection in a horizontal layer of Maxwellian visco-elastic nanofluid is studied in the presence of both magnetic field and rotation. The flux of volume fraction of nanoparticles is taken to be zero on the isothermal boundaries and the eigen value problem is solved using the Galerkin residual method. The results have been presented both analytically and graphically.

The main conclusions derived from the present chapter are as follows:

1. The instability purely phenomenon due to buoyancy coupled with the conservation of nanoparticle and is independent of the contribution of Brownian motion and thermophoresis.
2. For stationary convection Maxwellian visco-elastic nanofluid behaves like an ordinary Newtonian nanofluid.
3. Rotation and magnetic field stabilizes the stationary convection while Lewis number Le, modified diffusivity ratio N_A and nanoparticles Rayleigh number Rn destabilizes the stationary convection.

REFERENCES

Buongiorno, J. (2006). Convective transport in nanofluids. *ASME Journal of Heat Transfer*, *128*(3), 240–250. doi:10.1115/1.2150834

Chand, R. (2013b). On the onset of Rayleigh-Bénard convection in a layer of nanofluid in Hydromagnetics. *Int. J. of Nanoscience*, *12*(6), 1350038. doi:10.1142/S0219581X13500385

Chand, R., & Rana, G. C. (2012b). On the onset of thermal convection in rotating nanofluid layer saturating a Darcy-Brinkman porous medium. *International Journal of Heat and Mass Transfer*, *55*(21-22), 5417–5424. doi:10.1016/j.ijheatmasstransfer.2012.04.043

Chand, R., & Rana, G. C. (2014d). Thermal instability in a Brinkman porous medium saturated by nanofluid with no nanoparticle flux on boundaries. *Special Topics & Reviews in Porous Media: An International Journal*, *5*(4), 277–286. doi:10.1615/SpecialTopicsRevPorousMedia.v5.i4.10

Chand, R., & Rana, G. C. (2015a). Magneto convection in a layer of nanofluid in porous medium- a more realistic approach. *Journal of Nanofluids*, *4*(2), 196–202. doi:10.1166/jon.2015.1142

Chandrasekhar, S. (1961). *Hydrodynamic and Hydromagnetic Stability*. New York: Dover Publication.

Mahajan, A., & Arora, M. (2013). Convection in rotating magnetic nanofluids. *Applied Mathematics and Computation*, *219*(11), 6284–6296. doi:10.1016/j.amc.2012.12.012

Parekh, K., & Lee, H. S. (2011). Experimental investigation of thermal conductivity of magnetic nanofluids. In *Proceedings of the 56th DAE Solid State Physics Symposium 2011,AIP Conference Proceedings*, (vol. 1447, pp. 385–386).

Patel, R. (2012). Effective viscosity of magnetic nanofluids through capillaries. *Physical Review E: Statistical, Nonlinear, and Soft Matter Physics*, *85*(2), 026316. doi:10.1103/PhysRevE.85.026316 PMID:22463326

Chapter 5
Thermal Convection in a Layer of Maxwellian Visco-Elastic Nanofluid in a Porous Medium:
Darcy Model

INTRODUCTION

Thermal instability in a porous medium has many applications in geophysics, food processing, oil reservoir modeling, petroleum industry, bio-mechanics, building of thermal insulations and nuclear reactors. Many researchers have investigated thermal instability problems by taking different types of fluids. When a fluid permeates through an isotropic and homogeneous porous medium, the gross effect is represented by Darcy's law. According to which when fluid slowly percolates through the pores of the rock, the gross effect is represented by the Darcy's law. According to which the resistance term $-\frac{\mu}{k_1}q$ will replace the usual viscous term in the equation of motion, μ is

DOI: 10.4018/978-1-68318-006-7.ch005

viscosity of the fluid, k_1 is permeability of medium and **q** is the seepage velocity of fluid. The study of a horizontal layer of fluid heated from below in a porous media is of importance in geophysics, soil sciences, ground water hydrology and astrophysics. Lapwood (1948) and Wooding (1960) considered the stability of flow of a fluid through a porous medium taking into account the Darcy's law. A good account of convection problems in a porous medium is given by Vafai and Hadim (2000), Ingham and Pop (1981), Nield and Bejan (2013).

The thermal instability nanofluid in a porous medium has been a topic interest due to its applications in fields of food and chemical process, petroleum industry, bio-mechanics and geophysical problems. Owing the applications of the nanofluid and porous media theory in chemical engineering to study theory in drying and freezing of food, in cooling of microchips in computers by use of metal foams and their use in heat pipes etc. study of nanofluid in porous medium turns to be important to the researchers. Thermal instability of nanofluid in a porous medium has been studied by Nield and Kuznetsov (2009a, 2009b), Nield and Kuznetsov (2010b, 2011b), Kuznetsov and Nield (2010a, 2010b, 2010c), Chand and Rana (2012b).

In this chapter we studied the thermal instability of Maxwellian visco-elastic nanofluid in a porous medium for more realistic boundary condition. For porous medium Darcy model has been used and it is assumed that nanoparticles are suspended in the nanofluid using either surfactant or surface charge technology. This prevents the particles from agglomeration and deposition on the porous matrix.

MATHEMATICAL FORMULATIONS OF THE PROBLEM

The Physical Problem

In this chapter we shall investigate the thermal convection in a horizontal layer Maxwellian visco-elastic nanofluid. The physical configuration of the problem to be considered as:

An infinite horizontal layer of Maxwellian visco-elastic nanofluid of thickness'd' bounded by plane $z = 0$ and $z = d$, heated from below in a porous medium of medium permeability k_1 and porosity ε. Fluid layer is acted upon by a gravity force $\mathbf{g}(0,0,-g)$ and is heated from below in such a way that horizontal boundaries $z = 0$ and $z = d$ respectively maintained at a uniform temperature T_0 and T_1 ($T_0 > T_1$) as shown is Figure 1. The normal component of the nanoparticles flux has to vanish at an impermeable boundaries and

Figure 1. Physical configuration of the problem

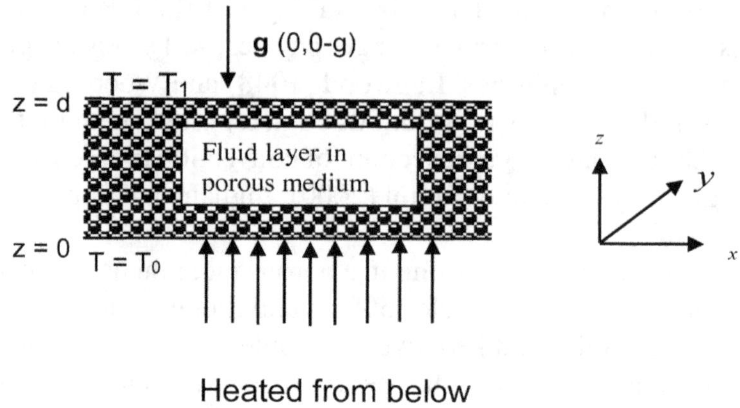

the reference scale for temperature and nanoparticles fraction is taken to be T_1 and φ_0 respectively It is assumed that nanoparticles are suspended in the nanofluid using either surfactant or surface charge technology. This prevents the particles from agglomeration and deposition on the porous matrix.

Assumptions

The mathematical equations describing the physical model are based upon the following assumptions:

1. Thermophysical properties of fluid expect for density in the buoyancy force (Boussinesq Hypothesis) are constant,
2. The fluid phase and nanoparticles are in thermal equilibrium state and thus, the heat flow has been described using one equation model,
3. Dilute mixture,
4. Nanoparticles are spherical,
5. Nanoparticles are non-magnetic,
6. No chemical reactions take place in fluid layer,
7. Negligible viscous dissipation,
8. Radiative heat transfer between the sides of wall is negligible when compared with other modes of the heat transfer,
9. Nanofluid is incompressible, electrically conducting, Newtonian and laminar flow,
10. Each boundary wall is assumed to be impermeable and perfectly thermal conducting,
11. Nanoparticles are suspended in the nanofluid using either surfactant or surface charge technology.

GOVERNING EQUATIONS

The governing equations for Maxwellian visco-elastic nanofluid in a porous medium under the Boussinesq approximation are given as

$$\nabla \cdot \mathbf{q} = 0, \tag{5.1}$$

$$\frac{\rho}{\varepsilon}\left(1+\lambda\frac{\partial}{\partial t}\right)\frac{d\mathbf{q}}{dt} = \left(1+\lambda\frac{\partial}{\partial t}\right)\left(-\nabla p+\rho g\right) - \frac{\mu}{k_1}\mathbf{q}, \tag{5.2}$$

where **q(u, v, w)** is the Darcy velocity vector, p is the hydrostatic pressure, μ is viscosity, α is the coefficient of thermal expansion, λ is the relaxation time, φ is the volume fraction of the nanoparticles, k_1 is the medium permeability, ε is the porosity parameter, ρ_p density of nanoparticles and ρ_f density of base fluid and $\frac{d}{dt} = \frac{\partial}{\partial t} + \frac{1}{\varepsilon}(\mathbf{q}\cdot\nabla)$ is stands for convection derivative.

The equation of energy for nanofluid in porous medium is

$$(\rho c)_m \frac{\partial T}{\partial t} + (\rho c)_f\, \mathbf{q}\cdot\nabla T = k_m \nabla^2 T + \varepsilon(\rho c)_p\left(D_B \nabla\varphi\cdot\nabla T + \frac{D_T}{T_1}\nabla T\cdot\nabla T\right), \tag{5.3}$$

where $(\rho c)_m$ is effective heat capacity of fluid, $(\rho c)_p$ is heat capacity of nanoparticles and km is effective thermal conductivity of the porous medium.

The continuity equation for the nanoparticles is

$$\frac{\partial\varphi}{\partial t} + \frac{1}{\mu}\mathbf{q}\cdot\nabla\varphi = D_B\nabla^2\varphi + \frac{D_T}{T_1}\nabla^2 T, \tag{5.4}$$

Where D_B is the Brownian diffusion coefficient, given by Einstein-Stokes equation and D_T is the thermoporetic diffusion coefficient of the nanoparticles.

We assume that the temperature is constant and nanoparticles flux is zero on the boundaries. Thus boundary conditions are

$$w = 0,\ T = T_0,\ D_B\frac{\partial\phi}{\partial z} + \frac{D_T}{T_1}\frac{\partial T}{\partial z} = 0 \ \text{at } z = 0$$

and

$$w = 0, T = T_1, \; D_B \frac{\partial \phi}{\partial z} + \frac{D_T}{T_1} \frac{\partial T}{\partial z} = 0 \;\text{ at }\; z = d. \tag{5.5}$$

Introducing non-dimensional variables as

$$(x', y', z') = \left(\frac{x,y,z}{d}\right), \; v'(u', v', w') = \upsilon\left(\frac{u,v,w}{\kappa}\right) d, \; t' = \frac{t^o}{\sigma d^2}, \; p' = \frac{p k_1}{\mu \kappa},$$

$$\varphi' = \frac{(\varphi - \varphi_0)}{\varphi_0}, \; T' = \frac{(T - T_1)}{(T_0 - T_1)},$$

where $\kappa = \dfrac{k_m}{(\rho c_p)_f}$ is thermal diffusivity of the fluid.

Equations (5.1) - (5.5) in non-dimensional form can be written as

$$\nabla' \cdot \mathbf{q}' = 0, \tag{5.6}$$

$$\left(1 + F\frac{\partial}{\partial t'}\right)\frac{1}{\Pr}\frac{\partial \mathbf{q}'}{\partial t'} = \left(1 + F\frac{\partial}{\partial t'}\right)\left(-\nabla' p' \text{-} Rm\hat{e}_z + RaT'\hat{e}_z - Rn\varphi'\hat{e}_z\right) - \mathbf{q}', \tag{5.7}$$

$$\frac{1}{\sigma}\frac{\partial \varphi'}{\partial t'} + \frac{1}{\varepsilon}\mathbf{q}' \cdot \nabla' \varphi' = \frac{1}{Le}\nabla'^2 \varphi' + \frac{N_A}{Le}\nabla'^2 T', \tag{5.8}$$

$$\frac{\partial T'}{\partial t'} + \mathbf{q}' \cdot \nabla' T' = \nabla'^2 T' + \frac{N_B}{Le}\nabla' \varphi' \cdot \nabla' T' + \frac{N_A N_B}{Le}\nabla' T' \cdot \nabla' T'. \tag{5.9}$$

Here the non-dimensional parameters are given as follows:

$Da = \dfrac{k_1}{d^2}$ is the Darcy number,

$\Pr = \dfrac{\mu}{\rho \kappa}$ is the Prandtl number,

$Le = \dfrac{\kappa}{D_B}$ is the Lewis number,

$F = \dfrac{\kappa \lambda}{d^2}$ is the stress relaxation parameter,

$$\mathrm{Ra} = \frac{\rho g \alpha d k_1 \left(T_0 - T_1\right)}{\mu \kappa} \text{ is the Rayleigh Darcy number,}$$

$$\mathrm{Rm} = \frac{\left(\rho_p \varphi_0 + \rho\left(1\text{-}\varphi_0\right)\right) g d k_1}{\mu \kappa} \text{ is the density Rayleigh Darcy number,}$$

$$\mathrm{Rn} = \frac{\left(\rho_p - \rho\right) \varphi_0 g d k_1}{\mu \kappa} \text{ is the nanoparticles Rayleigh Darcy number,}$$

$$N_A = \frac{D_T \left(T_0\text{-}T_1\right)}{D_B T_1 \varphi_0} \text{ is the modified diffusivity ratio,}$$

$$N_B = \frac{(\rho c)_p \varphi_0}{(\rho c)_f} \text{ is the modified particle-density increment.}$$

In spirit of Oberbeck-Boussinesq approximation, equation (5.7) has been linearized by the neglect of a term proportional to the product of $Æ_0$ and T. This approximation is valid in the case of small temperature gradients in a dilute suspension of nanoparticles.

The dimensionless boundary conditions are

$$w' = 0, \quad T' = 1, \quad \frac{\partial \phi'}{\partial z'} + N_A \frac{\partial T'}{\partial z'} = 0 \text{ at } \quad z' = 0$$

and

$$w' = 0, \ T' = 0, \ \frac{\partial \phi'}{\partial z'} + N_A \frac{\partial T'}{\partial z'} = 0 \text{ at } z' = 1. \tag{5.10}$$

THE BASIC STATE AND ITS SOLUTIONS

The basic state was assumed to be quiescent and is given by

$$q' \left(u', v', w'\right) = 0,$$

$$p' = p_b(z),$$

$$T' = T_b(z), \tag{5.11}$$

$$\varphi' = \varphi_b(z) \text{ and}$$

$$\rho = \rho_0\left(1+\alpha\left(T\text{-}T_0\right)\right).$$

Equations (5.6) – (5.9) reduce to

$$0 = -\frac{dp_b}{dz'}\text{-}Rm + RaT_b - Rn\text{Æ}_b, \tag{5.12}$$

$$\frac{d^2T_b}{dz'^2} + \frac{N_B}{Le}\frac{d\varphi_b}{dz'}\frac{dT_b}{dz'} + \frac{N_A N_B}{Le}\left(\frac{dT_b}{dz'}\right)^2 = 0, \tag{5.13}$$

$$\frac{d^2\varphi_b}{dz'^2} + N_A\frac{d^2T_b}{dz'^2} = 0, \tag{5.14}$$

Using boundary conditions in (5.10), equation (5.14) gives

$$\varphi_b = -N_A T_b + \left(1-N_A\right)z' + N_A \tag{5.15}$$

On substituting the value of the φ_b from equation (5.14) in equation (5.13), we get

$$\frac{d^2T_b}{dz'} + \frac{\left(1\text{-}N_A\right)N_B}{Le}\frac{dT_b}{dz'} = 0 \tag{5.16}$$

On integrating equation (5.16) with respect to z' and using boundary conditions (5.10), we get

$$T_b = \frac{1-e^{-\left(1-N_A\right)N_B\left(1-z'\right)/Le}}{1-e^{-\left(1-N_A\right)N_B/Le}}. \tag{5.17}$$

For most nanofluid investigated so far Le is large, is of order 10^2 - 10^3, while N_A is no greater than about 10. Then, the exponents in equation (5.17) are small. By expanding the exponential function into the power series and retaining up to the first order is negligible and so as to get good approximation for the solution as

$$T_b = 1 - z' \text{ and}$$

$$\varphi_b = \phi_0 + N_A z'$$

PERTURBATION SOLUTIONS

Let the initial basic state described by (5.11) is slightly perturbed so that perturbed state is given by

$$\begin{aligned} &\mathbf{q}'(u',v',w') = 0 + \mathbf{q}''(u'',v'',w''), \\ &T' = T_b + T'', \\ &\varphi' = \varphi_b + \varphi'', \\ &p' = p_b + p'', \end{aligned} \tag{5.18}$$

where $T_b = 1 - z'$, $\varphi_b = \phi_0 + N_A z'$ and (u'',v'',w''), T'', φ'' and p'' respectively the perturbations in initial velocity, temperature, volume fraction of the nanoparticles and pressure.

By substituting (5.18) in equations (5.6)–(5.9) and linearize by neglecting the product of the prime quantities, we obtained following equations

$$\nabla \cdot \mathbf{q} = 0, \tag{5.19}$$

$$\left(1 + F\frac{\partial}{\partial t}\right)\frac{1}{\Pr}\frac{\partial \mathbf{q}}{\partial t} = \left(1 + F\frac{\partial}{\partial t}\right)\left(-\nabla p + \mathrm{Ra}T\hat{e}_z - \mathrm{Rn}Æ\hat{e}_z\right) - \mathbf{q}, \tag{5.20}$$

$$\frac{1}{\sigma}\frac{\partial \varphi}{\partial t} + \frac{1}{\varepsilon} w N_A = \frac{1}{\mathrm{Le}}\nabla^2\varphi + \frac{N_A}{\mathrm{Le}}\nabla^2 T, \tag{5.21}$$

$$\frac{\partial T}{\partial t} - w = \nabla^2 T + \frac{N_B}{\mathrm{Le}}\left(N_A\frac{\partial T}{\partial z} - \frac{\partial \phi}{\partial z}\right) - \frac{2N_A N_B}{\mathrm{Le}}\frac{\partial T}{\partial z}. \tag{5.22}$$

Boundary conditions are

$$w = 0, T = 0, \quad \frac{\partial \phi}{\partial z} + N_A \frac{\partial T}{\partial z} = 0 \text{ at } z = 0, 1. \tag{5.23}$$

[Dashes (") have been suppressed for convenience]

Eliminating pressure term p from equation (5.20), by operating with curl twice on it, we have

$$\left(1 + F\frac{\partial}{\partial t}\right)\left(\frac{1}{\Pr}\frac{\partial}{\partial t}\left(\nabla^2 w\right) - \mathrm{Ra}\nabla_H^2 T + \mathrm{Rn}\nabla_H^2 \varphi\right) - \nabla^2 w = 0, \tag{5.24}$$

where ∇_H^2, is two-dimensional Laplacian operator on a horizontal plane.

NORMAL MODE ANALYSIS

We shall now analyze an arbitrary perturbation into a complete set of normal modes and then examine the stability of each of those modes individually. For the system of equations (5.24), (5.21) and (5.22) the analysis can be made in terms of two dimensional periodic wave numbers. Thus, assuming that the perturbed quantities are of the form

$$\left[w, T, \varphi\right] = \left[W(z), \Theta(z), \Phi(z)\right] \exp\left(ik_x x + ik_y y + nt\right), \tag{5.25}$$

where k_x, k_y are wave numbers in x and y direction and n is growth rate of disturbances.

Using equation (5.25), equations (5.24), (5.21) and (5.22) become

$$\left(D^2 - a^2 - \frac{n\left(1 + nF\right)}{Va}\right)W - \left(1 + nF\right)\left(a^2 \mathrm{Ra}\Theta - a^2 \mathrm{Rn}\Phi\right) = 0, \tag{5.27}$$

$$\frac{1}{\varepsilon}N_A W - \frac{N_A}{Le}\left(D^2 - a^2\right)\Theta - \left(\frac{1}{Le}\left(D^2 - a^2\right) - \frac{n}{\sigma}\right)\Phi = 0, \tag{5.28}$$

$$W + \left(D^2 - a^2 - n - \frac{N_A N_B}{Le}D\right)\Theta + \frac{N_B}{Le}D\Phi = 0, \tag{5.29}$$

where $D \equiv \frac{d}{dz}$ and $a = \sqrt{k_x^2 + k_y^2}$ is the dimensionless resultant wave number.

The boundary conditions of the problem in view of normal mode analysis are

$$W = 0, D^2W = 0, \Theta = 0, D\Phi + N_A D\Theta = 0 \text{ at } z = 0,1. \tag{5.30}$$

METHOD OF SOLUTION

The Galerkin weighted residuals method is used to obtain an approximate solution to the system of equations (5.27) – (5.29) with the corresponding boundary conditions (5.30). In this method, the test functions are the same as the base (trial) functions. Accordingly W, Θ and Φ are taken as

$$W = \sum_{p=1}^{N} A_p W_p, \Theta = \sum_{p=1}^{N} B_p \Theta_p, \Phi = \sum_{p=1}^{N} C_p \Phi_p, \tag{5.31}$$

where $W_p = \Theta_p = \sin p\pi z, \Phi_p = -N_A \sin p\pi z$, A_p, B_p and C_p are unknown coefficients, p = 1, 2, 3,..., N and the base functions W_p, Θ_p and Φ_p satisfying the boundary conditions (5.30). Using expression for W, Θ and Φ in equations (5.27) – (5.29) and multiplying the first equation by W_p the second equation by Θ_p and third equation by Φ_p and then integrating in the limits from zero to unity, we obtain a set of 3N linear homogeneous equations with 3N unknown A_p, B_p and C_p; p = 1, 2, 3,..., N. For existing of nontrivial solution, the vanishing of the determinant of coefficients produces the characteristics equation of the system in term of Rayleigh number Ra.

LINEAR STABILITY ANALYSIS

For the present formulation, we have considered the which system of equations (5.27) - (5.29) together with the boundary conditions (5.30) constitute a linear eigenvalue problem with variable coefficient for the growth rate of disturbance n of the system. Substituting equation (5.31) into the system of equations (5.27) -(5.29) and multiplying the first equation by W_p the second equation by Θ_p and third equation by Φ_p and then integrating in the limits from zero to unity and performing some integration by parts, one obtains the following matrix equation

$$\begin{bmatrix} \dfrac{\left(\pi^2+a^2\right)}{\left(1+nF\right)}+\dfrac{n}{\Pr}\left(\pi^2+a^2\right) & \text{-a}^2\text{Ra} & \text{-a}^2\text{N}_\text{A}\text{Rn} \\ 1 & -\left(\pi^2+a^2+n\right) & 0 \\ \dfrac{1}{\varepsilon} & \dfrac{1}{\text{Le}}\left(\pi^2+a^2\right) & -\left(\dfrac{1}{\text{Le}}\left(\pi^2+a^2\right)+\dfrac{\text{n}}{\sigma}\right) \end{bmatrix} \begin{bmatrix} W_0 \\ \Theta_0 \\ \Phi_0 \end{bmatrix} = \begin{bmatrix} 0 \\ 0 \\ 0 \end{bmatrix}. \tag{5.32}$$

The non-trivial solution of the above matrix requires that

$$\text{Ra} = \frac{1}{\text{a}^2}\left(\frac{\left(\pi^2+a^2\right)}{\left(1+nF\right)}+\frac{\text{n}\left(\pi^2+a^2\right)}{\Pr}\right)\left(\pi^2+a^2+n\right)-\frac{\left(\pi^2+a^2\right)+\dfrac{Le}{\varepsilon}\left(\pi^2+a^2+n\right)}{\left(\pi^2+a^2\right)+\text{n}\dfrac{\text{Le}}{\sigma}}\text{N}_\text{A}\text{Rn}. \tag{5.33}$$

The growth rate n is in general a complex quantity such that $\text{n}=\omega_r+i\omega$, the system with $\omega_r<0$ is always stable, while for $\omega_r>0$ it will become unstable. For neutral stability, the real part of n is zero. Hence, we now write $\text{n}=i\omega$, (where ω is real and is a dimensionless frequency) in equation (5.33), we have

$$\text{Ra} = \Delta_1 + i\omega\Delta_2, \tag{5.34}$$

where

$$''_1 = \frac{\left(\pi^2+a^2\right)}{\text{a}^2}\left(\frac{\left(\pi^2+a^2\right)}{1+\omega^2F^2}-\omega^2\left(\frac{1}{\Pr}-\frac{F}{1+\omega^2F^2}\right)\right)-\frac{\left(\pi^2+a^2\right)^2\left(\dfrac{Le}{\varepsilon}+1\right)+Le\left(\left(\pi^2+a^2\right)\right)+\omega^2Le^2}{\left(\pi^2+a^2\right)^2+\omega^2\dfrac{Le^2}{\sigma^2}}\text{N}_\text{A}\text{Rn} \tag{5.35}$$

and

$$''_2 = \frac{\left(\pi^2+a^2\right)}{\text{a}^2}\left(\frac{1}{1+\omega^2F^2}+\left(\pi^2+a^2\right)\left(\frac{1}{\Pr}-\frac{F}{1+\omega^2F^2}\right)\right)+\frac{\dfrac{Le}{\varepsilon}\left(\pi^2+a^2\right)-\left(\pi^2+a^2\right)\left(1+\dfrac{Le}{\varepsilon}\right)\dfrac{Le}{\sigma}}{\left(\pi^2+a^2\right)^2+\omega^2\dfrac{Le^2}{\sigma^2}}\text{N}_\text{A}\text{Rn}. \tag{5.36}$$

Since Ra is a physical quantity, so it must be real. Hence, it follows from the equation (5.34) that either $\omega = 0$ (exchange of stability, steady state) or $\Delta_2 = 0$ ($\omega \neq 0$ overstability or oscillatory onset).

STATIONARY CONVECTION

For the case of stationary convection [$n = \omega = 0$], equation (5.33) reduces to

$$(\mathrm{Ra})_s = \frac{(\pi^2 + a^2)^2}{a^2} - \left(1 + \frac{Le}{\varepsilon}\right) N_A \mathrm{Rn}. \tag{5.37}$$

It is observed that stationary Rayleigh number Ra is function of the Lewis number Le, the modified diffusivity ratio N_A, the nanoparticles Rayleigh Rn and porosity parameter ε but independent of visco- elastic parameter F, Prandtl number Pr and modified particle- density increment N_B. Thus Maxwellian visco-elastic nanofluid behaves like an ordinary Newtonian nanofluid and instability is purely a phenomenon due to buoyancy coupled with the conservation of nanoparticles.

To find the critical value of $(\mathrm{Ra})_s$, equation (5.37) is differentiated with respect to 'a^2' and then equated to zero. The minimum of first term of right-hand side of equation (5.37) is attained at $a_c = \sqrt{\pi}$ and minimum value found to $4\pi^2$ so the corresponding critical Rayleigh number given by

$$(\mathrm{Ra})_c = 4\pi^2 - \left(1 + \frac{Le}{\varepsilon}\right) N_A \mathrm{Rn}\,. \tag{5.38}$$

In the absence of nanoparticles ($\mathrm{Rn} = \mathrm{Le} = N_A = 0$), one recovers the well- known results that the critical Rayleigh-Darcy number is equal to $(\mathrm{Ra})_c = 4\pi^2$.

This is good agreement of the result obtained by Nield and Kuznetsov (2009a).

Thus presence of the nanoparticles lowers the value of the critical Rayleigh number by usually by substantial amount. Also parameter N_B does not appear in the equation (5.38); thus instability is purely phenomena due to buoyancy coupled with conservation of nanoparticles. Thus average contribution of nanoparticles flux in the thermal energy equation is zero with one-term Galerkin approximation.

OSCILLATORY CONVECTION

For oscillatory convection $\Delta_2 = 0$ and $\omega \neq 0$, thus equation (5.36) gives a dispersion relation of the form

$$a_2\left(\omega^2\right)^2 + a_1\left(\omega^2\right) + a_0 = 0; \tag{5.39}$$

where

$$a_2 = \frac{\left(\pi^2 + a^2\right)^2}{a^2}\frac{Le^2 F^2}{\sigma^2 \Pr},$$

$$a_1 = \frac{\left(\pi^2 + a^2\right)^3 F^2}{a^2 \Pr} + F^2\left(\left(\pi^2 + a^2\right)\frac{Le}{\varepsilon} - \left(\pi^2 + a^2\right)\left(1 + \frac{Le}{\varepsilon}\right)\frac{Le}{\sigma}\right)N_A Rn,$$

$$a_0 = \frac{\left(\pi^2 + a^2\right)^3}{a^2 \Pr}\left(1 + \left(\pi^2 + a^2\right)\left(1 - F\Pr\right)\right) + \left(\left(\pi^2 + a^2\right)\frac{Le}{\varepsilon} - \left(\pi^2 + a^2\right)\left(1 + \frac{Le}{\varepsilon}\right)\frac{Le}{\sigma}\right)N_A Rn.$$

Then equation (5.34) with $\Delta_2 = 0$ gives oscillatory Rayleigh number at the margin of stability as

$$\left(\mathrm{Ra}\right)_{osc} = \frac{\left(\pi^2 + a^2\right)}{a^2}\left(\frac{\left(\pi^2 + a^2\right)}{1 + \omega^2 F^2} - \omega^2\left(\frac{1}{\Pr} - \frac{F}{1 + \omega^2 F^2}\right)\right)$$
$$- \frac{\left(\pi^2 + a^2\right)^2\left(\frac{Le}{\varepsilon} + 1\right) + Le\left(\left(\pi^2 + a^2\right)\right) + \omega^2 Le^2}{\left(\pi^2 + a^2\right)^2 + \omega^2\frac{Le^2}{\sigma^2}}N_A \mathrm{Rn}. \tag{5.40}$$

For the oscillatory convection to occur, ω^2 must be positive. If there are no positive roots of ω^2 in equation (5.39), then oscillatory convection is not possible. If there are positive roots of ω^2, the critical Rayleigh number for oscillatory convection can be obtained numerically minimizing equation (5.40) with respect to wave number, after substituting various values of physical parameters for ω^2 of equation (5.39) to determine the various effect of different parameter on the onset of oscillatory convection.

RESULTS AND DISCUSSION

To study the effect of Lewis number Le, modified diffusivity ratio N_A and nanoparticles Rayleigh number Rn on stationary convection, we examine the behavior of $\frac{\partial(\mathrm{R}a)_s}{\partial Le}, \frac{\partial(\mathrm{R}a)_s}{\partial N_A}, \frac{\partial(\mathrm{R}a)_s}{\partial Rn}$ and $\frac{\partial(\mathrm{R}a)_s}{\partial \varepsilon}$ analytically.

From equation (5.37), we have

1. $\frac{(\partial \mathrm{R}a)_s}{\partial Le} < 0$,
2. $\frac{(\partial \mathrm{R}a)_s}{\partial N_A} < 0$,
3. $\frac{(\partial \mathrm{R}a)_s}{\partial Rn} < 0$,
4. $\frac{(\partial \mathrm{R}a)_s}{\partial \varepsilon} > 0$.

These inequalities imply that Lewis number Le, modified diffusivity ratio N_A and nanoparticles Rayleigh number Rn destabilize while porosity parameter stabilizes the fluid layer.

Expression for stationary Rayleigh number, which characterizes the stability of the system, is found to be depend upon Lewis number, modified diffusivity ratio and nanoparticles Rayleigh number. The computations are carried out for different values of parameters considered in the range $-10^2 \leq Ra \leq 10^5$ (thermal Rayleigh number), $1 \leq N_A \leq 10$ (modified diffusivity ratio), $10^2 \leq Le \leq 10^4$ (Lewis number), $10^{-1} \leq Rn \leq 10^1$ (nanoparticles Rayleigh number), $10^{-2} \leq \varepsilon \leq 1$ (porosity parameter).

The variations of the stationary thermal Rayleigh number with wave number have been plotted graphically as shown in Figures 2 – 5.

Figures 2 – 5 demonstrate the neutral curve on the $((\mathrm{Ra})_s, a)$ plane for different values of the Lewis number Le, the modified diffusivity ratio N_A, the nanoparticles Rayleigh Rn and porosity parameter ε.

Figure 2 shows the variation of stationary Rayleigh number with wave number for different value of Lewis number with the fixed value of other parameters. It is found that stationary Rayliegh number decreases as the value of Lewis number increases, indicating that Lewis number destabilize the stationary convection. It is due to the fact that thermophoresis at a higher value of thermophoretic diffusivity is more supportable to the disturbance

Figure 2. Variation of the stationary Rayleigh number with wave number for different value of Lewis number

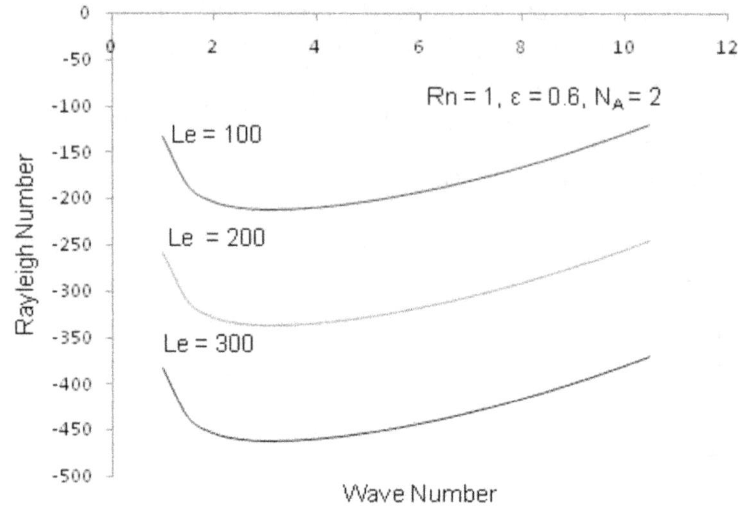

in nanofluids, while both thermophoresis and Brownian motion are driving forces in favor of the motion of nanoparticles.

Figure 3 shows the variation of stationary Rayleigh number with wave number for different value of modified diffusivity ratio with fixed value of other parameters and it is found that stationary Rayliegh number decreases with an increase in the value of modified diffusivity ratio, which indicate that modified diffusivity ratio destabilize the stationary convection. This may lead to an increase in volumetric fraction, which shows that Brownian motion of the nanoparticles will also increase, which may cause destabilizing effect. This is good agrement of the result obtained by Chand and Rana (2014d, 2015a).

Figure 4 shows the variation of stationary Rayleigh number with wave number for different values of nanoparticle Rayleigh number with fixed value of other parameters and it is found that stationary Rayliegh number decreases as the value of the nanoparticles Rayleigh number increases, which mean that nanoparticle Rayleigh number has destabilizing effect on the stationary convection. It has destabilizing effect because the heavier nanoparticles moving through the base fluid makes more strong disturbances as compared with the lighter nanoparticles.

Figure 3. Variation of the stationary Rayleigh number with wave number for different value of modified diffusivity ratio

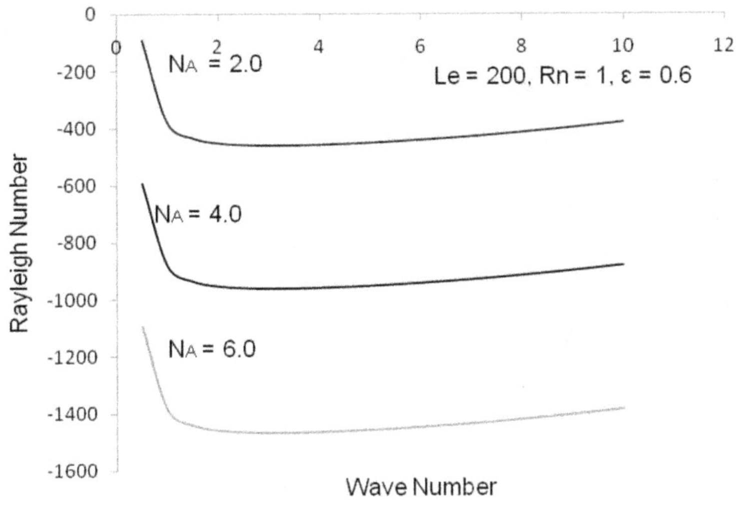

Figure 4. Variation of the stationary Rayleigh number with wave number for different value of modified diffusivity ratio

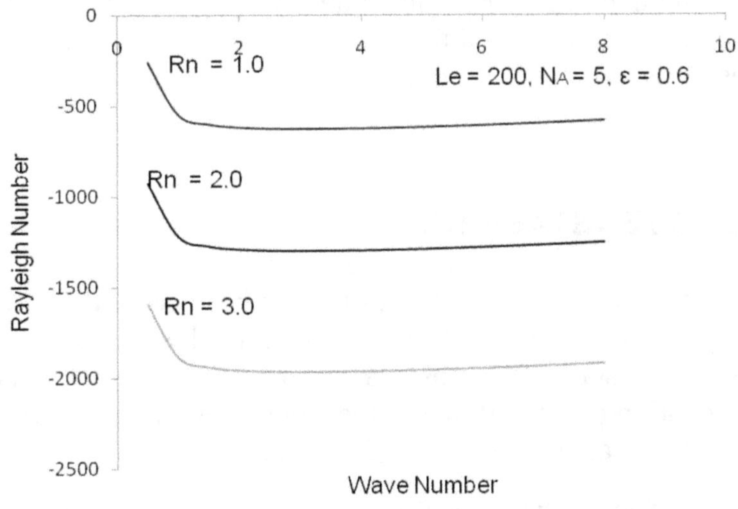

Figure 5. Variation of the stationary Rayleigh number with wave number for different value of porosity parameter

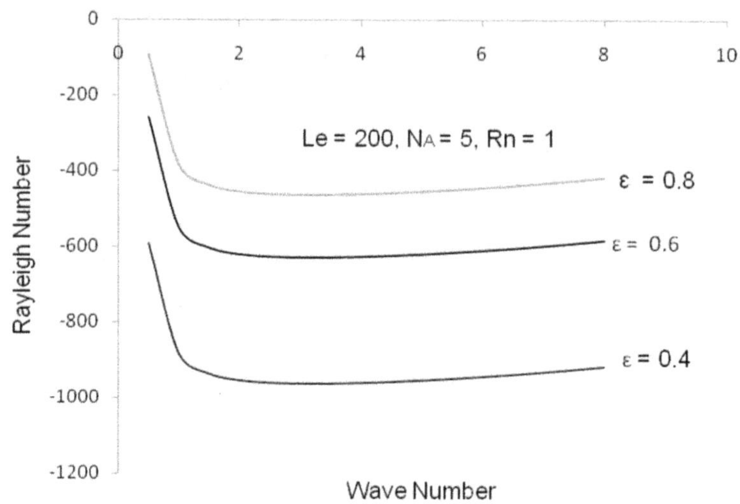

To assess the effect of porous medium on the stability of the system, the variation of the stationary Rayleigh number is shown in Figure 5 as a function of wave number a for different values of the porosity ε. We found that with an increase in the value of the porosity ε, the stationary Rayleigh number increases, indicating that it delays the onset of convection in nanofluid saturated in porous medium.

CASE OF OVERSTABILITY

Here we examine the possibility of as to whether instability may occur as overstability. Since we wish to determine the Rayleigh number for the onset of instability via a state of pure oscillation, it suffices to find the conditions for which equation (5.39) will admit the solution with real values of ω. All the coefficients a_0, a_1, a_2 in equation (5.39) are real.

Now the product of the roots of equation (5.39) $= \left(\frac{a_0}{a_2}\right)$ is positive.

a_2 is always positive and

a_0 is negative if $1 < FP,\ \ 1 < \left(1 + \frac{Le}{\varepsilon}\right)\frac{\varepsilon}{\sigma}$ (5.40)

Thus inequality $1 < FP,\quad 1 < \left(1 + \frac{Le}{\varepsilon}\right)\frac{\varepsilon}{\sigma}$ is sufficient condition for the non-existence of overstability, the violation of which does not necessarily imply the occurrence of overstability.

CONCLUSION

Thermal convection in a horizontal layer of Maxwellian visco-elastic nanofluid in porous medium is studied. Darcy model is used for porous medium. The flux of volume fraction of nanoparticles is taken to be zero on the isothermal boundaries and the eigen value problem is solved using the Galerkin residual method. The results have been presented both analytically and graphically.

The main conclusions derived from the present chapter are as follows

1. The instability purely phenomenon due to buoyancy coupled with the conservation of nanoparticle and is independent of the contribution of Brownian motion and thermophoresis.
2. For stationary convection Maxwellian visco-elastic nanofluid behaves like an ordinary Newtonian nanofluid.
3. Lewis number Le, modified diffusivity ratio N_A and nanoparticles Rayleigh number Rn destabilizes while porosity parameter stabilize the stationary convection.
4. Sufficient condition for the non-existence of overstability is $1 < FP,\quad 1 < \left(1 + \frac{Le}{\varepsilon}\right)\frac{\varepsilon}{\sigma}$.

REFERENCES

Chand, R., & Rana, G. C. (2012b). On the onset of thermal convection in rotating nanofluid layer saturating a Darcy-Brinkman porous medium. *International Journal of Heat and Mass Transfer*, *55*(21-22), 5417–5424. doi:10.1016/j.ijheatmasstransfer.2012.04.043

Chand, R., & Rana, G. C. (2014d). Thermal instability in a Brinkman porous medium saturated by nanofluid with no nanoparticle flux on boundaries. *Special Topics & Reviews in Porous Media: An International Journal*, *5*(4), 277–286. doi:10.1615/SpecialTopicsRevPorousMedia.v5.i4.10

Chand, R., & Rana, G. C. (2015a). Magneto convection in a layer of nanofluid in porous medium- a more realistic approach. *Journal of Nanofluids*, *4*(2), 196–202. doi:10.1166/jon.2015.1142

Ingham, D. B., & Pop, I. (1981). *Transport Phenomena in Porous Media.* New York: Elsvier.

Kuznetsov, A. V., & Nield, D. A. (2010a). Effect of local thermal non-equilibrium on the onset of convection in a porous medium layer saturated by a nanofluid. *Transport in Porous Media*, *83*(2), 425–436. doi:10.1007/s11242-009-9452-8

Kuznetsov, A. V., & Nield, D. A. (2010b). Thermal instability in a porous medium layer saturated by a nanofluid: Brinkman Model. *Transport in Porous Media*, *81*(3), 409–422. doi:10.1007/s11242-009-9413-2

Kuznetsov, A. V., & Nield, D. A. (2010c). The onset of double-diffusive nanofluid convection in a layer of a saturated porous medium. *Transport in Porous Media*, *85*(3), 941–951. doi:10.1007/s11242-010-9600-1

Lapwood, E. R. (1948). Convection of a fluid in porous medium. *Proceedings of the Cambridge Philosophical Society*, *44*(04), 508–519. doi:10.1017/S030500410002452X

Nield, D. A., & Bejan, A. (2013). *Convection in porous media.* New York: Springer-Verlag. doi:10.1007/978-1-4614-5541-7

Nield, D. A., & Kuznetsov, A. V. (2009a). Thermal instability in a porous medium layer saturated by a nanofluid. Int. *J. Heat Mass Transf.*, *52*(25-26), 5796–5801. doi:10.1016/j.ijheatmasstransfer.2009.07.023

Nield, D. A., & Kuznetsov, A. V. (2009b). The Cheng–Minkowycz problem for natural convective boundary-layer flow in a porous medium saturated by a nanofluid. *International Journal of Heat and Mass Transfer*, *52*(25-26), 5792–5795. doi:10.1016/j.ijheatmasstransfer.2009.07.024

Nield, D. A., & Kuznetsov, A. V. (2010b). The onset of convection in a layer of cellular porous material: Effect of temperature-dependent conductivity arising from radiative transfer. *Journal of Heat Transfer*, *132*(7), 074503. doi:10.1115/1.4001125

Nield, D. A., & Kuznetsov, A. V. (2011b). The effect of vertical through flow on thermal Instability in a porous medium layer saturated by a nanofluid. *Transport in Porous Media*, *87*(3), 765–775. doi:10.1007/s11242-011-9717-x

Vafai, K., & Hadim, H. A. (2000). *Handbook of Porous Media.* New York: Decker.

Wooding, R. A. (1960). Rayleigh instability of a thermal boundary layer in flow through a porous medium. *Journal of Fluid Mechanics*, *9*(02), 183–192. doi:10.1017/S0022112060001031

Chapter 6
Thermal Convection in a Layer of Maxwellian Visco-Elastic Nanofluid in a Porous Medium:
Brinkman Model

INTRODUCTION

Thermal instability in a porous medium is now regarded as a classical problem due to its wide range of applications in geothermal reservoirs, agricultural product storage, enhanced oil recovery, packed-bed catalytic reactors and the pollutant transport in underground. It has many applications in geophysics, food processing, oil reservoir modeling, building of thermal insulations and nuclear reactors. Lapwood (1948) has studied the convective flow in a porous medium using linearized stability theory. The Rayleigh instability of a thermal boundary layer in flow through a porous medium has been considered by Wooding (1960). The investigation in porous media has been started with the simple Darcy model and gradually was extended to Darcy-Brinkman model. A good account of convection problems in a porous medium is given by Vafai and Hadim (2000), Ingham and Pop (1981) and Nield and Bejan (2013).

DOI: 10.4018/978-1-68318-006-7.ch006

The study of flow, heat and mass transfer about natural convection of non-Newtonian fluids in porous media has gained much attention from the researchers because of its engineering and industrial applications. These applications include design of chemical processing equipment, formation and dispersion of fog, distributions of temperature and moisture over agricultural fields and groves of fruit trees and damage of crops due to freezing and pollution of the environment etc. In the recent years, considerable interest has been evinced in the study Rayleigh-Bénard convection problem for a nanofluid having relevance in engineering, automotive industries and biomedical engineering etc. Nanofluids have novel properties that make them potentially useful in wide range of engineering applications where cooling is of primary concern. Nanofluid used as heat transfer, chemical nanofluids, smart fluids, bio nanofluids, medical nanofluids (drug delivery and functional tissue cell interaction) etc. in many industrial applications.

Thermal instability of nanofluid in a Brinkman porous medium has been studied by Kuznetsov and Nield (2010b), Chand and Rana (2012b).

Due to importance of non-Newtonian nanofluids in Brinkman porous medium an attempt has been made in this chapter to study the thermal instability of a horizontal layer of Maxwellian visco-elastic nanofluids for more realistic boundary conditions in Brinkman porous medium.

MATHEMATICAL FORMULATIONS OF THE PROBLEM

The Physical Problem

In this chapter we shall investigate the thermal convection in a horizontal layer Maxwellian visco-elastic nanofluid in Brinkman porous medium. The physical configuration of the problem to be considered as:

An infinite horizontal layer of Maxwellian visco-elastic nanofluid of thickness'd' bounded by plane $z = 0$ and $z = d$, heated from below in a porous medium of medium permeability k_1 and porosity ε. Fluid layer is acted upon by a gravity force $\mathbf{g}(0,0,-g)$ and is heated from below in such a way that horizontal boundaries $z = 0$ and $z = d$ respectively maintained at a uniform temperature T_0 and T_1 ($T_0 > T_1$) as shown is Figure 1. The normal component of the nanoparticles flux has to vanish at an impermeable boundaries and the reference scale for temperature and nanoparticles fraction is taken to be T_1 and φ_0 respectively. It is assumed that nanoparticles are suspended in the nanofluid using either surfactant or surface charge technology. This prevents the particles from agglomeration and deposition on the porous

Figure 1. Physical configuration of the problem

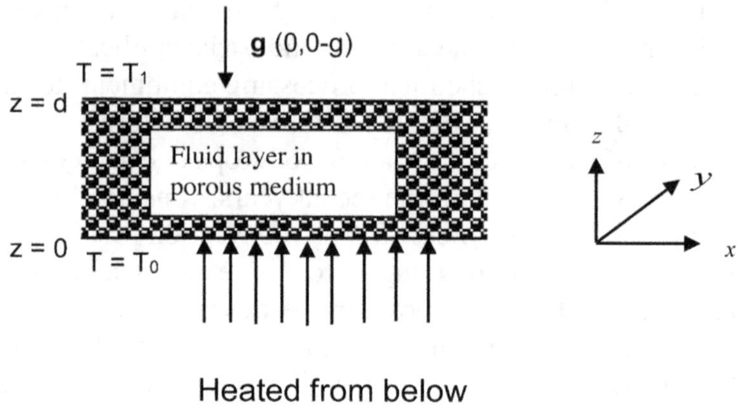

Assumptions

The mathematical equations describing the physical model are based upon the following assumptions:

1. Thermophysical properties of fluid expect for density in the buoyancy force (Boussinesq Hypothesis) are constant,
2. The fluid phase and nanoparticles are in thermal equilibrium state and thus, the heat flow has been described using one equation model,
3. Dilute mixture,
4. Nanoparticles are spherical,
5. Nanoparticles are non-magnetic,
6. No chemical reactions take place in fluid layer,
7. Negligible viscous dissipation,
8. Radiative heat transfer between the sides of wall is negligible when compared with other modes of the heat transfer,
9. Nanofluid is incompressible, electrically conducting, Newtonian and laminar flow,
10. Each boundary wall is assumed to be impermeable and perfectly thermal conducting,
11. Nanoparticles are suspended in the nanofluid using either surfactant or surface charge technology.

GOVERNING EQUATIONS

The governing equations for Maxwellian visco-elastic nanofluid in Brinkman porous medium under the Boussinesq approximation are given as

$$\nabla \cdot \mathbf{q} = 0, \tag{6.1}$$

$$\frac{\rho}{\varepsilon}\left(1+\lambda\frac{\partial}{\partial t}\right)\frac{d\mathbf{q}}{dt} = \left(1+\lambda\frac{\partial}{\partial t}\right)\left(-\nabla p + \rho g\right) + \tilde{\mu}\,\nabla^2\mathbf{q} - \frac{\mu}{k_1}\mathbf{q}, \tag{6.2}$$

where **q(u, v, w)** is the Darcy velocity vector, p is the hydrostatic pressure, μ the is viscosity, $\tilde{\mu}$ is the effective viscosity, λ is the relaxation time, k_1 is the medium permeability, ε is the porosity parameter, α is the coefficient of thermal expansion, φ is the volume fraction of the nanoparticles, ρ_p density of nanoparticles and ρ_f density of base fluid and $\frac{d}{dt} = \frac{\partial}{\partial t} + \frac{1}{\varepsilon}(\mathbf{q}\cdot\nabla)$ is stands for convection derivative.

The equation of energy for nanofluid in porous medium is

$$(\rho c)_m \frac{\partial T}{\partial t} + (\rho c)_f\, \mathbf{q}\cdot\nabla T = k_m \nabla^2 T + \varepsilon (\rho c)_p \left[D_B \nabla\varphi\cdot\nabla T + \frac{D_T}{T_1}\nabla T\cdot\nabla T\right], \tag{6.3}$$

where $(\rho c)_m$ is effective heat capacity of fluid, $(\rho c)_p$ is heat capacity of nanoparticles and k_m is effective thermal conductivity of the porous medium.

The continuity equation for the nanoparticles is

$$\frac{\partial\varphi}{\partial t} + \frac{1}{\varepsilon}\mathbf{q}\cdot\nabla\varphi = D_B \nabla^2\varphi + \frac{D_T}{T_1}\nabla^2 T, \tag{6.4}$$

where D_B is the Brownian diffusion coefficient, given by Einstein-Stokes equation and D_T is the thermoporetic diffusion coefficient of the nanoparticles.

We assume that the temperature is constant and nanoparticles flux is zero on the boundaries. Thus boundary conditions are

$$w = 0,\ T = T_0,\ D_B\frac{\partial\phi}{\partial z} + \frac{D_T}{T_1}\frac{\partial T}{\partial z} = 0 \text{ at } z = 0 \text{ and}$$

$$w = 0,\ T = T_1,\ D_B\frac{\partial\phi}{\partial z} + \frac{D_T}{T_1}\frac{\partial T}{\partial z} = 0 \text{ at } z = d. \tag{6.5}$$

Introducing non-dimensional variables as

$$(x',y',z') = \left(\frac{x,y,z}{d}\right),\ q'(u',v',w',) = q\left(\frac{u,v,w}{\kappa}\right)d,\ t' = \frac{t\kappa}{d^2},\ p' = \frac{pk_1}{\mu\kappa},$$

$$\varphi' = \frac{(\varphi - \varphi_0)}{\varphi_0},\ T' = \frac{(T - T_1)}{(T_0 - T_1)},$$

where

$$\kappa = \frac{k_m}{(\rho c)_p},\ \sigma = \frac{(\rho c)_m}{(\rho c)_f}.$$

Equations (6.1) - (6.5) in non-dimensional form can be written as

$$\nabla' \cdot \mathbf{q}' = 0, \tag{6.6}$$

$$\left(1 + F\frac{\partial}{\partial t'}\right)\frac{1}{Va}\frac{\partial \mathbf{q}'}{\partial t'} = \left(1 + F\frac{\partial}{\partial t'}\right)\left(-\nabla' p' \text{-} Rm\hat{e}_z + RaT'\hat{e}_z - Rn\varphi'\hat{e}_z\right) + \tilde{D}a\nabla'^2\mathbf{q}' - \mathbf{q}\,, \tag{6.7}$$

$$\frac{1}{\sigma}\frac{\partial \varphi'}{\partial t'} + \frac{1}{\varepsilon}\mathbf{q}' \cdot \nabla'\varphi' = \frac{1}{Le}\nabla'^2\varphi' + \frac{N_A}{Le}\nabla'^2 T', \tag{6.8}$$

$$\frac{\partial T'}{\partial t'} + \mathbf{q}' \cdot \nabla' T' = \nabla'^2 T' + \frac{N_B}{Le}\nabla'\varphi' \cdot \nabla' T' + \frac{N_A N_B}{Le}\nabla' T' \cdot \nabla' T', \tag{6.9}$$

here non-dimensional parameters are given as

$Va = \dfrac{\varepsilon Pr}{Da}$ is the Prandtl- Darcy Number (Vadasz Number),

$Da = \dfrac{k_1}{d^2}$ is the Darcy number,

$Pr = \dfrac{\mu}{\rho\kappa}$ is the Prandtl number,

$\tilde{Da} = \dfrac{\tilde{\mu} k_1}{\mu d^2}$ is the Brinkman-Darcy number,

$\mathrm{Le} = \dfrac{\kappa}{\mathrm{D_B}}$ is the Lewis number,

$F = \dfrac{\kappa \lambda}{d^2}$ is the stress relaxation parameter,

$\mathrm{Ra} = \dfrac{\rho g \alpha d k_1 \left(T_0 - T_1\right)}{\mu \kappa}$ is the Rayleigh Darcy number,

$\mathrm{Rm} = \dfrac{\left(\rho_p \varphi_0 + \rho\left(1\text{-}\varphi_0\right)\right) g d k_1}{\mu \kappa}$ is the density Rayleigh Darcy number,

$\mathrm{Rn} = \dfrac{\left(\rho_p - \rho\right) \varphi_0 g d k_1}{\mu \kappa}$ is the nanoparticles Rayleigh Darcy number,

$\mathrm{N_A} = \dfrac{\mathrm{D_T}\left(T_0\text{-}T_1\right)}{\mathrm{D_B} T_1 \varphi_0}$ is the modified diffusivity ratio,

$\mathrm{N_B} = \dfrac{\left(\rho c\right)_p \varphi_0}{\left(\rho c\right)_f}$ is the modified particle-density increment.

In spirit of Oberbeck-Boussinesq approximation, equation (6.7) has been linearized by the neglect of a term proportional to the product of φ_0 and T. This approximation is valid in the case of small temperature gradients in a dilute suspension of nanoparticles.

The dimensionless boundary conditions are

$$\mathrm{w}' = 0, \quad \mathrm{T}' = 1, \quad \frac{\partial \phi'}{\partial \mathrm{z}'} + N_A \frac{\partial T'}{\partial \mathrm{z}'} = 0 \text{ at } \quad \mathrm{z}' = 0$$

and

$$w' = 0,\ T' = 0,\ \frac{\partial \phi'}{\partial z'} + N_A \frac{\partial T'}{\partial z'} = 0 \text{ at } z' = 1. \tag{6.10}$$

THE BASIC STATE AND ITS SOLUTIONS

The basic state was assumed to be quiescent and is given by

$$q'\left(u', v', w'\right) = 0,$$

$$p' = p_b(z),$$

$$T' = T_b(z), \tag{6.11}$$

$$\varphi' = \varphi_b(z) \text{ and}$$

$$\rho = \rho_0\left(1 + \alpha\left(T\text{-}T_0\right)\right).$$

Equations (6.6) – (6.9) reduce to

$$0 = -\frac{dp_b}{dz'}\text{-}Rm + RaT_b - Rn\varphi_b, \tag{6.12}$$

$$\frac{d^2T_b}{dz'^2} + \frac{N_B}{Le}\frac{d\varphi_b}{dz'}\frac{dT_b}{dz'} + \frac{N_A N_B}{Le}\left(\frac{dT_b}{dz'}\right)^2 = 0, \tag{6.13}$$

$$\frac{d^2\varphi_b}{dz'^2} + N_A\frac{d^2T_b}{dz'^2} = 0, \tag{6.14}$$

Using boundary conditions in (6.10), equation (6.14) gives

$$\varphi_b = -N_A T_b + \left(1 - N_A\right)z' + N_A. \tag{6.15}$$

On substituting the value of the φ_b from equation (6.15) in equation (6.14), we get

$$\frac{d^2T_b}{dz'} + \frac{(1\text{-}N_A)N_B}{Le}\frac{dT_b}{dz'} = 0 \tag{6.16}$$

On integrating equation (6.16) with respect to z' and using boundary conditions (6.10), we get

$$T_b = \frac{1 - e^{-(1-N_{z'})N_B(1-z')/Le}}{1 - e^{-(1-N_A)N_B/Le}}. \tag{6.17}$$

For most nanofluid investigated so far Le is large, is of order 10^2 - 10^3, while N_A is no greater than about 10. Then, the exponents in equation (6.17) are small. By expanding the exponential function into the power series and retaining up to the first order is negligible and hence an approximate solution for basic state is given by

$T_b = 1 - z'$ and

$\varphi_b = \phi_0 + N_A z'$

PERTURBATION SOLUTIONS

Let the initial basic state described by (6.11) is slightly perturbed so that perturbed state is given by

$$\mathbf{q}'(u',v',w') = 0 + \mathbf{q}''(u'',v'',w''),$$

$$T' = T_b + T'', \tag{6.18}$$

$$\varphi' = \varphi_b + \varphi'',$$

$$p' = p_b + p'',$$

where $T_b = 1 - z'$, $\varphi_b = \phi_0 + N_A z$ and (u'',v'',w''), T'', φ'' and p'' respectively the perturbations in initial velocity, temperature, volume fraction of the nanoparticles and pressure.

By substituting (6.18) in equations (6.6) – (6.9) and linearize by neglecting the product of the prime quantities, we obtained following equations

$$\nabla \cdot \mathbf{q} = 0, \tag{6.19}$$

$$\left(1+F\frac{\partial}{\partial t}\right)\frac{1}{Va}\frac{\partial \mathbf{q}}{\partial t} = \left(1+F\frac{\partial}{\partial t}\right)\left(-\nabla p + \mathrm{RaT}\hat{e}_z - \mathrm{Rn}\varphi\hat{e}_z\right) + \tilde{\mathrm{D}}\mathrm{a}\nabla^2\mathbf{q} - \mathbf{q}, \tag{6.20}$$

$$\frac{1}{\sigma}\frac{\partial \varphi}{\partial t} + \frac{1}{\varepsilon}wN_A = \frac{1}{\mathrm{Le}}\nabla^2\varphi + \frac{\mathrm{N_A}}{\mathrm{Le}}\nabla^2\mathrm{T}, \tag{6.21}$$

$$\frac{\partial \mathrm{T}}{\partial t} - w = \nabla^2\mathrm{T} + \frac{\mathrm{N_B}}{\mathrm{Le}}\left(N_A\frac{\partial T}{\partial z} - \frac{\partial \phi}{\partial z}\right) - \frac{2\mathrm{N_A N_B}}{\mathrm{Le}}\frac{\partial T}{\partial z}. \tag{6.22}$$

Boundary conditions are

$$\mathrm{w} = 0, \mathrm{T} = 0, \quad \frac{\partial \phi}{\partial z} + N_A\frac{\partial T}{\partial z} = 0 \ \text{at}\ z = 0, 1. \tag{6.23}$$

[Dashes (") have been suppressed for convenience]

Eliminating pressure term p from equation (6.20), we have

$$\left(1+F\frac{\partial}{\partial t}\right)\left(\frac{1}{\mathrm{Va}}\frac{\partial}{\partial t}\left(\nabla^2 \mathrm{w}\right) - \mathrm{Ra}\nabla_H^2\mathrm{T} + \mathrm{Rn}\nabla_H^2\varphi\right) - \tilde{\mathrm{D}}\mathrm{a}\nabla^4\mathrm{w} + \nabla^2\mathrm{w} = 0, \tag{6.24}$$

where ∇_{H}^2, is two-dimensional Laplacian operator on a horizontal plane.

NORMAL MODE ANALYSIS

We shall now analyze an arbitrary perturbation into a complete set of normal modes and then examine the stability of each of those modes individually. For the system of equations (6.24), (6.21) and (6.22) the analysis can be made in terms of two dimensional periodic wave numbers. Thus, assuming that the perturbed quantities are of the form

$$\left[\mathrm{w},\mathrm{T},\varphi\right] = \left[\mathrm{W(z)},\Theta(\mathrm{z}),\Phi(\mathrm{z})\right]\exp\left(\mathrm{i}k_x\mathrm{x} + \mathrm{i}k_y\mathrm{y} + \mathrm{nt}\right), \tag{6.25}$$

where k_x, k_y are wave numbers in x and y direction and n is growth rate of disturbances.

Using equation (6.25), equations (6.24), (6.21) and (6.22) become

$$\left(D^2-a^2\right)\left[\tilde{D}a\left(D^2-a^2\right)-1-\frac{\left(1+nF\right)n}{Va}\right]W-\left(1+nF\right)\left(a^2Ra\Theta-a^2Rn\Phi\right)=0, \tag{6.27}$$

$$\frac{1}{\varepsilon}N_AW-\frac{N_A}{Le}\left(D^2-a^2\right)\Theta-\left(\frac{1}{Le}\left(D^2-a^2\right)-\frac{n}{\sigma}\right)\Phi=0, \tag{6.28}$$

$$W+\left(D^2-a^2-n-\frac{N_AN_B}{Le}D\right)\Theta+\frac{N_B}{Le}D\Phi=0, \tag{6.29}$$

where $D\equiv\frac{d}{dz}$ and $a=\sqrt{k_x^2+k_y^2}$ is the dimensionless resultant wave number.

The boundary conditions of the problem in view of normal mode analysis are

$$W=0,\ D^2W=0, \Theta=0,\ D\Phi+N_AD\Theta=0 \ \text{at} \quad z=0,1. \tag{6.30}$$

Method of Solution

The Galerkin weighted residuals method is used to obtain an approximate solution to the system of equations (6.27) – (6.29) with the corresponding boundary conditions (6.30). In this method, the test functions are the same as the base (trial) functions. Accordingly W, Θ and Φ are taken as

$$W=\sum_{p=1}^{N}A_pW_p,\ \Theta=\sum_{p=1}^{N}B_p\Theta_p,\ \Phi=\sum_{p=1}^{N}C_p\Phi_p, \tag{6.31}$$

where $W_p=\Theta_p=\sin p\pi z, \Phi_p=-N_A\sin p\pi z$, A_p, B_p and C_p are unknown coefficients, p = 1, 2, 3,..., N and the base functions W_p, Θ_p and Φ_p satisfying the boundary conditions (6.30). Using expression for W, Θ and Φ in equations (6.27) – (6.29) and multiplying the first equation by W_p the second equation by Θ_p and third equation by Φ_p and then integrating in the limits from zero to unity, we obtain a set of 3N linear homogeneous equations with

3N unknown A_p, B_p and C_p; p = 1, 2, 3,..., N. For existing of nontrivial solution, the vanishing of the determinant of coefficients produces the characteristics equation of the system in term of Rayleigh number Ra.

LINEAR STABILITY ANALYSIS

For the present formulation, we have considered the which system of equations (6.27) - (6.29) together with the boundary conditions (6.30) constitute a linear eigenvalue problem with variable coefficient for the growth rate of disturbance n of the system. Substituting equation (6.31) into the system of equations (6.27) -(6.29) and multiplying the first equation by W_p the second equation by Θ_p and third equation by Φ_p and then integrating in the limits from zero to unity and performing some integration by parts, one obtains the following matrix equation

$$\begin{bmatrix} \dfrac{\tilde{D}a\left(\pi^2+a^2\right)^2+\left(\pi^2+a^2\right)}{\left(1+nF\right)}+\dfrac{n\left(\pi^2+a^2\right)}{\mathrm{Va}} & -a^2\mathrm{Ra} & -a^2N_A\mathrm{Rn} \\ 1 & -\left(\pi^2+a^2+n\right) & 0 \\ \dfrac{1}{\varepsilon} & \dfrac{1}{\mathrm{Le}}\left(\pi^2+a^2\right) & -\left(\dfrac{1}{\mathrm{Le}}\left(\pi^2+a^2\right)+\dfrac{n}{\sigma}\right) \end{bmatrix}\begin{bmatrix} W_0 \\ \Theta_0 \\ \Phi_0 \end{bmatrix}=\begin{bmatrix} 0 \\ 0 \\ 0 \end{bmatrix}. \tag{6.32}$$

The non-trivial solution of the above matrix requires that

$$\mathrm{Ra}=\frac{1}{a^2}\left(\frac{\tilde{D}a\left(\pi^2+a^2\right)^2+\left(\pi^2+a^2\right)}{\left(1+nF\right)}+\frac{n\left(\pi^2+a^2\right)}{\mathrm{Va}}\right)\left(\pi^2+a^2+n\right)$$

$$-\frac{\left(\pi^2+a^2\right)+\dfrac{Le}{\varepsilon}\left(\pi^2+a^2+n\right)}{\left(\pi^2+a^2\right)+\dfrac{n\mathrm{Le}}{\sigma}}N_A\mathrm{Rn}. \tag{6.33}$$

The growth rate n is in general a complex quantity such that $n = \omega_r + i\omega$, the system with $\omega_r < 0$ is always stable, while for $\omega_r > 0$ it will become unstable. For neutral stability, the real part of n is zero. Hence, we now write $n = i\omega$, (where ω is real and is a dimensionless frequency) in equation (6.33), we have

$$\mathrm{Ra} = \Delta_1 + i\omega\Delta_2, \tag{6.34}$$

where

$$\Delta_1 = \frac{1}{\mathrm{a}^2}\left(\frac{\tilde{D}a\left(\pi^2+a^2\right)^3+\left(\pi^2+a^2\right)^2}{1+\omega^2F^2} - \omega^2\left(\frac{\left(\pi^2+a^2\right)}{Va} - \frac{\tilde{D}a\left(\pi^2+a^2\right)^2+\left(\pi^2+a^2\right)}{1+\omega^2F^2}\right)\right) - \frac{\left(\pi^2+a^2\right)^2\left(\frac{Le}{\varepsilon}+1\right)+Le\left(\left(\pi^2+a^2\right)\right)+\omega^2Le^2}{\left(\pi^2+a^2\right)^2+\omega^2\frac{Le^2}{\sigma^2}}\mathrm{N_A Rn} \tag{6.35}$$

and

$$\Delta_2 = \frac{1}{\mathrm{a}^2}\left(\frac{\tilde{D}a\left(\pi^2+a^2\right)^2+\left(\pi^2+a^2\right)}{1+\omega^2F^2} + \left(\frac{\left(\pi^2+a^2\right)^2}{Va} - \frac{\tilde{D}a\left(\pi^2+a^2\right)^2+\left(\pi^2+a^2\right)}{1+\omega^2F^2}\right)\right) + \frac{\frac{Le}{\varepsilon}\left(\pi^2+a^2\right)-\left(\pi^2+a^2\right)\left(1+\frac{Le}{\varepsilon}\right)\frac{Le}{\sigma}}{\left(\pi^2+a^2\right)^2+\omega^2\frac{Le^2}{\sigma^2}}\mathrm{N_A Rn}. \tag{6.36}$$

Since Ra is a physical quantity, so it must be real. Hence, it follows from the equation (6.34) that either $\omega = 0$ (exchange of stability, steady state) or $\Delta_2 = 0$ ($\omega \neq 0$ overstability or oscillatory onset).

STATIONARY CONVECTION

For the case of stationary (non- oscillatory) convection $[\mathrm{n} = \omega = 0]$, equation (6.33) reduces to

$$\left(\mathrm{Ra}\right)_s = \frac{\tilde{D}a\left(\pi^2+a^2\right)^3+\left(\pi^2+a^2\right)^2}{\mathrm{a}^2} - \left(1+\frac{Le}{\varepsilon}\right)\mathrm{N_A Rn}. \tag{6.37}$$

It is observed that stationary Rayleigh number Ra is function of the Lewis number Le, the modified diffusivity ratio N_A and the nanoparticles Rayleigh Rn but independent of visco- elastic parameter F, Vadasz number Va and modified particle- density increment N_B. Thus Maxwellian visco-elastic nanofluid behaves like an ordinary Newtonian nanofluid and instability is purely a phenomenon due to buoyancy coupled with the conservation of nanoparticles.

For the case when $\tilde{D}a = 0$, minimum of first term of right- hand side of equation (6.37) is attained at $a = \pi$ and minimum value found to $4\pi^2$, so the corresponding critical Rayleigh number given by $(\mathrm{Ra})_c = 4\pi^2 - \left(N_A + \frac{Le}{\varepsilon}\right)\mathrm{Rn}$.

This is same result which was derived by Kuznetsov and Nield (2014).

For the case when $\tilde{D}a$ is large as compared to unity, minimum value of first term of right- hand side of equation (6.37) is attained at $a_c = \frac{\pi}{\sqrt{2}}$ and minimum value found to $\frac{27\pi^4}{4}\tilde{D}a$, so the corresponding critical Rayleigh number given by

$$(\mathrm{Ra})_c = \frac{27\pi^4}{4}\tilde{D}a - \left(N_A + \frac{Le}{\varepsilon}\right)\mathrm{Rn}. \tag{6.38}$$

This is well known result derived by Kuznetsov and Nield (2010b).

OSCILLATORY CONVECTION

For oscillatory convection $\Delta_2 = 0$ and $\omega \neq 0$, thus equation (6.36) gives a dispersion relation of the form

$$a_2(\omega^2)^2 + a_1(\omega^2) + a_0 = 0; \tag{6.39}$$

where

$$a_2 = \frac{(\pi^2 + a^2)^2}{a^2}\frac{F^2}{Va},$$

$$a_1 = \frac{Le^2}{a^2\sigma^2}\left(\tilde{D}a\left(\pi^2+a^2\right)^2+\left(\pi^2+a^2\right)-\tilde{D}a\left(\pi^2+a^2\right)^3+\left(\pi^2+a^2\right)^2\right)+\frac{\left(\pi^2+a^2\right)^5F^2}{a^2Va}$$
$$+F^2\left(\left(\pi^2+a^2\right)\frac{Le}{\varepsilon}-\left(\left(\pi^2+a^2\right)\left(1+\frac{Le}{\varepsilon}\right)\frac{Le}{\sigma}\right)\right)N_ARn,$$

$$a_0 = \frac{1}{a^2}\left[\tilde{D}a\left(\pi^2+a^2\right)^4+\left(\pi^2+a^2\right)^3+\frac{\left(\pi^2+a^2\right)^3}{Va}-\tilde{D}a\left(\pi^2+a^2\right)^2+\left(\pi^2+a^2\right)\right]$$
$$+\left(\left(\pi^2+a^2\right)\frac{Le}{\varepsilon}-\left(\pi^2+a^2\right)\left(1+\frac{Le}{\varepsilon}\right)\frac{Le}{\sigma}\right)N_ARn.$$

Now equation (6.34) together with $\Delta_2 = 0$ gives oscillatory Rayleigh number at the margin of stability as

$$(\mathrm{Ra})_{osc} = \frac{1}{\omega^2}\left(\frac{\tilde{D}a\left(\pi^2+a^2\right)^3+\left(\pi^2+a^2\right)^2}{1+\omega^2F^2}-\omega^2\left(\frac{\left(\pi^2+a^2\right)}{Va}-\frac{\tilde{D}a\left(\pi^2+a^2\right)^2+\left(\pi^2+a^2\right)}{1+\omega^2F^2}\right)\right)$$
$$-\frac{\left(\pi^2+a^2\right)^2\left(\frac{Le}{\varepsilon}+1\right)+Le\left(\left(\pi^2+a^2\right)\right)+\omega^2Le^2}{\left(\pi^2+a^2\right)^2+\omega^2\frac{Le^2}{\sigma^2}}\mathrm{N_A Rn}. \tag{6.40}$$

For the oscillatory convection to occur, ω^2 must be positive. If there are no positive roots of ω^2 in equation (6.39), then oscillatory convection is not possible. If there are positive roots of ω^2, the critical Rayleigh number for oscillatory convection can be obtained numerically minimizing equation (6.40) with respect to wave number, after substituting various values of physical parameters for ω^2 of equation (6.39) to determine the various effect of different parameter on the onset of oscillatory convection.

RESULTS AND DISCUSSION

To study the effect of Brinkmanship-Darcy number, Lewis number Le, modified diffusivity ratio N_A and nanoparticles Rayleigh number Rn and porosity parameter on stationary convection, we examine the behavior of $\frac{\partial(\mathrm{Ra})_s}{\partial \tilde{D}a}$, $\frac{\partial(\mathrm{Ra})_s}{\partial Le}$, $\frac{\partial(\mathrm{Ra})_s}{\partial N_A}$, $\frac{\partial(\mathrm{Ra})_s}{\partial Rn}$ and $\frac{\partial(\mathrm{Ra})_s}{\partial \varepsilon}$ analytically.

From equation (6.37), we have

1. $\frac{(\partial Ra)_s}{\partial \tilde{D}a} > 0,$
2. $\frac{(\partial Ra)_s}{\partial Le} < 0,$
3. $\frac{(\partial Ra)_s}{\partial N_A} < 0,$
4. $\frac{(\partial Ra)_s}{\partial Rn} < 0,$
5. $\frac{(\partial Ra)_s}{\partial \varepsilon} > 0.$

These inequalities imply that Lewis number Le, modified diffusivity ratio N_A and nanoparticles Rayleigh number Rn have destabilizing effect while Brinkman-Darcy number and porosity parameter have stabilizing effect on the stationary convection.

Expression for stationary Rayleigh number, which characterizes the stability of the system, is found to be depend upon Lewis number, modified diffusivity ratio and nanoparticles Rayleigh number.

The computations are carried out for different values of parameters considered in the range $-10^2 \leq Ra \leq 10^5$ (thermal Rayleigh number), $1 \leq N_A \leq 10$ (modified diffusivity ratio), $10^2 \leq Le \leq 10^4$ (Lewis number), $10^{-1} \leq Rn \leq 10^1$ (nanoparticles Rayleigh number), $10^{-4} \leq \tilde{D}a < 1$ (Brinkman-Darcy number) and $10^{-2} \leq \varepsilon \leq 1$ (porosity parameter).

The variations of the stationary thermal Rayleigh number with wave number have been plotted graphically as shown in Figures. 2 – 6.

Figures 2 – 6 demonstrate the neutral curve on the $((Ra)_s, a)$ plane for different values of the Lewis number Le, the modified diffusivity ratio N_A, the nanoparticles Rayleigh Rn and porosity parameter ε.

Figure 2 shows the variation of stationary Rayleigh number with wave number for different values of the Brinkman-Darcy number. We found that with an increase in the value of the Brinkman-Darcy number, the stationary Rayleigh number increases, indicating that it delays the onset of convection in Maxwellian visco elastic nanofluid saturated in Brinkman porous medium. This is because; increase in the value of Brinkman-Darcy number is related to increase in the value of effective viscosity and which has the tendency to retard the fluid flow and hence higher heating is required for the onset of convection in a nanofluid-saturated porous medium.

Figure 2. Variation of the stationary Rayleigh number with wave number for different value of Brinkman- Darcy number

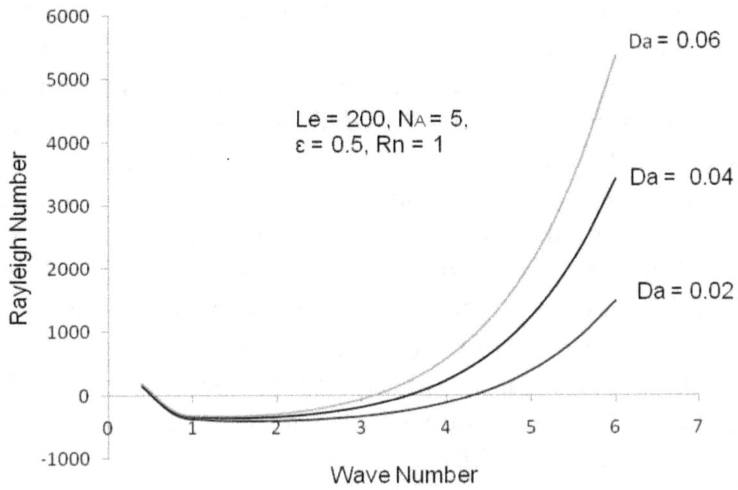

Figure 3 shows the variation of stationary Rayleigh number with wave number for different value of Lewis number with the fixed value of other parameters. It is found that stationary Rayliegh number decreases as the value of Lewis number increases, indicating that Lewis number destabilize the stationary convection. It is due to the fact that thermophoresis at a higher value of thermophoretic diffusivity is more supportable to the disturbance

Figure 3. Variation of the stationary Rayleigh number with wave number for different value of Lewis number

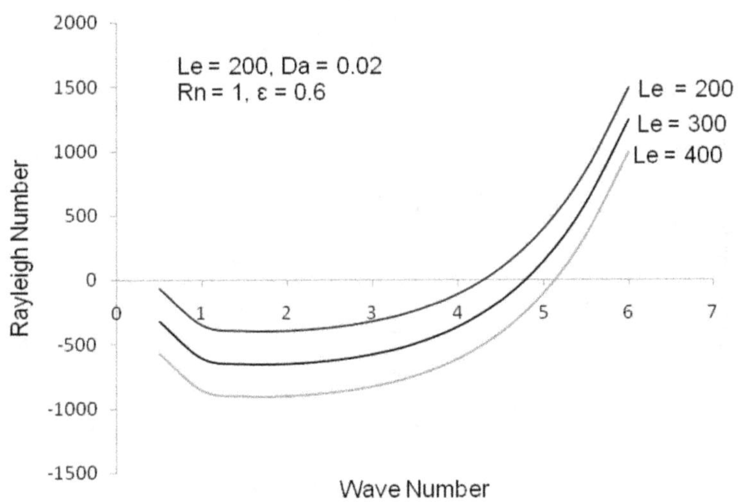

in nanofluids, while both thermophoresis and Brownian motion are driving forces in favor of the motion of nanoparticles.

Figure 4 shows the variation of stationary Rayleigh number with wave number for different value of modified diffusivity ratio with fixed value of other parameters and it is found that stationary Rayliegh number decreases with an increase in the value of modified diffusivity ratio, which indicate that modified diffusivity ratio destabilize the stationary convection. This may lead to an increase in volumetric fraction, which shows that Brownian motion of the nanoparticles will also increase, which may cause destabilizing effect.

Figure 5 shows the variation of stationary Rayleigh number with wave number for different values of nanoparticle Rayleigh number with fixed value of other parameters and it is found that stationary Rayliegh number decreases as the value of the nanoparticles Rayleigh number increases, which mean that nanoparticle Rayleigh number has destabilizing effect on the stationary convection. It has destabilizing effect because the heavier nanoparticles moving through the base fluid makes more strong disturbances as compared with the lighter nanoparticles.

To assess the effect of porous medium on the stability of the system, the variation of the stationary Rayleigh number is shown in Fig. 6 as a function of wave number a for different values of the porosity ε. We found that with an increase in the value of the porosity ε, the stationary Rayleigh number increases, indicating that it delays the onset of convection in nanofluid saturated in porous medium.

Figure 4. Variation of the stationary Rayleigh number with wave number for different value of modified diffusivity ratio

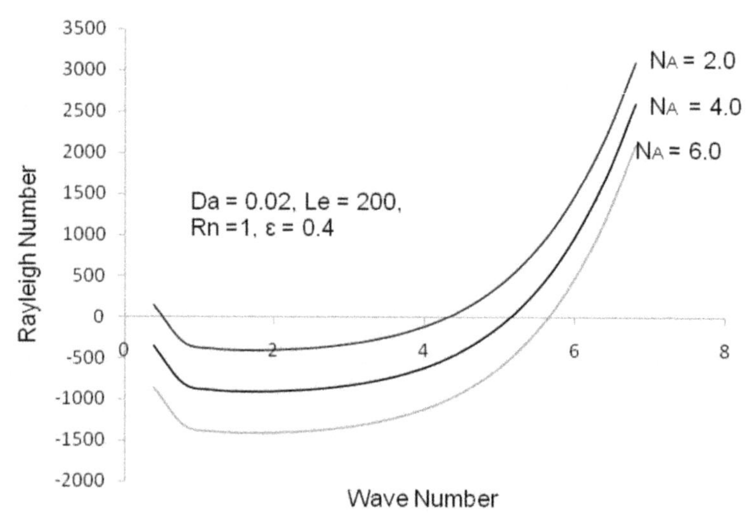

Figure 5. Variation of the stationary Rayleigh number with wave number for different value of modified diffusivity ratio

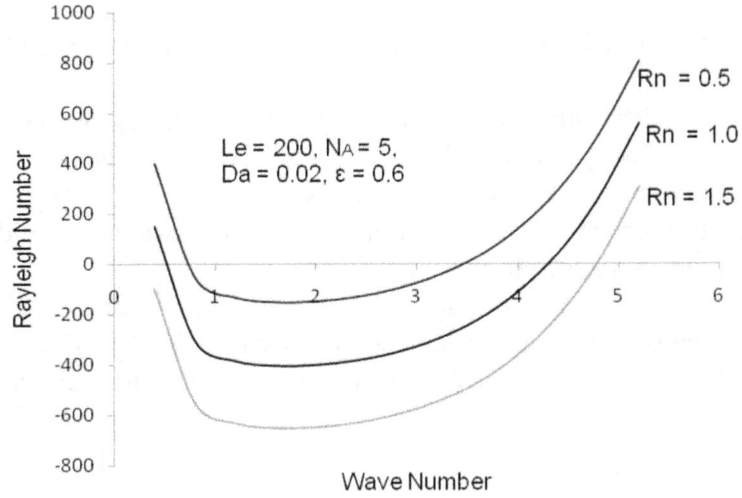

Figure 6. Variation of the stationary Rayleigh number with wave number for different value of porosity parameter

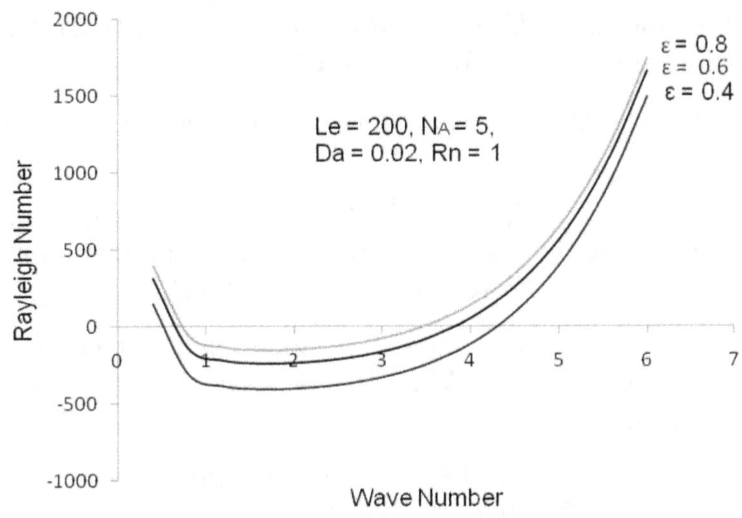

CONCLUSION

Thermal convection in a horizontal layer of Maxwellian visco-elastic nanofluid in a porous medium is studied. Brinkman model is used for porous medium. The flux of volume fraction of nanoparticles is taken to be zero on the isothermal boundaries and the eigen value problem is solved using the Galerkin residual method. The results have been presented both analytically and graphically.

The main conclusions derived from the present chapter are as follows:

1. The instability purely phenomenon due to buoyancy coupled with the conservation of nanoparticles and is independent of the contribution of Brownian motion and thermophoresis.
2. For stationary convection Maxwellian visco-elastic nanofluid behaves like an ordinary Newtonian nanofluid.
3. Brinkman- Darcy number and porosity parameter have stabilizing effect while Lewis number Le, modified diffusivity ratio N_A and nanoparticles Rayleigh number Rn have destabilizing effect the stationary convection.

REFERENCES

Chand, R., & Rana, G. C. (2012b). On the onset of thermal convection in rotating nanofluid layer saturating a Darcy-Brinkman porous medium. *International Journal of Heat and Mass Transfer*, *55*(21-22), 5417–5424. doi:10.1016/j.ijheatmasstransfer.2012.04.043

Ingham, D. B., & Pop, I. (1981). *Transport Phenomena in Porous Media*. New York: Elsvier.

Kuznetsov, A. V., & Nield, D. A. (2010b). Thermal instability in a porous medium layer saturated by a nanofluid: Brinkman Model. *Transport in Porous Media*, *81*(3), 409–422. doi:10.1007/s11242-009-9413-2

Lapwood, E. R. (1948). Convection of a fluid in porous medium. *Proceedings of the Cambridge Philosophical Society*, *44*(04), 508–519. doi:10.1017/S030500410002452X

Nield, D. A., & Bejan, A. (2013). *Convection in porous media*. New York: Springer-Verlag. doi:10.1007/978-1-4614-5541-7

Nield, D. A., & Kuznetsov, A. V. (2014). Thermal instability in a porous medium layer saturated by a nanofluid: A revised model, Int. *J. Heat and Mass Transf.*, *68*, 211–214. doi:10.1016/j.ijheatmasstransfer.2013.09.026

Vafai, K., & Hadim, H. A. (2000). *Handbook of Porous Media. M.* New York: Decker.

Wooding, R. A. (1960). Rayleigh instability of a thermal boundary layer in flow through a porous medium. *Journal of Fluid Mechanics*, *9*(02), 183–192. doi:10.1017/S0022112060001031

Chapter 7
Effect of Variable Gravity on the Thermal Convection in a Layer of Maxwellian Visco-Elastic Nanofluid

INTRODUCTION

The idealization of uniform gravity assumed in theoretical investigations, although valid for laboratory purposes can scarcely be justified for large-scale convection phenomena occurring in atmosphere, the ocean or mantle of the Earth. It then becomes imperative to consider gravity as variable quantity varying with distance from surface or reference point. Pradhan and Samal (1987), Pradhan et al. (1989) studied the thermal instability of a fluid layer in a variable gravitational field while Alex et al. (2001), Alex and Prabhamani (2001) studied the variable gravity effects on the thermal instability in a porous medium with internal heat source and inclined temperature gradient. Rionero and Strughan (1990) discuss the various type of variable gravity parameter on the stability convection and recently Chand (2010, 2011, 2013c) studied the variable gravity effects on the thermal instability in fluid layer. Effect of variable gravity in layer of nanofluid in a porous medium is studied by Chand et al. (2013a) and found that gravity parameter play significant role on the stability of fluid.

DOI: 10.4018/978-1-68318-006-7.ch007

In this chapter an attempt has been made to study the thermal instability in a horizontal layer of Maxwellian visco-elastic nanofluids in the presence of variable gravity for more realistic boundary conditions.

MATHEMATICAL FORMULATIONS OF THE PROBLEM

The Physical Problem

In this chapter we shall investigate effect of variable gravity on the onset of thermal convection in a horizontal layer of Maxwellian visco-elastic nanofluid. The physical configuration of the problem to be considered as:

An infinite horizontal layer of Maxwell visco-elastic nanofluid of thickness 'd' bounded by horizontal boundaries z = 0 and z = d. Fluid layer is acted upon by a gravity force **g**(0,0,-g) and and it is assumed that gravity force vector is varies linearly with z i.e. $\mathbf{g} = \left(1 + \delta\, h(z)\right) g$, where $\delta h(z)$ is the variable gravity parameter. Fluid layer is heated from below in such a way that horizontal boundaries z = 0 and z = d respectively maintained at a uniform temperature T_0 and T_1 ($T_0 > T_1$) as shown is Figure 1. The normal component of the nanoparticles flux has to vanish at an impermeable boundaries and the reference scale for temperature and nanoparticles fraction is taken to be T_1 and φ_0 respectively.

Figure 1. Physical configuration of the problem

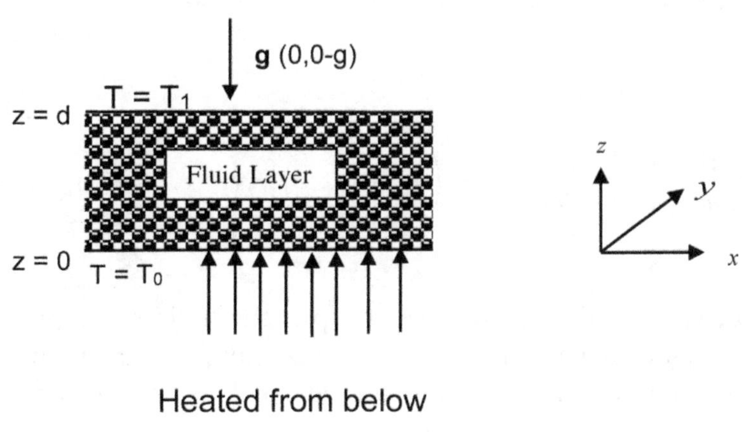

Assumptions

The mathematical equations describing the physical model are based upon the following assumptions:

1. Thermophysical properties of fluid expect for density in the buoyancy force (Boussinesq Hypothesis) are constant,
2. The fluid phase and nanoparticles are in thermal equilibrium state and thus, the heat flow has been described using one equation model,
3. Dilute mixture,
4. Nanoparticles are spherical,
5. Nanoparticles are non-magnetic,
6. No chemical reactions take place in fluid layer,
7. Negligible viscous dissipation,
8. Radiative heat transfer between the sides of wall is negligible when compared with other modes of the heat transfer,
9. Nanofluid is incompressible, electrically conducting, Newtonian and laminar flow,
10. Each boundary wall is assumed to be impermeable and perfectly thermal conducting.

GOVERNING EQUATIONS

The governing equations for Maxwellian visco-elastic nanofluid in the presence of variable gravity under the Boussinesq approximation are

$$\nabla \cdot \mathbf{v} = 0, \tag{7.1}$$

$$\rho\left(1+\lambda\frac{\partial}{\partial t}\right)\frac{d\mathbf{v}}{dt} = \left(1+\lambda\frac{\partial}{\partial t}\right)\left(-\nabla p + \left(\varphi\rho_p + (1-\varphi)\left\{\rho_f\left(1-\alpha\left(T-T_0\right)\right)\right\}\right)\mathbf{g}\right) + \mu\nabla^2\mathbf{v}\,, \tag{7.2}$$

where $\mathbf{v(u, v, w)}$ is the velocity vector, p is the hydrostatic pressure, μ is viscosity, α is the coefficient of thermal expansion, λ is the relaxation time, φ is the volume fraction of the nanoparticles, ρ_p density of nanoparticles and ρ_f density of base fluid and $\frac{d}{dt} = \frac{\partial}{\partial t} + (\mathbf{v}\cdot\nabla)$ is stands for convection derivative.

Equation of energy for Maxwellian visco-elastic nanofluid is given by

$$\rho c\frac{\partial T}{\partial t}+\mathbf{v}\cdot\nabla T=k_m\nabla^2 T+(\rho c)_p\left[D_B\nabla\varphi\cdot\nabla T+\frac{D_T}{T_1}\nabla T\cdot\nabla T\right], \tag{7.3}$$

where ρc is heat capacity of fluid, $(\rho c)_p$ is heat capacity of nanoparticles, T_1 is the temperature of the fluid layer at $z = d$, and k_m is thermal conductivity.

Equation of continuity for the nanoparticles is

$$\frac{\partial\varphi}{\partial t}+\mathbf{v}\cdot\nabla\varphi=D_B\nabla^2\varphi+\frac{D_T}{T_1}\nabla^2 T. \tag{7.4}$$

We assume that the temperature is constant and nanoparticles flux is zero on the boundaries. Thus boundary conditions are

$$w=0,\ T=T_0,\ D_B\frac{\partial\phi}{\partial z}+\frac{D_T}{T_1}\frac{\partial T}{\partial z}=0 \ \text{ at } z=0$$

and

$$w=0,\ T=T_1,\ D_B\frac{\partial\phi}{\partial z}+\frac{D_T}{T_1}\frac{\partial T}{\partial z}=0 \ \text{ at } z=d. \tag{7.5}$$

Introducing non-dimensional variables as

$$(x',y',z',)=\left(\frac{x,y,z}{d}\right),\ v'(u',v',w')=\upsilon\left(\frac{u,v,w}{\kappa}\right)d,\ t'=\frac{t^o}{d^2},\ p'=\frac{pd^2}{\mu\kappa},$$

$$\varphi'=\frac{(\varphi-\varphi_0)}{\varphi_0},\ T'=\frac{(T-T_1)}{(T_0-T_1)},$$

where $\kappa=\dfrac{k_m}{\rho c}$ is thermal diffusivity of the fluid.

Equations (7.1) - (7.4) in non-dimensional form can be written as

$$\nabla'\cdot\mathbf{v}'=0, \tag{7.6}$$

$$\left(1+F\frac{\partial}{\partial t'}\right)\frac{1}{\Pr}\frac{\partial \mathbf{v}'}{\partial t'}=\left(1+F\frac{\partial}{\partial t'}\right)\begin{pmatrix}-\nabla' p'\text{-Rm}\left(1+\delta\, h(z)\right)\hat{e}_z+\text{Ra}\left(1+\delta\, h(z)\right)T'\hat{e}_z\\ -\text{Rn}\left(1+\delta\, h(z)\right)\varphi'\hat{e}_z\end{pmatrix}+\nabla'^2\mathbf{v}', \tag{7.7}$$

$$\frac{\partial \varphi'}{\partial t'}+\mathbf{v}'\cdot\nabla'\varphi'=\frac{1}{\text{Le}}\nabla'^2\varphi'+\frac{N_A}{\text{Le}}\nabla'^2 T', \tag{7.8}$$

$$\frac{\partial T'}{\partial t'}+\mathbf{v}'\cdot\nabla' T'=\nabla'^2 T'+\frac{N_B}{\text{Le}}\nabla'\varphi'\cdot\nabla' T'+\frac{N_A N_B}{\text{Le}}\nabla' T'\cdot\nabla' T'. \tag{7.9}$$

Here the non-dimensional parameters are given as follows:

$\Pr=\frac{\mu}{\rho\kappa}$ is the Prandtl number,

$Le=\frac{\kappa}{D_B}$ is the Lewis number,

$F=\frac{\kappa\lambda}{d^2}$ is the stress relaxation parameter,

$\text{Ra}=\frac{\rho g\alpha d^3\left(T_0-T_1\right)}{\mu\kappa}$ is the Rayleigh number,

$\text{Rm}=\frac{\left(\acute{A}_p\varphi_0+\acute{A}(1\text{-}\varphi_0)\right)gd^3}{\mu\kappa}$ is the density Rayleigh number,

$\text{Rn}=\frac{\left(\rho_p-\rho\right)\varphi_0 gd^3}{\mu\kappa}$ is the nanoparticles Rayleigh number,

$N_A=\frac{D_T\left(T_0\text{-}T_1\right)}{D_B T_1\varphi_0}$ is the modified diffusivity ratio,

$N_B=\frac{(\rho c)_p\,\varphi_0}{(\rho c)_f}$ is the modified particle-density increment.

In spirit of Oberbeck-Boussinesq approximation, equation (7.7) has been linearized by the neglect of a term proportional to the product of $Æ_0$ and T.

This approximation is valid in the case of small temperature gradients in a dilute suspension of nanoparticles.

The dimensionless boundary conditions are

$$w' = 0, \quad T' = 1, \quad \frac{\partial \phi'}{\partial z'} + N_A \frac{\partial T'}{\partial z'} = 0 \text{ at } z' = 0$$

and

$$w' = 0, \; T' = 0, \quad \frac{\partial \phi'}{\partial z'} + N_A \frac{\partial T'}{\partial z'} = 0 \text{ at } z' = 1. \tag{7.10}$$

THE BASIC STATE AND ITS SOLUTIONS

The basic state was assumed to be quiescent and is given by

$$v_i' \left(u', v', w'\right) = 0,$$

$$p' = p_b(z),$$

$$T' = T_b(z), \tag{7.11}$$

$$\varphi' = \varphi_b(z) \text{ and}$$

$$\rho = \rho_0 \left(1 + \alpha\left(T\text{-}T_0\right)\right).$$

Equations (7.6) – (7.9) reduce to

$$0 = -\frac{dp_b}{dz'}\text{-}Rm + RaT_b - Rn\varphi_b, \tag{7.12}$$

$$\frac{d^2T_b}{dz'^2} + \frac{N_B}{Le}\frac{d\varphi_b}{dz'}\frac{dT_b}{dz'} + \frac{N_A N_B}{Le}\left(\frac{dT_b}{dz'}\right)^2 = 0, \tag{7.13}$$

$$\frac{d^2\varphi_b}{dz'^2} + N_A \frac{d^2T_b}{dz'^2} = 0, \tag{7.14}$$

Using boundary conditions in (7.10), equation (7.14) gives

$$\varphi_b = -N_A T_b + (1 - N_A) z' + N_A. \tag{7.15}$$

On substituting the value of the φ_b from equation (7.15) in equation (7.13), we have

$$\frac{d^2 T_b}{dz'} + \frac{(1\text{-}N_A) N_B}{Le} \frac{dT_b}{dz'} = 0 \tag{7.16}$$

On integrating equation (7.16) with respect to z' and using boundary conditions (7.10), we get

$$T_b = \frac{1 - e^{-(1-N_A)N_B(1-z')/Le}}{1 - e^{-(1-N_A)N_B/Le}} \tag{7.17}$$

For the most of nanofluid investigated so far Le is large, is of order 10^2 - 10^3, while N_A is no greater than about 10. Then, the exponents in equation (7.17) are small. By expanding the exponential function into the power series and retaining up to the first order is negligible and so to a good approximation for the solution of basic state is given by

$$T_b = 1 - z' \text{ and}$$

$$\varphi_b = \phi_0 + N_A z'$$

PERTURBATION SOLUTIONS

Let the initial basic state described by (7.11) is slightly perturbed so that perturbed state is given by

$$(u', v', w') = 0 + (u'', v'', w''),$$

$$T' = T_b + T'', \tag{7.18}$$

$$\varphi' = \varphi_b + \varphi'',$$

$$p' = p_b + p'',$$

where $T_b = 1 - z'$, $\varphi_b = \phi_0 + N_A z'$ and (u'',v'',w''), T'', φ'' and p'' respectively the perturbations in initial velocity, temperature, volume fraction of the nanoparticles and pressure.

By substituting (7.18) in equations (7.6) – (7.9) and linearize by neglecting the product of the prime quantities, we obtained following equations

$$\nabla \cdot \mathbf{v} = 0, \tag{7.19}$$

$$\left(1 + F\frac{\partial}{\partial t}\right)\frac{1}{\Pr}\frac{\partial \mathbf{v}}{\partial t} = \left(1 + F\frac{\partial}{\partial t}\right)\left(-\nabla p + \mathrm{Ra}\left(1 + \delta\, h(z)\right)T\hat{e}_z - \mathrm{Rn}\left(1 + \delta\, h(z)\right)\varphi\hat{e}_z\right) + \nabla^2 \mathbf{v}, \tag{7.20}$$

$$\frac{\partial \varphi}{\partial t} + wN_A = \frac{1}{\mathrm{Le}}\nabla^2\varphi + \frac{N_A}{\mathrm{Le}}\nabla^2 T, \tag{7.21}$$

$$\frac{\partial T}{\partial t} - w = \nabla^2 T + \frac{N_B}{\mathrm{Le}}\left(N_A\frac{\partial T}{\partial z} - \frac{\partial \phi}{\partial z}\right) - \frac{2N_A N_B}{\mathrm{Le}}\frac{\partial T}{\partial z}. \tag{7.22}$$

Boundary conditions are

$$w = 0,\ T = 0,\quad \frac{\partial \phi}{\partial z} + N_A\frac{\partial T}{\partial z} = 0 \ \text{ at } z = 0, 1. \tag{7.23}$$

[Dashes (") have been suppressed for convenience]

Eliminating pressure term p from equation (7.20), we have

$$\left(1 + F\frac{\partial}{\partial t}\right)\left(\frac{1}{\Pr}\frac{\partial}{\partial t}\left(\nabla^2 w\right) - \mathrm{Ra}\left(1 + \delta\, h(z)\right)\nabla_H^2 T + \mathrm{Rn}\left(1 + \delta\, h(z)\right)\nabla_H^2\varphi\right) - \nabla^4 w = 0, \tag{7.24}$$

where ∇_H^2, is two-dimensional Laplacian operator on a horizontal plane.

NORMAL MODE ANALYSIS

We shall now analyze an arbitrary perturbation into a complete set of normal modes and then examine the stability of each of those modes individually. For the system of equations (7.24), (7.22) - (7.23) the analysis can be made in terms of two dimensional periodic wave numbers. Thus, assuming that the perturbed quantities are of the form

$$[w,T,\varphi]=[W(z),\Theta(z),\Phi(z)]\exp\left(ik_x x+ik_y y+nt\right), \tag{7.25}$$

where k_x, k_y are wave numbers in x and y direction and n is growth rate of disturbances.

Using equation (7.25), equations (7.24), (7.21) and (7.22) become

$$\left(D^2-a^2\right)\left(D^2-a^2-\frac{n\left(1+nF\right)}{Pr}\right)W-\left(1+nF\right)\left(a^2Ra\left(1+\delta\, h(z)\right)\Theta-a^2Rn\left(1+\delta\, h(z)\right)\Phi\right)=0, \tag{7.26}$$

$$N_A W-\frac{N_A}{Le}\left(D^2-a^2\right)\Theta-\left(\frac{1}{Le}\left(D^2-a^2\right)-n\right)\Phi=0, \tag{7.27}$$

$$W+\left(D^2-a^2-n-\frac{N_A N_B}{Le}D\right)\Theta-\frac{N_B}{Le}D\Phi=0, \tag{7.28}$$

where $D\equiv\frac{d}{dz}$ and $a=\sqrt{k_x^2+k_y^2}$ is the dimensionless resultant wave number.

The boundary conditions of the problem in view of normal mode analysis are

$$W=0,\ D^2W=0, \Theta=0,\ D\Phi+N_A D\Theta=0\ \text{at}\quad z=0,1. \tag{7.29}$$

METHOD OF SOLUTION

The Galerkin weighted residuals method is used to obtain an approximate solution to the system of equations (7.26) - (7.28) with the corresponding boundary conditions (7.29). In this method, the test functions are the same as the base (trial) functions. Accordingly W, **Θ** and **Φ** are taken as

$$W=\sum_{p=1}^{N}A_pW_p,\ \Theta=\sum_{p=1}^{N}B_p\Theta_p,\ \ =\sum_{p=1}^{N}C_p\Phi_p, \tag{7.30}$$

where $W_p=\Theta_p=\sin p\pi z, \Phi_p=-N_A\sin p\pi z$, A_p, B_p and C_p are unknown coefficients, p = 1, 2, 3,..., N and the base functions W_p, Θ_p, and Φ_p satisfying the boundary conditions (7.29). Using expression for W, Θ and Φ in equations (7.26) - (7.28) and multiplying the first equation by W_p the second equation by Θ_p and third equation by Φ_p and then integrating in the limits from zero to unity, we obtain a set of 3N linear homogeneous equations with 3N unknown A_p, B_p and C_p; p = 1, 2, 3,..., N. For existing of nontrivial solution, the vanishing of the determinant of coefficients produces the characteristics equation of the system in term of Rayleigh number Ra.

LINEAR STABILITY ANALYSIS

For the present formulation, we have considered the which system of equations (7.26) - (7.28) together with the boundary conditions (7.29) constitute a linear eigenvalue problem with variable coefficient for the growth rate of disturbance n of the system. Substituting equation (7.30) into the system of equations (7.26) - (7.28) and multiplying the first equation by W_p the second equation by Θ_p and third equation by Φ_p and then integrating in the limits from zero to unity and performing some integration by parts, one obtains the following matrix equation

$$\begin{bmatrix} \dfrac{(\pi^2+a^2)^2}{(1+nF)}+\dfrac{n}{Pr}(\pi^2+a^2) & -a^2Ra(1+\delta h(z)) & -a^2N_ARn(1+\delta h(z)) \\ 1 & -(\pi^2+a^2+n) & 0 \\ 1 & \dfrac{1}{Le}(\pi^2+a^2) & -\left(\dfrac{1}{Le}(\pi^2+a^2)+n\right) \end{bmatrix}\begin{bmatrix} W_0 \\ \Theta_0 \\ \Phi_0 \end{bmatrix}=\begin{bmatrix} 0 \\ 0 \\ 0 \end{bmatrix}. \tag{7.31}$$

The non-trivial solution of the above matrix requires that

$$Ra=\frac{(\pi^2+a^2)}{a^2(1+nF)(1+\delta h(z))}\left((\pi^2+a^2)+\frac{n(1+nF)}{Pr}\right)(\pi^2+a^2+n)$$

$$-\frac{\left(\pi^2+a^2\right)+Le\left(\pi^2+a^2+n\right)}{\left(\pi^2+a^2\right)+\mathrm{nLe}}\mathrm{N_ARn}. \tag{7.32}$$

The growth rate n is in general a complex quantity such that $n=\omega_r+i\omega$, the system with $\omega_r<0$ is always stable, while for $\omega_r>0$ it will become unstable. For neutral stability, the real part of n is zero. Hence, we now write $n=i\omega$, (where ω is real and is a dimensionless frequency) in equation (7.32), we have

$$\mathrm{Ra}=\Delta_1+i\omega\Delta_2\,, \tag{7.33}$$

where

$$''_1=\frac{\left(\pi^2+a^2\right)}{\mathrm{a}^2\left(1+\delta\,\mathrm{h}\left(\mathrm{z}\right)\right)}\left(\frac{\left(\pi^2+a^2\right)^2+\acute{E}^2F\left(\pi^2+a^2\right)}{1+\acute{E}^2F^2}-\frac{\acute{E}^2}{\mathrm{Pr}}\right)$$

$$-\frac{\left(\pi^2+a^2\right)^2\left(Le+1\right)+Le\left(\left(\pi^2+a^2\right)\right)+\acute{E}^2Le^2}{\left(\pi^2+a^2\right)^2+\acute{E}^2Le^2}\mathrm{N_ARn} \tag{7.34}$$

and

$$\Delta_2=\frac{\left(\pi^2+a^2\right)^2}{\mathrm{a}^2\left(1+\delta\,\mathrm{h}\left(\mathrm{z}\right)\right)}\left(\frac{1-F\left(\pi^2+a^2\right)}{1+\omega^2F^2}+\frac{1}{\mathrm{Pr}}\right)+\frac{Le^2\left(\pi^2+a^2\right)}{\left(\pi^2+a^2\right)^2+\omega^2Le^2}\mathrm{N_ARn}. \tag{7.35}$$

Since Ra is a physical quantity, so it must be real. Hence, it follows from the equation (7.33) that either $\omega=0$ (exchange of stability, steady state) or $\Delta_2=0$ ($\omega\neq 0$ overstability or oscillatory onset).

STATIONARY CONVECTION

For the case of stationary convection $[n=\omega=0]$, equation (7.32) reduces to

$$\left(\mathrm{Ra}\right)_s = \frac{\left(\pi^2 + a^2\right)^3}{\mathrm{a}^2\left(1 + \delta\, \mathrm{h}\left(\mathrm{z}\right)\right)} - \left(1 + Le\right)\mathrm{N}_\mathrm{A}\mathrm{Rn}. \tag{7.36}$$

It is observed that stationary Rayleigh number Ra is function of the Lewis number Le, the modified diffusivity ratio N_A, the nanoparticles Rayleigh Rn and variable gravity parameter $\delta h(z)$ but independent of visco- elastic parameter F, Prandtl number Pr and modified particle-density increment N_B. Thus Maxwell visco-elastic nanofluid behaves like an ordinary Newtonian nanofluid and instability is purely a phenomenon due to buoyancy coupled with the conservation of nanoparticles.

If gravity is constant ($\delta h(z) = 0$), then equation (7.36) reduces to

$$\left(\mathrm{Ra}\right)_s = \frac{\left(\pi^2 + a^2\right)^3}{\mathrm{a}^2} - \left(1 + Le\right)\mathrm{N}_\mathrm{A}\mathrm{Rn}. \tag{7.37}$$

This is the good agreement of the result (1.47) obtained in Chapter 1.

To find the critical value of $\left(\mathrm{Ra}\right)_\mathrm{s}$, equation (7.36) is differentiated with respect to 'a^2' and then equated to zero. The minimum of first term of right-hand side of equation (7.36) is attained at $\mathrm{a}_\mathrm{c} = \frac{\pi}{\sqrt{2}}$ and minimum value found to $\frac{27\pi^4}{4}$ so the corresponding critical Rayleigh number given by

$$\left(\mathrm{Ra}\right)_c = \frac{27\pi^4}{4} - \left(1 + Le\right)\mathrm{N}_\mathrm{A}\mathrm{Rn}.$$

In the absence of nanoparticles ($\mathrm{Rn} = \mathrm{Le} = \mathrm{N}_\mathrm{A} = 0$), one recovers the well- known results that the critical Rayleigh number is equal to $\left(\mathrm{Ra}\right)_c = \frac{27\pi^4}{4}$.

Thus presence of the nanoparticles lowers the value of the critical Rayleigh number by usually by substantial amount. Also parameter N_B does not appear in the equation (7.36); thus instability is purely phenomena due to buoyancy coupled with conservation of nanoparticles. Thus average contribution of nanoparticles flux in the thermal energy equation is zero with one-term Galerkin approximation.

OSCILLATORY CONVECTION

For oscillatory convection $\Delta_2 = 0$ and $\omega \neq 0$, thus equation (7.35) gives a dispersion relation of the form

$$a_2\left(\omega^2\right)^2 + a_1\left(\omega^2\right) + a_0 = 0; \tag{7.38}$$

where

$$a_2 = \frac{Le^2F^2}{\Pr},$$

$$a_1 = \frac{a^2\left(1+\delta\,h(z)\right)Le^2F^2}{\pi^2 + a^2} N_A Rn + Le^2\left(1 - F\left(\pi^2 + a^2\right)\right) + F\left(\pi^2 + a^2\right) + Le^2,$$

$$a_0 = \frac{a^2\left(1+\delta\,h(z)\right)Le^2}{\pi^2 + a^2} N_A Rn + \left(\pi^2 + a^2\right)\left(1 - F\left(\pi^2 + a^2\right)\right) + \frac{\left(\pi^2 + a^2\right)}{\Pr}.$$

Then equation (7.33) with $\Delta_2 = 0$ gives oscillatory Rayleigh number at the margin of stability as

$$(\mathrm{Ra})_{osc} = \frac{\left(\pi^2 + a^2\right)^2}{a^2\left(1+\delta\,h(z)\right)}\left(\frac{\left(\pi^2 + a^2\right) + \omega^2 F}{1 + \omega^2F^2} - \frac{\omega^2}{\Pr}\right) - \frac{\left(\pi^2 + a^2\right)^2\left(Le+1\right) + Le\left(\pi^2 + a^2\right) + \omega^2Le^2}{\left(\pi^2 + a^2\right)^2 + \omega^2Le^2} N_A \mathrm{Rn} \cdot \tag{7.39}$$

For the oscillatory convection to occur, ω^2 must be positive. If there are no positive roots of ω^2 in equation (7.38), then oscillatory convection is not possible. If there are positive roots of ω^2, the critical Rayleigh number for oscillatory convection can be obtained numerically minimizing equation (7.39) with respect to wave number, after substituting various values of physical parameters for ω^2 of equation (7.38) to determine the various effect of different parameter on the onset of oscillatory convection.

RESULTS AND DISCUSSION

To study the effect of Lewis number Le, modified diffusivity ratio N_A and nanoparticles Rayleigh number Rn on stationary convection, we examine the behavior of $\frac{\partial(\mathrm{R}a)_s}{\partial Le}$, $\frac{\partial(\mathrm{R}a)_s}{\partial N_A}$ and $\frac{\partial(\mathrm{R}a)_s}{\partial Rn}$ analytically.

From equation (7.47), we have

1. $\frac{(\partial \mathrm{R}a)_s}{\partial Le} < 0$,
2. $\frac{(\partial \mathrm{R}a)_s}{\partial N_A} < 0$,
3. $\frac{(\partial \mathrm{R}a)_s}{\partial Rn} < 0$.

These inequalities show that Lewis number Le, modified diffusivity ratio N_A and nanoparticles Rayleigh number Rn destabilizes the stationary convection.

Expression for stationary Rayleigh number, which characterizes the stability of the system, is found to be depend upon Lewis number, modified diffusivity ratio and nanoparticles Rayleigh number.

The computations are carried out for different values of parameters considered in the range $10^2 \leq Ra \leq 10^5$ (thermal Rayleigh number), $1 \leq N_A \leq 10$ (modified diffusivity ratio), $10^2 \leq Le \leq 10^4$ (Lewis number), $10^{-1} \leq Rn \leq 10^1$ (nanoparticles Rayleigh number).

The effects of variable gravity parameter on stationary convection have been presented graphically. Stability curves for variable gravity parameter are shown in Figures 2 – 3.

Figure 2 indicates the effect of variable gravity parameter on the stationary convection and it is found that fluid layer has stabilizing effect when the gravity parameter varies as h(z) = z^2- 2z, h(z) = -z, h(z) = -z^2 and has destabilizing effect when gravity parameter is h(z) = z. These results are good agreements of the results obtained by Rionero and Straughan (1990) and Chand et al. (2013a).

Figure 3 indicates the effect of Lewis number on stationary convection and it is found that the critical Rayleigh number increases with an increase in the value of Lewis number, indicating that the effect of Lewis number is to inhibit the onset of convection. Also it is found that the value of stationary Rayleigh number increases when we taken decreasing gravity profile i.e. when variable gravity parameter is h(z) = - z, while the value of stationary

Figure 2. Stability curve for different values of variable gravity parameter

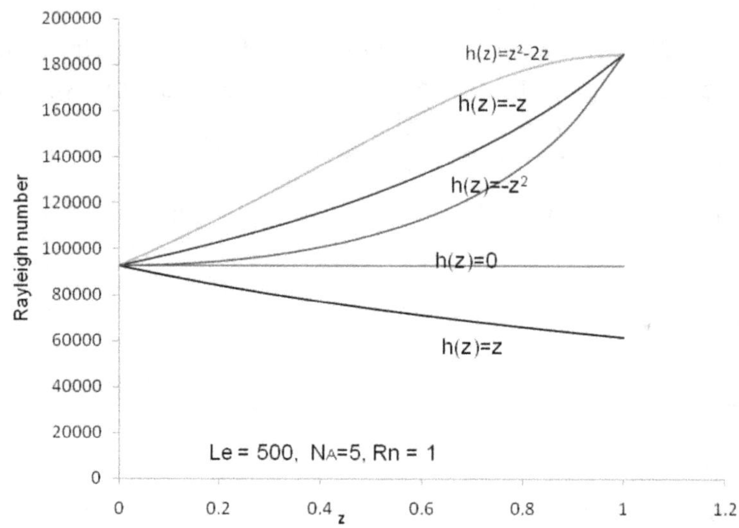

Figure 3. Stability curve for different values of Lewis number

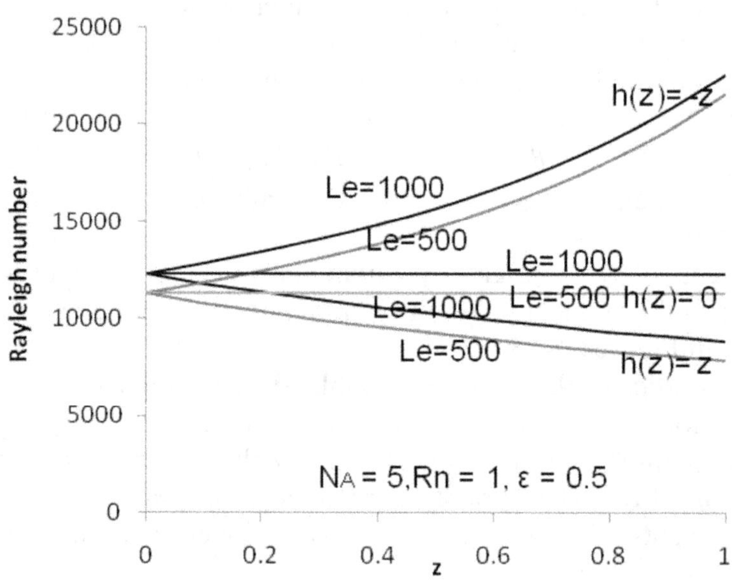

Rayleigh number decreases when we take increasing gravity profile i.e. when variable gravity parameter is $h(z) = z$. Thus decreasing gravity parameter has stabilizing effect while increasing gravity parameter has destabilizing effect on the stationary convection.

Figure 3 shows the variation of thermal Rayleigh number for different value of Lewis number Le and for the fixed value of other parameters. It is found that stationary Rayliegh number decreases as the values of Lewis number increases, indicating that Lewis number destabilizes the stationary convection.

CONCLUSION

Effect of variable gravity on the thermal instability in a horizontal layer of Maxwellian visco-elastic nanofluid is studied. The flux of volume fraction of nanoparticles is taken to be zero on the isothermal boundaries and the eigen value problem is solved using the Galerkin residual method. The results have been presented both analytically and graphically.

The main conclusions derived from the chapter are as follows:

1. The instability purely phenomenon due to buoyancy coupled with the conservation of nanoparticles and is independent of the contribution of Brownian motion and thermophoresis.
2. For stationary convection Maxwellian visco-elastic nanofluid behaves like an ordinary Newtonian nanofluid.
3. Stationary convection has stabilizing effect when the gravity parameter varies as $h(z) = z^2 - 2z$, $h(z) = -z$, $h(z) = -z^2$ and has destabilizing effect when gravity parameter varies as $h(z) = z$. In other word decreasing gravity parameter has stabilizing effect while increasing gravity parameter has destabilizing effect on the stationary convection.
4. Lewis number Le, modified diffusivity ratio N_A and nanoparticles Rayleigh number Rn have destabilizing effect on the stationary convection.

REFERENCES

Alex, S. M., & Prabhamani, R. P. (2001). Effect of variable gravity field on Soret driven thermosolutal convection in a porous medium. *International Communications in Heat and Mass Transfer*, *28*(4), 509–518. doi:10.1016/S0735-1933(01)00255-X

Alex, S. M., Prabhamani, R. P., & Vankatakrishan, K. S. (2001). Variable gravity effects on thermal instability in a porous medium with internal heat source and inclined temperature gradient. *Fluid Dynamics Research*, *29*(1), 1–6. doi:10.1016/S0169-5983(01)00016-8

Chand, R. (2010). Gravitational effect on thermal instability of Maxwell visco-elastic fluid in porous medium. *Ganita Sandesh*, *24*(2), 166–170.

Chand, R. (2011). Effect of suspended particles on thermal instability of Maxwell visco-elastic fluid with variable gravity in porous medium. *Antarctica Journal of Mathematics*, *8*(6), 487–497.

Chand, R. (2013c). Thermal Instability of rotating Maxwell visco-elastic fluid with variable gravity in porous medium. *Journal of the Indian Math. Soc., 80*(2), 23-31.

Chand, R., Rana, G. C., & Kumar, S. (2013a). Variable gravity effects on thermal instability of Nanofluid in anisotropic porous medium. Int. *J. of Applied Mechanics and Engineering*, *18*(3), 631–642.

Pradhan, G. K., & Samal, P. C. (1987). Thermal stability of a fluid layer under variable body forces. *Journal of Mathematical Analysis and Applications*, *122*(2), 487–495. doi:10.1016/0022-247X(87)90280-0

Pradhan, G. K., Samal, P. C., & Tripathy, U. K. (1989). Thermal stability of a fluid layer in a variable gravitational field. *Indian J. Pure Appl. Math.*, *20*(7), 736–745.

Rionero, S., & Straughan, B. (1990). Convection in a porous medium with internal heat source and variable gravity effects. *International Journal of Engineering Science*, *28*(6), 497–503. doi:10.1016/0020-7225(90)90052-K

Chapter 8 Thermal Convection in a Layer of Maxwellian Visco-Elastic Nanofluid with Hall Current

INTRODUCTION

Magnetic fluids (ferromagnetic fluid) are kinds of special nanofluids. They are stable colloidal suspensions of small magnetic particles such as magnetite (Fe_3O_4). The properties of the magnetic nanoparticles, the magnetic component of magnetic nanofluids, may be tailored by varying their size and adapting their surface coating in order to meet the requirements of colloidal stability of magnetic nanofluids with non-polar and polar carrier liquids. Recently, the study of magnetohydrodynamics (MHD) became important in engineering applications, such as in designing cooling system with liquid metals, MHD generator and other devices in the petroleum industry, materials processing, Plasma studies, nuclear reactors, geophysics, astrophysics, aeronautics and aerodynamics Chandrasekhar (1961) and Kent (1966). If an electric field is applied right angle to magnetic field, the whole current will not flow along the electric field. The tendency of the electric current of

DOI: 10.4018/978-1-68318-006-7.ch008

flow across an electric field in the presence of magnetic field is called 'Hall effect'. Researchers Sherman and Sutton (1962), Oberoi and Devanathan (1963), Gupta (1967), Sharma et al. (2000), Sharma and Kumar (1996b) have studied the Hall effect in thermal instability of different types of Newtonian and non-Newtonian fluids. The Hall effect is likely to be important in geophysical and astrophysical situation. The study of MHD flows with Hall currents has important engineering applications in MHD generators, Hall accelerators, refrigeration coils, electric transformers etc. The uncommon properties of nanofluids as such as thermal transfer fluids for instance, these fluids can be used in a plethora of engineering applications ranging from use in the automotive industry to the medical arena to use in power plant cooling systems as well as computers.

Due to importance Hall effect on the onset of thermal convection of Maxwellian visco-elastic nanofluids, in this chapter an attempt has been made to study the thermal convection in a horizontal layer of Maxwellian visco-elastic nanofluid in the presence of Hall effect for more realistic boundary conditions. Stability is discussed analytically as well as numerically using Galerkin-type weighted residuals method. It has been observed that the Hall effect parameter destabilize the fluid layer.

MATHEMATICAL FORMULATIONS OF THE PROBLEM

The Physical Problem

The physical configuration of the problem to be considered as:

An infinite horizontal layer of Maxwellian visco-elastic nanofluid of thickness 'd' bounded by horizontal boundaries z = 0 and z = d. A Cartesian coordinate system (x, y, z) is chosen with the origin at the bottom of the fluid layer and the z- axis normal to the fluid layer. Fluid layer is acted upon by gravity force **g**(0, 0,-g) and a uniform vertical magnetic field **H** (0, 0, H). Fluid layer is heated from below in such a way that horizontal boundaries z = 0 and z = d respectively maintained at a uniform temperature T_0 and T_1 ($T_0 > T_1$). The normal component of the nanoparticles flux has to vanish at an impermeable boundary and the temperature T is taken to be T_0 at z = 0 and T_1 at z = d, ($T_0 > T_1$) as shown in Figure 1. The reference scale for temperature and nanoparticles fraction is taken to be T_1 and φ_0 respectively.

Figure 1. Physical configuration of the problem

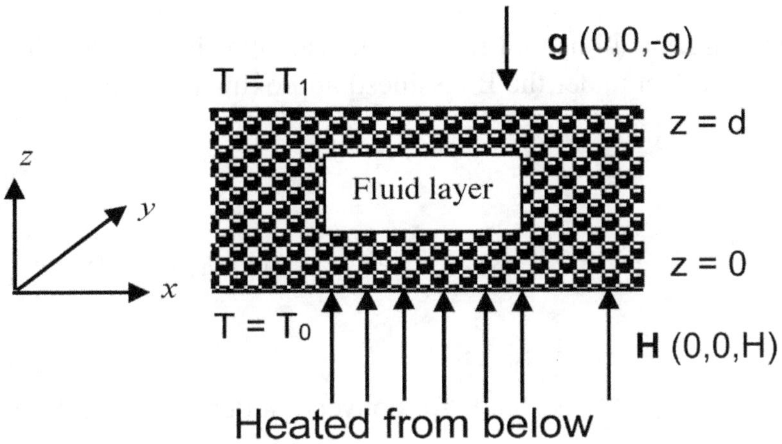

Assumptions

The mathematical equations describing the physical model are based upon the following assumptions:

1. Thermophysical properties of fluid expect for density in the buoyancy force (Boussinesq Hypothesis) are constant,
2. The fluid phase and nanoparticles are in thermal equilibrium state and thus, the heat flow has been described using one equation model,
3. Dilute mixture,
4. Nanoparticles are spherical,
5. Nanoparticles are non-magnetic,
6. No chemical reactions take place in fluid layer,
7. Negligible viscous dissipation,
8. Radiative heat transfer between the sides of wall is negligible when compared with other modes of the heat transfer
9. Nanofluid is incompressible, electrically conducting, Newtonian and laminar flow,
10. Each boundary wall is assumed to be impermeable and perfectly thermal conducting.

GOVERNING EQUATIONS

The governing equations for Maxwellian visco-elastic nanofluid in the presence of Hall effect under the Boussinesq approximation are

$$\nabla \cdot \mathbf{v} = 0\,, \tag{8.1}$$

$$\rho\left(1+\lambda\frac{\partial}{\partial t}\right)\frac{d\mathbf{v}}{dt} = \left(1+\lambda\frac{\partial}{\partial t}\right)\left(-\nabla p + \left(\varphi\rho_p + (1-\varphi)\left\{\acute{A}_f\left(1-\alpha\left(T-T_0\right)\right)\right\}\right)\mathbf{g}\right) + \mu\nabla^2\mathbf{v} + \frac{\mu_e}{4\pi}\left(\mathbf{H}\cdot\nabla\right)\mathbf{H}\,, \tag{8.2}$$

where $\frac{d}{dt} = \frac{\partial}{\partial t} + (\mathbf{v}\cdot\nabla)$ is stands for convection derivative while H, v, p, ρ, ρ_0, μ, μ_e, g and α stands for magnetic field, fluid velocity, hydrostatic pressure, density of nanofluid, density of the nanofluid at reference temperature viscosity, magnetic permeability, acceleration due to gravity and the coefficient of thermal expansion respectively.

Equation of energy for Maxwellian visco-elastic nanofluid is given by

$$\rho c\frac{\partial T}{\partial t} + \mathbf{v}\cdot\nabla T = k_m\nabla^2 T + (\rho c)_p\left(D_B\nabla\varphi\cdot\nabla T + \frac{D_T}{T_1}\nabla T\cdot\nabla T\right), \tag{8.3}$$

where ρc is heat capacity of fluid, $(\rho c)_p$ is heat capacity of nanoparticles, T_1 is the temperature of the fluid layer at z = d and k_m is thermal conductivity.

Equation of continuity for the nanoparticles is given by

$$\frac{\partial\varphi}{\partial t} + \mathbf{v}\cdot\nabla\varphi = D_B\nabla^2\varphi + \frac{D_T}{T_1}\nabla^2 T\,, \tag{8.4}$$

where D_B is the Brownian diffusion coefficient, given by Einstein-Stokes equation and D_T is the thermoporetic diffusion coefficient of the nanoparticles.

Maxwell equations are

$$\frac{d\mathbf{H}}{dt} = (\mathbf{H}\cdot\nabla)\mathbf{v} + \eta\nabla^2 H - \frac{C}{4\pi Ne}\nabla\times\left((\nabla\times H)\times H\right), \tag{8.5}$$

$$\nabla\cdot\mathbf{H} = 0, \tag{8.6}$$

where η, C, N, e, stand for the electrical resitivity, speed of light, electron number density and charge of electron respectively.

We assume that the temperature is constant and nanoparticles flux is zero on the boundaries. Thus boundary conditions are

$$w = 0, T = T_0, \ D_B \frac{\partial \phi}{\partial z} + \frac{D_T}{T_1} \frac{\partial T}{\partial z} = 0 \ \text{at } z = 0$$

and

$$w = 0, T = T_1, \ D_B \frac{\partial \phi}{\partial z} + \frac{D_T}{T_1} \frac{\partial T}{\partial z} = 0 \ \text{at } z = d. \tag{8.7}$$

Introducing non-dimensional variables as

$$(x',y',z') = \left(\frac{x,y,z}{d}\right), \ v'(u',v',w') = v\left(\frac{u,v,w}{\kappa}\right)d, \ t' = \frac{t\kappa}{d^2}, \ p' = \frac{pd^2}{\mu\kappa},$$

$$\varphi' = \frac{(\varphi - \varphi_0)}{\varphi_0}, \ T' = \frac{(T - T_1)}{(T_0 - T_1)}, \ H' = \frac{H}{H_0},$$

where

$\kappa = \dfrac{k_m}{\rho c}$ is thermal diffusivity of the fluid.

Equations (8.1) - (8.7) in non-dimensional form can be written as

$$\nabla' \cdot \mathbf{v}' = 0, \tag{8.8}$$

$$\left(1 + F\frac{\partial}{\partial t'}\right)\frac{1}{\Pr}\frac{\partial \mathbf{v}'}{\partial t'} = \left(1 + F\frac{\partial}{\partial t'}\right)\left(-\nabla' p' \text{-} Rm\hat{e}_z + RaT'\hat{e}_z - Rn\varphi'\hat{e}_z\right) + \nabla'^2 \mathbf{v}' + \frac{\Pr}{\Pr_M} Q\left(\mathbf{H}' \cdot \nabla'\right)\mathbf{H}', \tag{8.9}$$

$$\frac{\partial \varphi'}{\partial t'} + \mathbf{v}' \cdot \nabla' \varphi' = \frac{1}{Le}\nabla'^2 \varphi' + \frac{N_A}{Le}\nabla'^2 T', \tag{8.10}$$

$$\frac{\partial T'}{\partial t'} + \mathbf{v}' \cdot \nabla' T' = \nabla'^2 T' + \frac{N_B}{Le} \nabla' \varphi' \cdot \nabla' T' + \frac{N_A N_B}{Le} \nabla' T' \cdot \nabla' T'. \qquad (8.11)$$

$$\frac{d\mathbf{H}'}{dt'} = (\mathbf{H}' \cdot \nabla')\mathbf{v}' + \frac{Pr}{Pr_M} \nabla'^2 H' - \frac{Pr}{Pr_M} \sqrt{M} \nabla' \times \left[(\nabla' \times H') \times H' \right], \qquad (8.12)$$

$$\nabla' \cdot \mathbf{H}' = 0. \qquad (8.13)$$

The dimensionless boundary conditions are

$$w' = 0, \quad T' = 1, \quad \frac{\partial \phi'}{\partial z'} + N_A \frac{\partial T'}{\partial z'} = 0 \ \text{at} \ \ z' = 0$$

and

$$w' = 0, \ T' = 0, \quad \frac{\partial \phi'}{\partial z'} + N_A \frac{\partial T'}{\partial z'} = 0 \ \text{at} \ z' = 1. \qquad (8.14)$$

Here the non-dimensional parameters are given as follows:

$Pr = \frac{\mu}{\rho \kappa}$ is the Prandtl number,

$Pr_M = \frac{\mu}{\rho \eta}$ is the magnetic Prandtl number,

$Le = \frac{\kappa}{D_B}$ is the Lewis number,

$F = \frac{\kappa \lambda}{d^2}$ is the stress relaxation parameter,

$Ra = \frac{\rho g \alpha d^3 (T_0 - T_1)}{\mu \kappa}$ is the Rayleigh number,

$Rm = \frac{(\rho_p \varphi_0 + \acute{A}(1-\varphi_0)) g d^3}{\mu \kappa}$ is the density Rayleigh number,

$$Rn = \frac{(\rho_p - \rho)\varphi_0 g d^3}{\mu\kappa}$$ is the nanoparticles Rayleigh number,

$$Q = \frac{\mu_e H^2 d^2}{4\pi\rho_0 \upsilon\eta}$$ is the nanofluid magnetic number,

$$M = \left(\frac{CH_0^2}{4\pi Ne\eta}\right)^2$$ is Hall effect parameter,

$$N_A = \frac{D_T (T_0\text{-}T_1)}{D_B T_1 \varphi_0}$$ is the modified diffusivity ratio,

$$N_B = \frac{(\rho c)_p \varphi_0}{(\rho c)_f}$$ is the modified particle-density increment.

Equation (8.9) has been linearized by the neglect of a term proportional to the product of φ_0 and T. This approximation is valid in the case of small temperature gradients in a dilute suspension of nanoparticles.

THE BASIC STATE AND ITS SOLUTIONS

The basic state was assumed to be quiescent and is given by

$$v_i' \left(u',v',w'\right) = 0,$$

$$p' = p_b(z),$$

$$T' = T_b(z),$$

$$H' = H'(z), \tag{8.15}$$

$$\varphi' = \varphi_b(z) \text{ and}$$

$$\rho = \rho_0\left(1 + \alpha\left(T\text{-}T_0\right)\right).$$

Equations (8.8) – (8.13) reduce to

$$0 = -\frac{dp_b}{dz'}\text{-Rm} + RaT_b - Rn\varphi_b, \tag{8.16}$$

$$\frac{d^2T_b}{dz'^2} + \frac{N_B}{Le}\frac{d\varphi_b}{dz'}\frac{dT_b}{dz'} + \frac{N_A N_B}{Le}\left(\frac{dT_b}{dz'}\right)^2 = 0, \tag{8.17}$$

$$\frac{d^2\varphi_b}{dz'^2} + N_A\frac{d^2T_b}{dz'^2} = 0, \tag{8.18}$$

Using boundary conditions in (8.14), equation (8.18) gives

$$\varphi_b = -N_A T_b + (1 - N_A)z' + N_A \tag{8.19}$$

On substituting the value of the φ_b from equation (8.19) in equation (8.17), we get

$$\frac{d^2T_b}{dz'} + \frac{(1\text{-}N_A)N_B}{Le}\frac{dT_b}{dz'} = 0. \tag{8.20}$$

On integrating equation (8.20) with respect to z' and using boundary conditions (8.14), we get

$$T_b = \frac{1 - e^{-(1-N_A)N_B(1-z')/Le}}{1 - e^{-(1-N_A)N_B/Le}}. \tag{8.21}$$

The nanofluid investigated so far Le is large, is of order 10^2-10^3, while N_A is no greater than about 10. Then, the exponents in equation (8.21) are small. By expanding the exponential function into the power series and retaining up to the first order is negligible and hence an approximate solution is given by $T_b = 1 - z'$ and $\varphi_b = \phi_0 + N_A z'$

PERTURBATION SOLUTIONS

Let the initial basic state described by (8.15) is slightly perturbed so that perturbed state is given by

$$(u',v',w') = 0 + (u'',v'',w''),$$

$$T' = T_b + T'', \tag{8.22}$$

$$\varphi' = \varphi_b + \varphi'',$$

$$p' = p_b + p'',$$

$$H' = H_b + h(h_x,h_y,h_z)$$

where $T_b = 1 - z'$, $\varphi_b = \phi_0 + N_A z$ and (u'',v'',w''), T'', φ'' and p'' respectively the perturbations in the initial velocity, temperature, volume fraction of the nanoparticles and pressure.

By substituting (8.22) in equations (8.8) – (8.13) and linearize by neglecting the product of the prime quantities, we obtained following equations

$$\nabla \cdot \mathbf{v} = 0, \tag{8.23}$$

$$\left(1 + F\frac{\partial}{\partial t}\right)\frac{1}{\Pr}\frac{\partial \mathbf{v}}{\partial t} = \left(1 + F\frac{\partial}{\partial t}\right)\left(-\nabla p + \mathrm{Ra}T\hat{e}_z - \mathrm{Rn}\varphi\hat{e}_z\right) + \nabla^2\mathbf{v} + Q\frac{\Pr}{\Pr_M}\frac{\partial h}{\partial z}, \tag{8.24}$$

$$\frac{\partial \varphi}{\partial t} + wN_A = \frac{1}{\mathrm{Le}}\nabla^2\varphi + \frac{N_A}{\mathrm{Le}}\nabla^2 T, \tag{8.25}$$

$$\frac{\partial T}{\partial t} - w = \nabla^2 T + \frac{N_B}{\mathrm{Le}}\left(N_A\frac{\partial T}{\partial z} - \frac{\partial \phi}{\partial z}\right) - \frac{2N_A N_B}{\mathrm{Le}}\frac{\partial T}{\partial z}, \tag{8.26}$$

$$\frac{\partial h}{\partial t} = \frac{\partial w}{\partial z} + \frac{\Pr}{\Pr_M}\nabla^2 h - \frac{\Pr}{\Pr_M}\sqrt{M}\nabla \times \frac{\partial h}{\partial z}, \tag{8.27}$$

$$\nabla \cdot h = 0. \tag{8.28}$$

Boundary conditions are

$$w = 0, T = 0, \quad \frac{\partial \phi}{\partial z} + N_A \frac{\partial T}{\partial z} = 0 \text{ at } z = 0, 1. \tag{8.29}$$

[Dashes (") have been suppressed for convenience]

Eliminating pressure term p from equation (8.24), (8.27) by making use of equations (8.23) and (8.28), we have

$$\begin{aligned} &\left(\left(\frac{\Pr}{\Pr_M}\nabla^2 - \frac{\partial}{\partial t}\right)\left(\nabla^2 - \left(1 + F\frac{\partial}{\partial t}\right)\frac{1}{\Pr}\frac{\partial}{\partial t}\right) - Q\frac{\Pr}{\Pr_M}D^2\right)\nabla^2 w \\ &+ \left(\frac{\Pr}{\Pr_M}\nabla^2 - \frac{\partial}{\partial t}\right)\left(1 + F\frac{\partial}{\partial t}\right)\left(Ra\nabla_H^2 T - \mathrm{Rn}\nabla_H^2 \varphi\right) - \left(\frac{\Pr}{\Pr_M}\right)^2 Q\sqrt{M}\nabla^2 D^2 \xi = 0, \end{aligned} \tag{8.30}$$

$$\left(\left(1 + F\frac{\partial}{\partial t}\right)\frac{1}{\Pr}\frac{\partial}{\partial t} - \nabla^2\right)\varsigma - \frac{\Pr}{\Pr_M}Q\frac{\partial}{\partial z}\xi = 0, \tag{8.31}$$

$$\left(\frac{\Pr}{\Pr_M}\nabla^2 - \frac{\partial}{\partial z}\right)\xi + \frac{\partial \varsigma}{\partial z} - \frac{\Pr}{\Pr_M}\sqrt{M}\nabla^2\frac{\partial hz}{\partial z} = 0, \tag{8.32}$$

$$\left(\frac{\Pr}{\Pr_M}\nabla^2 - \frac{\partial}{\partial t}\right)h_z - \frac{\Pr}{\Pr_M}\sqrt{M}\frac{\partial \xi}{\partial z} + \frac{\partial w}{\partial z} = 0, \tag{8.33}$$

where $\varsigma = \frac{\partial v}{\partial x} - \frac{\partial u}{\partial y}$ and $\xi = \frac{\partial h_y}{\partial x} - \frac{\partial h_x}{\partial y}$ stand for the z-components of vorticity and current density respectively.

Eliminating ς, ξ and h_z from equations (8.30) - (8.33), we have

$$\left[\begin{array}{l} \left(\left(1 + F\frac{\partial}{\partial t}\right)\left(\frac{1}{\Pr}\frac{\partial}{\partial t} - \nabla^2\right)\left(\frac{\Pr}{\Pr_M}\nabla^2 - \frac{\partial}{\partial t}\right)^2 + \frac{\Pr}{\Pr_M}Q\left(\frac{\Pr}{\Pr_M}\nabla^2 - \frac{\partial}{\partial t}\right)D^2 \right. \\ \left. + \left(\frac{\Pr}{\Pr_M}\right)^2 M\left(\left(1 + F\frac{\partial}{\partial t}\right)\frac{1}{\Pr}\frac{\partial}{\partial t} - \nabla^2\right)\nabla^2 D^2 \right) \\ \left(\left(\frac{\Pr}{\Pr_M}\nabla^2 - \frac{\partial}{\partial t}\right)\left(\nabla^2 - \left(1 + F\frac{\partial}{\partial t}\right)\frac{1}{\Pr}\frac{\partial}{\partial t}\right) - \frac{\Pr}{\Pr_M}QD^2\right) + \left(\frac{\Pr}{\Pr_M}\right)^3 MQ\nabla^2 D^4 \end{array}\right]\nabla^2 w$$

$$+\left(\begin{array}{l}\left(1+F\frac{\partial}{\partial t}\right)\left(\frac{1}{\Pr}\frac{\partial}{\partial t}-\nabla^2\right)\left(\frac{\Pr}{\Pr_M}\nabla^2-\frac{\partial}{\partial t}\right)^2+\frac{\Pr}{\Pr_M}Q\left(\frac{\Pr}{\Pr_M}\nabla^2-\frac{\partial}{\partial t}\right)D^2\\+\left(\frac{\Pr}{\Pr_M}\right)^2M\left(\left(1+F\frac{\partial}{\partial t}\right)\frac{1}{\Pr}\frac{\partial}{\partial t}-\nabla^2\right)\nabla^2D^2\end{array}\right)$$

$$\times\left(\frac{\Pr}{\Pr_M}\nabla^2-\frac{\partial}{\partial t}\right)\left(1+F\frac{\partial}{\partial t}\right)\left(Ra\nabla_H^2\mathrm{T}-\mathrm{Rn}\nabla_H^2\varphi\right)=0. \tag{8.34}$$

where $\nabla_{\mathrm{H}}^2=\frac{\partial^2}{\partial x^2}+\frac{\partial^2}{\partial y^2}$ is the two-dimensional Laplacian operator on the horizontal plane.

NORMAL MODE ANALYSIS

Analyzing the disturbances into the normal modes and assuming that the perturbed quantities are of the form

$$\left[\mathrm{w},\mathrm{T},\varphi\right]=\left[\mathrm{W(z)},\Theta(\mathrm{z}),\Phi(\mathrm{z})\right]\exp\left(\mathrm{i}k_x\mathrm{x}+\mathrm{i}k_y\mathrm{y}+\mathrm{nt}\right), \tag{8.35}$$

where k_x, k_y are wave numbers in x and y direction and n is growth rate of disturbances.

Using equation (8.35), equations (8.34), (8.25) and (8.26) become

$$\left(\begin{array}{l}\left(\begin{array}{l}(1+nF)\left(\frac{n}{\Pr}-(D^2-a^2)\right)\left(\frac{\Pr}{\Pr_M}(D^2-a^2)-n\right)^2+\frac{\Pr}{\Pr_M}Q\left(\frac{\Pr}{\Pr_M}(D^2-a^2)-n\right)D^2\\+\left(\frac{\Pr}{\Pr_M}\right)^2M\left((1+nF)\frac{n}{\Pr}-(D^2-a^2)\right)(D^2-a^2)D^2\end{array}\right)\\\left(\left(\frac{\Pr}{\Pr_M}(D^2-a^2)-n\right)\left((D^2-a^2)-(1+nF)\frac{n}{\Pr}\right)-\frac{\Pr}{\Pr_M}QD^2\right)\\+\left(\frac{\Pr}{\Pr_M}\right)^3MQ(D^2-a^2)D^4\end{array}\right)(D^2-a^2)w$$

$$+\left(\begin{array}{l}\left(1+nF\right)\left(\frac{n}{\Pr}-\left(D^2-a^2\right)\right)\left(\frac{\Pr}{\Pr_M}\left(D^2-a^2\right)-n\right)^2 \\ +\frac{\Pr}{\Pr_M}Q\left(\frac{\Pr}{\Pr_M}\left(D^2-a^2\right)-n\right)D^2 \\ +\left(\frac{\Pr}{\Pr_M}\right)^2 M\left(\left(1+nF\right)\frac{n}{\Pr}-\left(D^2-a^2\right)\right)\left(D^2-a^2\right)D^2\end{array}\right)$$

$$\times\left(\frac{\Pr}{\Pr_M}\left(D^2-a^2\right)-n\right)\left(1+nF\right)\left(-a^2Ra\nabla_H^2\mathrm{T}+a^2\mathrm{Rn}\nabla_H^2\varphi\right)=0. \quad (8.36)$$

$$\mathrm{N_A W}-\frac{N_A}{Le}\left(\mathrm{D}^2-\mathrm{a}^2\right)\Theta-\left(\frac{1}{Le}\left(\mathrm{D}^2-\mathrm{a}^2\right)-n\right)\Phi=0, \quad (8.37)$$

$$\mathrm{W}+\left(\mathrm{D}^2-\mathrm{a}^2-\mathrm{n}-\frac{\mathrm{N_A}N_B}{\mathrm{Le}}\mathrm{D}\right)\Theta\text{-}\frac{\mathrm{N_B}}{\mathrm{Le}}\mathrm{D}\Phi=0, \quad (8.38)$$

where $\mathrm{D}\equiv\frac{\mathrm{d}}{\mathrm{dz}}$ and $\mathrm{a}=\sqrt{k_x^2+k_y^2}$ is the dimensionless resultant wave number.

The boundary conditions of the problem in view of normal mode analysis are

$$\mathrm{W}=0,\ \mathrm{D^2W}=0,\tilde{}\ =0,\ \mathrm{D}\Phi+\mathrm{N_A D}\Theta=0 \ \text{ at } \ \mathrm{z}=0,1. \quad (8.39)$$

METHOD OF SOLUTION

The Galerkin weighted residuals method is used to obtain an approximate solution to the system of equations (8.36) – (8.38) with the corresponding boundary conditions (8.39). In this method, the test functions are the same as the base (trial) functions. Accordingly W, Θ and Φ are taken as

$$\mathrm{W}=\sum_{p=1}^{N}A_p\mathrm{W_p},\ \Theta=\sum_{p=1}^{N}B_p\Theta_p,\Phi=\sum_{p=1}^{N}C_p\Phi_p, \quad (8.40)$$

where $\mathrm{W_p}=\Theta_\mathrm{p}=\sin p\pi z, \Phi_\mathrm{p}=-N_A\sin p\pi z$, $\mathrm{A_p}$, $\mathrm{B_p}$ and $\mathrm{C_p}$ are unknown coefficients, p = 1, 2, 3,..., N and the base functions $\mathrm{W_p}$, Θ_p, and Φ_p satisfying the boundary conditions (8.39). Using expression for W, Θ and Φ in equations (8.36) – (8.38) and multiplying the first equation by W_p the second equation by Θ_p and third equation by Φ_p and then integrating in the limits from zero to unity, we obtain a set of 3N linear homogeneous equations with 3N unknown A_p, B_p and C_p; p = 1, 2, 3,..., N. For existing of nontrivial solution, the vanishing of the determinant of coefficients produces the characteristics equation of the system in term of Rayleigh number Ra.

Linear Stability Analysis

For the present formulation, we have considered the which system of equations (8.36) – (8.38) together with the boundary conditions (8.39) constitute a linear eigenvalue problem with variable coefficient for the growth rate of disturbance of the system. Substituting equation (8.40) into the system of equations (8.36) - (8.38) and multiplying the first equation by W_p the second equation by Θ_p and third equation by Φ_p and then integrating in the limits from zero to unity and performing some integration by parts, we obtain expression for stationary Rayleigh number Ra as

$$\mathrm{R}a = \frac{1}{a^2}\left(\left(\pi^2+a^2\right)^3 + \left(\pi^2+a^2\right)\pi^2 Q - \frac{MQ\pi^4\left(\pi^2+a^2\right)^2}{Q\pi^2 + \frac{\Pr}{\Pr_M} M\pi^2\left(\pi^2+a^2\right) + \left(\pi^2+a^2\right)^2} \right) - \left(1+Le\right)\mathrm{N_A Rn}. \tag{8.41}$$

Equation (8.41) expresses the thermal Rayleigh number Ra as a function of dimensionless wave number a, Hall effect parameter M, magnetic field Q, Lewis number Le, modified diffusivity N_A, nanoparticles Rayleigh number Rn. It is also noted that parameter N_B and stress relaxation parameter F does not appear in the equation, thus Maxwellian visco-elastic nanofluid behaves like an ordinary Newtonian nanofluid and instability is purely a phenomenon due to buoyancy coupled with the conservation of nanoparticles. It is independent of the contributions of Brownian motion and thermophoresis to the thermal energy equation. The parameter N_B drops out because of an orthogonal property of the first order trail functions and their first derivatives.

In the absence of Hall effect and magnetic field the Rayleigh number Ra for steady onset is

$$\mathrm{R}a = \frac{\left(\pi^2 + a^2\right)^3}{\mathrm{a}^2} - \left(1 + Le\right)\mathrm{N_A Rn}\,. \tag{8.42}$$

This is the good agreement of the result (1.47) obtained in Chapter 1.

To find the critical value of Ra equation (8.42) is differentiated with respect to 'a^2' and then equated to zero. The minimum of first term of right- hand side of equation (8.42) is attained at $\mathrm{a_c} = \frac{\pi}{\sqrt{2}}$ and minimum value found to $\frac{27\pi^4}{4}$ so the corresponding critical Rayleigh number given by

$$\left(\mathrm{Ra}\right)_c = \frac{27\pi^4}{4} - \left(1 + Le\right)\mathrm{N_A Rn}\,. \tag{8.43}$$

The interweaving behaviors' of Brownian motion and thermoporesis of nanoparticles evidently does not change the critical size of the Bénard cell at the onset of instability. As such, the critical size is not a function of any thermo physical properties of nanofluid.

In the absence of Hall effect, magnetic field magnetic field and nanoparticles i.e. for ordinary fluid, the critical Rayleigh number given by

$$\left(\mathrm{R}a\right)_c = \frac{27\pi^2}{4}\,.$$

This is well known result derived by Chandrasekhar (1961).

Thus presence of the nanoparticles lowers the value of the critical Rayleigh number by usually by substantial amount. Also parameter N_B does not appear in the equation (8.41); thus instability is purely phenomena due to buoyancy coupled with conservation of nanoparticles. Thus average contribution of nanoparticles flux in the thermal energy equation is zero with one-term Galerkin approximation.

RESULTS AND DISCUSSION

To study the effect of Hall effect parameter, Chandrasekhar number, Lewis number, modified diffusivity ratio and nanoparticles Rayleigh number on stationary convection, we examine the behavior of $\frac{\partial \mathrm{Ra}}{\partial \mathrm{M}}, \frac{\partial \mathrm{Ra}}{\partial \mathrm{Q}}, \frac{\partial \mathrm{Ra}}{\partial \mathrm{L}e}, \frac{\partial \mathrm{Ra}}{\partial \mathrm{N}_A}$ and $\frac{\partial Ra}{\partial Rn}$ analytically.

From equation (8.41), we have

$$\frac{\partial \mathrm{Ra}}{\partial \mathrm{M}} = -\frac{\left(\pi^2+a^2\right)Q^2\pi^6+\left(\pi^2+a^2\right)^3 Q\pi^4}{\left(\mathrm{Q}\pi^2+\frac{\mathrm{Pr}}{\mathrm{Pr}_M}M\pi^2\left(\pi^2+a^2\right)+\left(\pi^2+a^2\right)^2\right)^2}<0.$$

Thus Hall effect has a destabilizing effect on the layer of nanofluid fluid. Equation (8.41) also yields

$$\frac{\partial \mathrm{Ra}}{\partial \mathrm{Q}} = \frac{\left(\mathrm{Q}\pi^2+\frac{\mathrm{Pr}}{\mathrm{Pr}_M}M\pi^2\left(\pi^2+a^2\right)+\left(\pi^2+a^2\right)^2\right)^2-\left(\frac{\mathrm{Pr}}{\mathrm{Pr}_M}M\pi^2\left(\pi^2+a^2\right)^2+\left(\pi^2+a^2\right)^2\right)M\left(\pi^2+a^2\right)}{\left(\mathrm{Q}\pi^2+\frac{\mathrm{Pr}}{\mathrm{Pr}_M}M\pi^2\left(\pi^2+a^2\right)+\left(\pi^2+a^2\right)^2\right)^2}.$$

Since M<<1, thus $\frac{\partial \mathrm{Ra}}{\partial \mathrm{Q}}>0$.

Also we have $\frac{\partial \mathrm{Ra}}{\partial \mathrm{Le}}<0$, $\frac{\partial \mathrm{Ra}}{\partial \mathrm{N}_A}<0$ and $\frac{\partial \mathrm{Ra}}{\partial \mathrm{Rn}}<0$,

thus from these inequalities it is observed that magnetic field has stabilizing effect while Lewis number, modified diffusivity ratio and concentration Rayleigh number have destabilizing effects on the stationary convection.

The computations are carried out for different values of parameters considered in the range $10^2 \le Ra \le 10^5$ (thermal Rayleigh number), $10 \le M \le 10^2$ (Hall effect parameter, $10 \le Q \le 10^4$ (Chandrasekhar number), $1 \le N_A \le 10$ (modified diffusivity ratio), $10^2 \le Le \le 10^4$ (Lewis number), $10^{-1} \le Rn \le 10^1$ (nanoparticles Rayleigh number).

The variations of the stationary thermal Rayleigh number with wave number have been plotted graphically as shown in Figures 2 - 6.

The convection curves in (Ra, a) plane for various values of Hall effect parameter M and fixed values of other parameters is shown in Figure 2. It has been found that the Rayleigh number decreases with an increase in the value of Hall effect parameter M, thus Hall effect has destabilizing effect on fluid layer.

Figure 2. Variation of the Rayleigh number with wave number for different values of Hall effect parameter M

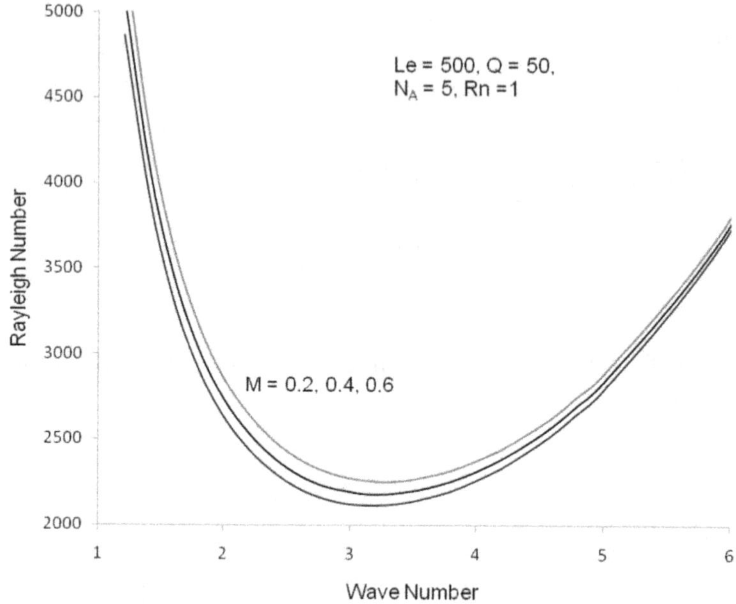

Figure 3 shows the variation of thermal Rayleigh number with wave number for different values of Nanofluid magnetic number Q. It is found that the thermal Rayliegh number increases as values of magnetic number Q increases. Thus, magnetic field has stabilizing effect on fluid layer.

Figure 4 shows the variation of thermal Rayleigh number with wave number for different values of Lewis number Le. It is found that the thermal Rayliegh number decreases as the value of Lewis number Le increases. Thus, Lewis number Le destabilizes the fluid layer. It is due to the fact that thermophoresis at a higher value of thermophoretic diffusivity is more supportable to the disturbance in nanofluids, while both thermophoresis and Brownian motion are driving forces in favor of the motion of nanoparticles. This is good agrement of the result obtained by Chand and Rana (2014d, 2015a).

Figure 5 shows the variation of thermal Rayleigh number with wave number for different values of modified diffusivity ratio N_A. It is found that the thermal Rayliegh number decreases as value of modified diffusivity ratio N_A increases. Hence, modified diffusivity ratio N_A has stabilizing effect on fluid layer. This may lead to an increase in volumetric fraction, which shows that Brownian motion of the nanoparticles will also increase, which may cause destabilizing effect. This is good agrement of the result obtained by Chand and Rana (2014d, 2015a).

Figure 3. Variation of the Rayleigh number with wave number for different values of Nanofluid magnetic number Q

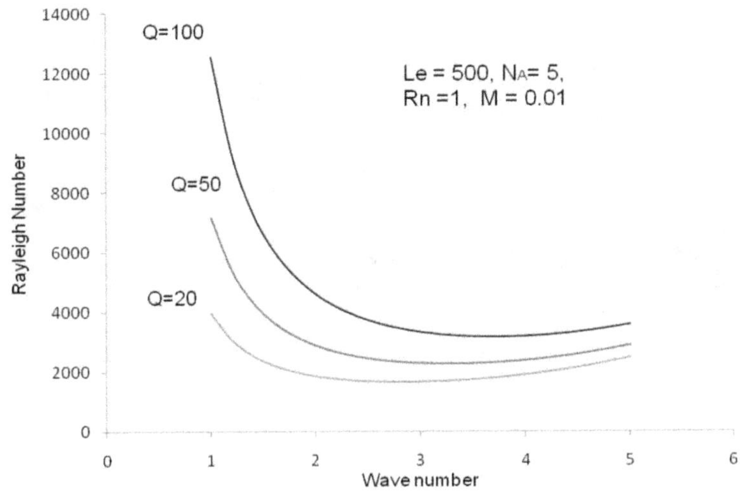

Figure 4. Variation of the Rayleigh number with wave number for different values of Lewis number

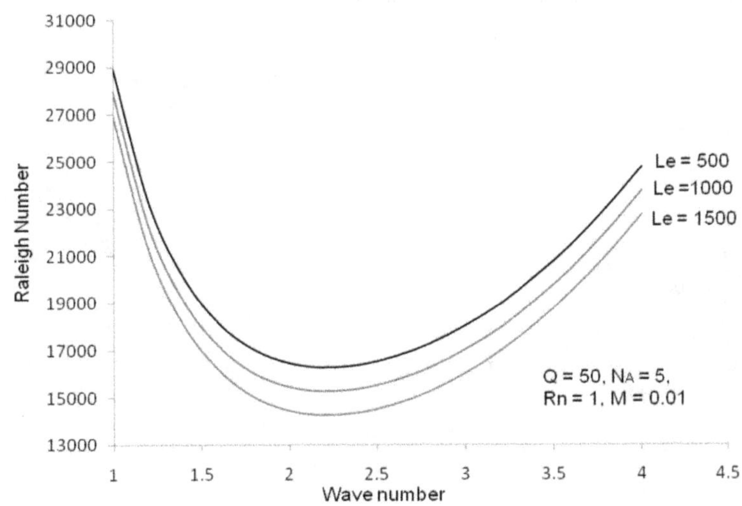

Figure 6 shows the variation of Rayleigh numbers with wave number for different values of the nanoparticles Rayleigh number Rn and for the fixed values of other parameters. It is found that thermal Rayleigh number decreases with an increase in value of the nanoparticles Rayleigh number Rn, which means that the nanoparticles Rayleigh number Rn has destabilizing effect

Figure 5. Variation of the stationary Rayleigh number with wave number for different value of modified diffusivity ratio

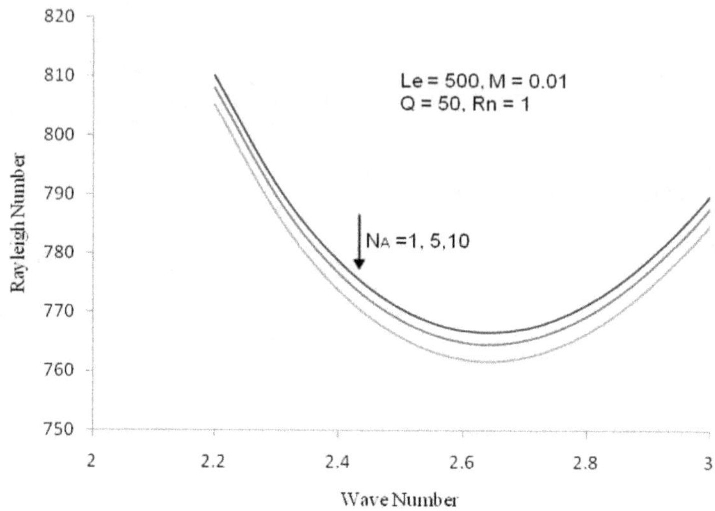

Figure 6. Variation of the stationary Rayleigh number with wave number for different value of nanoparticles Rayleigh number

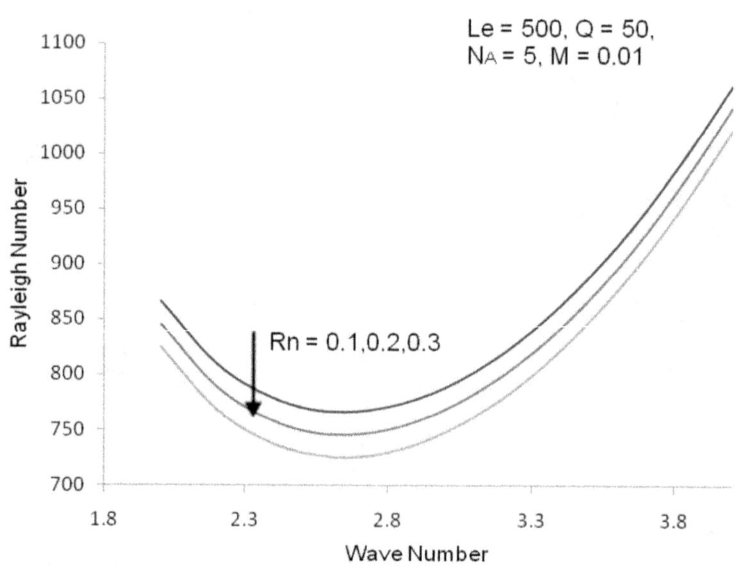

on fluid layer. It has destabilizing effect because the heavier nanoparticles moving through the base fluid makes more strong disturbances as compared with the lighter nanoparticles. This is good agrement of the result obtained by Chand and Rana (2014d, 2015a).

CONCLUSION

A linear stability analysis for a horizontal layer of Maxwellian visco-elastic in the presence of Hall effect is investigated. The flux of volume fraction of nanoparticles is taken to be zero on the isothermal boundaries and Galerkin residuals method is used for the stability analysis. Results has been depicted both analytically graphically. The main conclusions of the present chapter are summarized as follows:

1. The critical cell size is not a function of any thermo physical properties of nanofluid.
2. Instability is independent of the contributions of Brownian motion and is purely phenomenon due to buoyancy coupled with the conservation of nanoparticles.
3. For stationary convection Maxwellian visco-elastic nanofluid behaves like an ordinary Newtonian nanofluid.
4. For the case of stationary convection the Hall effect, the Lewis number, the modified diffusivity ratio and the concentration Rayleigh number have has destabilizing effect while the magnetic field has stabilizing effect on fluid layer.

REFERENCES

Chand, R., & Rana, G. C. (2014d). Thermal instability in a Brinkman porous medium saturated by nanofluid with no nanoparticle flux on boundaries. *Special Topics & Reviews in Porous Media: An International Journal*, *5*(4), 277–286. doi:10.1615/SpecialTopicsRevPorousMedia.v5.i4.10

Chand, R., & Rana, G. C. (2015a). Magneto convection in a layer of nanofluid in porous medium- a more realistic approach. *Journal of Nanofluids*, *4*(2), 196–202. doi:10.1166/jon.2015.1142

Chandrasekhar, S. (1961). *Hydrodynamic and Hydromagnetic Stability*. New York: Dover Publication.

Gupta, A. S. (1967). Hall effects on thermal instability. *Rev. Roumaine Math. Pures Appl.*, *12*, 665–677.

Kent, A. (1966). Instability of laminar flow of a magnetofluid. *Physics of Fluids*, *9*(7), 1286–1289. doi:10.1063/1.1761842

Oberoi, C., & Devanathan, C. (1963). *Proc. Summer Seminar in Magneto-hydrodynamics*. IIT Bangalore.

Sharma, R. C., Sunil, & Chand, S. (2000). Hall effect on thermal instability of Rivlin-Ericksen elastic viscous fluid. *Indian J. Pure Appl. Math.*, *31*(4), 49-59.

Sharma, R. C., & Kumar, P. (1996b). Hall effect on thermosolutal instability in Maxwellian visco-elastic fluid in porous medium. *Arch. Mech.*, *48*(1), 199–209.

Sherman, A., & Sutton, G. W. (1962). *Magnetohydrodynamics*. North-Western University Press.

Chapter 9
Effect of Internal Heat Source on the Onset of Thermal Convection in a Layer of Maxwellian Visco-Elastic Nanofluid

INTRODUCTION

The study of the flow and internal heat generation in nanofluids is of special interest and has many practical applications in manufacturing processes industry. Effect of heat generation/absorption in thermal convection is significant where there exists high temperature difference between the surface and the ambient fluid. Possible heat generation also alters the temperature distribution; consequently the particle deposition rate in nuclear reactors, electronic chips and semiconductor wafers. The effect of internal heat source on the onset of convection has been carried out for various types of fluids by many researchers. Thermal instability induced by internal heat sources has been widely studied because of its wide range of applications in astrophysics and geophysics. The buoyancy force is incremented due to heat source resulting in modification of heat/mass transfer characteristic in such type of fluid flow problems. The internal heat source effects on the onset of thermal

DOI: 10.4018/978-1-68318-006-7.ch009

convection in a horizontal layer of ordinary Newtonian fluid were studied by researchers Rionero and Straughan (1990), Char and Chiang (1994), Shivakumara and Suma (2000), Alex et al. (2001), Nanjundappa et al. (2011). The study was extended to the nanofluid by Nield and Kuznetsov (2013), Yadav et al. (2012c, 2015), Kiran et al. (2016) and they observed that presence of constant internal heating makes both basic temperature distribution and basic volumetric fraction of nanoparticles distribution to deviate from linear to nonlinear. Due to wide range of applications in industry an attempt has been made to investigate the effect of internal heat source on the thermal instability in a horizontal layer of Maxwellian visco-elastic nanofluids for more realistic boundary conditions.

MATHEMATICAL FORMULATIONS OF THE PROBLEM

The Physical Problem

In this chapter we shall investigate the effect of rotation on the onset of thermal convection in a horizontal layer Maxwellian visco-elastic nanofluid. The physical configuration of the problem to be considered as:

An infinite horizontal layer of Maxwellian visco-elastic nanofluid of thickness 'd' bounded by horizontal boundaries $z = 0$ and $z = d$ subjected to a uniformly internal heat sources Q_0 and heated from below, here, Q_0 is the overall uniformly distributed effective volumetric internal heat source. A Cartesian coordinate system (x, y, z) is chosen with the origin at the bottom of the fluid layer and the z- axis normal to the fluid layer. Fluid layer is acted upon by gravity force $\mathbf{g}(0, 0,-g)$. Fluid layer is heated from below in such a way that horizontal boundaries $z = 0$ and $z = d$ respectively maintained at a uniform temperature T_0 and T_1 ($T_0 > T_1$). The normal component of the nanoparticles flux has to vanish at an impermeable boundary and the temperature T is taken to be T_0 at $z = 0$ and T_1 at $z = d$, ($T_0 > T_1$) as shown in Figure 1. The reference scale for temperature and nanoparticles fraction is taken to be T_1 and φ_0 respectively.

Assumptions

The mathematical equations describing the physical model are based upon the following assumptions:

Figure 1. Physical configuration of the problem

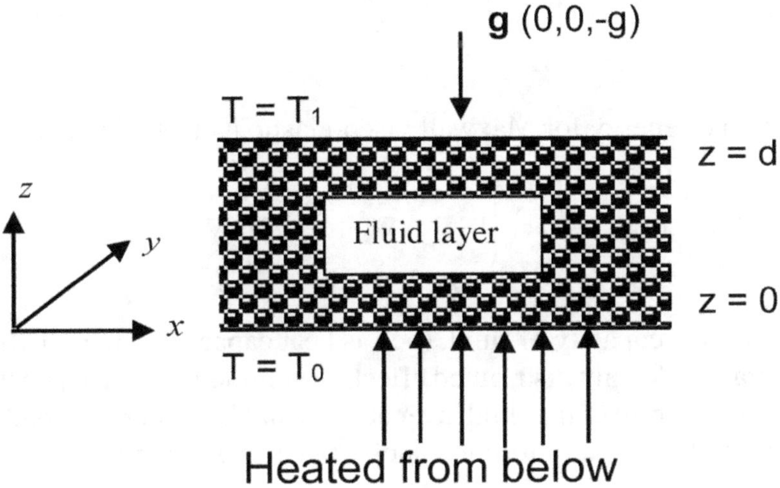

1. Thermophysical properties of fluid expect for density in the buoyancy force (Boussinesq Hypothesis) are constant,
2. The fluid phase and nanoparticles are in thermal equilibrium state and thus, the heat flow has been described using one equation model,
3. Dilute mixture,
4. Nanoparticles are spherical,
5. Nanoparticles are non-magnetic,
6. No chemical reactions take place in fluid layer,
7. Negligible viscous dissipation,
8. Radiative heat transfer between the sides of wall is negligible when compared with other modes of the heat transfer,
9. Nanofluid is incompressible, electrically conducting, Newtonian and laminar flow,
10. Each boundary wall is assumed to be impermeable and perfectly thermal conducting.

GOVERNING EQUATIONS

The governing equations for Maxwellian visco-elastic nanofluid in the internal heat under the Boussinesq approximation are given as:

$$\nabla \cdot \mathbf{v} = 0, \tag{9.1}$$

$$\rho\left(1+\lambda\frac{\partial}{\partial t}\right)\frac{d\mathbf{v}}{dt}=\left(1+\lambda\frac{\partial}{\partial t}\right)\left(-\nabla p+\left(\varphi\rho_p+(1-\varphi)\left\{\rho_f\left(1-\alpha\left(T-T_0\right)\right)\right\}\right)\mathbf{g}\right)+\mu\nabla^2\mathbf{v}\,. \tag{9.2}$$

Equation of energy for Maxwell visco-elastic nanofluid is given by

$$\rho c\frac{\partial T}{\partial t}+\mathbf{v}\cdot\nabla T=k_m\nabla^2 T+(\rho c)_p\left[D_B\nabla\varphi\cdot\nabla T+\frac{D_T}{T_1}\nabla T\cdot\nabla T\right]+Q_0\,, \tag{9.3}$$

where ρc is heat capacity of fluid, $(\rho c)_p$ is heat capacity of nanoparticles, Q_0 is the overall uniformly distributed effective volumetric internal heat source. T_1 is the temperature of the fluid layer at $z=d$ and k_m is thermal conductivity.

Equation of continuity for the nanoparticles is given by

$$\frac{\partial\varphi}{\partial t}+\mathbf{v}\cdot\nabla\varphi=D_B\nabla^2\varphi+\frac{D_T}{T_1}\nabla^2 T\,, \tag{9.4}$$

where DB is the Brownian diffusion coefficient, given by Einstein-Stokes equation and D_T is the thermoporetic diffusion coefficient of the nanoparticles.

We assume that the temperature is constant and nanoparticles flux is zero on the boundaries. Thus boundary conditions are

$$w=0,\ T=T_0,\ D_B\frac{\partial\phi}{\partial z}+\frac{D_T}{T_1}\frac{\partial T}{\partial z}=0\ \text{ at } z=0$$

and

$$w=0,\ T=T_1,\ D_B\frac{\partial\phi}{\partial z}+\frac{D_T}{T_1}\frac{\partial T}{\partial z}=0\ \text{ at } z=d. \tag{9.5}$$

Introducing non-dimensional variables as

$$(x',y',z')=\left(\frac{x,y,z}{d}\right),\ v'(u',v',w')=v\left(\frac{u,v,w}{\kappa}\right)d,\ t'=\frac{t\kappa}{d^2},\ p'=\frac{pd^2}{\mu\kappa},\ \varphi'=\frac{(\varphi-\varphi_0)}{\varphi_0},$$

$$T'=\frac{(T-T_1)}{(T_0-T_1)},$$

where

$\kappa = \frac{k_m}{\rho c}$ is thermal diffusivity of the fluid.

Equations (9.1) - (9.5) in non-dimensional form can be written as

$$\nabla' \cdot \mathbf{v}' = 0, \tag{9.6}$$

$$\left(1 + F\frac{\partial}{\partial t'}\right)\frac{1}{\mathrm{Pr}}\frac{\partial \mathbf{v}'}{\partial t'} = \left(1 + F\frac{\partial}{\partial t'}\right)\left(-\nabla' p' \text{-} \mathrm{Rm}\hat{e}_z + \mathrm{Ra}T'\hat{e}_z - \mathrm{Rn}\varphi'\hat{e}_z\right) + \nabla'^2 \mathbf{v}', \tag{9.7}$$

$$\frac{\partial \varphi'}{\partial t'} + \mathbf{v}' \cdot \nabla'\varphi' = \frac{1}{\mathrm{Le}}\nabla'^2\varphi' + \frac{N_A}{\mathrm{Le}}\nabla'^2 T' + \mathrm{Hs}, \tag{9.8}$$

$$\frac{\partial T'}{\partial t'} + \mathbf{v}' \cdot \nabla' T' = \nabla'^2 T' + \frac{N_B}{\mathrm{Le}}\nabla'\varphi' \cdot \nabla' T' + \frac{N_A N_B}{\mathrm{Le}}\nabla' T' \cdot \nabla' T'. \tag{9.9}$$

Here the non-dimensional parameters are given as follows:

$\mathrm{Pr} = \frac{\mu}{\rho\kappa}$ is the Prandtl number,

$\mathrm{Le} = \frac{\kappa}{D_B}$ is the Lewis number,

$F = \frac{\kappa\lambda}{d^2}$ is the stress relaxation parameter,

$\mathrm{Ra} = \frac{\rho g \alpha d^3 \left(T_0 - T_1\right)}{\mu\kappa}$ is the Rayleigh number,

$\mathrm{Rm} = \frac{\left(\rho_p \varphi_0 + \rho\left(1\text{-}\varphi_0\right)\right) g d^3}{\mu\kappa}$ is the density Rayleigh number,

$\mathrm{Rn} = \frac{\left(\acute{A}_p - \acute{A}\right)\varphi_0 g d^3}{\mu\kappa}$ is the nanoparticles Rayleigh number,

$\mathrm{Hs} = \dfrac{Q_0 d^2}{k_m (T_0 - T_1)}$ is the dimensionless constant of heat source strength,

$N_A = \dfrac{D_T (T_0 - T_1)}{D_B T_1 \varphi_0}$ is the modified diffusivity ratio,

$N_B = \dfrac{(\rho c)_p \varphi_0}{(\rho c)_f}$ is the modified particle-density increment.

Equation (9.7) has been linearized by the neglecting term proportional to the product of φ_0 and T. This approximation is valid in the case of small temperature gradients in a dilute suspension of nanoparticles.

The dimensionless boundary conditions are

$$w' = 0, \quad T' = 1, \quad \frac{\partial \phi'}{\partial z'} + N_A \frac{\partial T'}{\partial z'} = 0 \text{ at } z' = 0$$

and

$$w' = 0, \; T' = 0, \quad \frac{\partial \phi'}{\partial z'} + N_A \frac{\partial T'}{\partial z'} = 0 \text{ at } z' = 1. \tag{9.10}$$

THE BASIC STATE AND ITS SOLUTIONS

The basic state was assumed to be quiescent and is given by

$$v_i' (u', v', w') = 0,$$

$$p' = p_b(z),$$

$$T' = T_b(z), \tag{9.11}$$

$$\varphi' = \varphi_b(z) \text{ and}$$

$$\rho = \rho_0 \left(1 + \alpha (T - T_0)\right).$$

Equations (9.6) – (9.9) reduce to

$$0 = -\frac{dp_b}{dz'} \text{-Rm} + RaT_b - Rn\varphi_b, \tag{9.12}$$

$$\frac{d^2T_b}{dz'^2} + \frac{N_B}{Le}\frac{d\varphi_b}{dz'}\frac{dT_b}{dz'} + \frac{N_A N_B}{Le}\left(\frac{dT_b}{dz'}\right)^2 + Hs = 0, \tag{9.13}$$

$$\frac{d^2\varphi_b}{dz'^2} + N_A\frac{d^2T_b}{dz'^2} = 0, \tag{9.14}$$

Using boundary conditions in (9.10), equation (9.14) gives

$$\varphi_b = -N_A T_b + \phi_0 + N_A. \tag{9.15}$$

On substituting the value of the φ_b from equation (9.15) in equation (9.13), we get

$$\frac{d^2T_b}{dz'} + \frac{(1\text{-}N_A)N_B}{Le}\frac{dT_b}{dz'} + Hs = 0 \tag{9.16}$$

On integrating equation (9.16) with respect to z' and using boundary conditions (9.10), we get

$$T_b = \frac{e^{\frac{-(1-N_A)N_B}{Le}z}\left[-N_B\left(-1+N_A\right) - LeHs\right] - LeHs\left(-1+z'\right) + e^{\frac{-(1-N_A)N_B}{Le}\left[(-1+N_A)N_B + LeHsz'\right]}}{-1+e^{\frac{-(1-N_A)N_B}{Le}(-1+N_A)N_B}}. \tag{9.17}$$

$$\phi_b = z' + \frac{N_A\left[\left(-1+N_A\right)N_B - LeHs\right]\left(-1+e^{\frac{-(1-N_A)N_B}{Le}z} + z' - ze^{\frac{-(1-N_A)N_B}{Le}}\right)}{-1+e^{\frac{-(1-N_A)N_B}{Le}(-1+N_A)N_B}}. \tag{9.18}$$

According to Buongiorno (2006), for most nanofluid investigated so far Le is large, is of order 10^2-10^3, while N_A is no greater than about 10, N_B is of order 10^{-4} to 10^{-2}. Then, the exponents $\frac{-(1-N_A)N_B}{Le}$ in equations (9.17) - (9.18) are very small. By expanding the exponential function into the power series and retaining up to the first order is negligible and so to a good approximation for the solution of basic state is given by

$$T_b = \frac{1}{2}\left(2 - 2z' + Hsz'\text{-}Hsz'\right)$$

and

$$\varphi_b = \phi_0 + N_A\left(z - \frac{Hsz'}{2} + \frac{Hsz'}{2}\right).$$

In the absence of internal heat generation i.e. Hs = 0, then basic flow distributions for temperature and nanoparticles volume fraction are:

$$T_b = 1 - z' \text{ and } \varphi_b = \phi_0 + N_A z' \tag{9.19}$$

To see the effect of internal heat source strength Hs on the criterion for the onset of thermal convection in nanofluids, the distributions of dimensionless basic temperature and basic nanoparticles volumetric fraction are drawn in the Figure 2 for different values of Hs. The discrete values of Hs are purposely taken to see the behavior of both distributions. This plot shows the behavior of basic temperature distribution which is parabolic in positive direction and same behavior in negative direction for the basic nanoparticles distribution as internal heat source strength Hs increases. That is increase in the internal heat source strength Hs amounts to increase in energy supply to the system. This gives large deviations in these distributions which in turn improve the disturbances in the layer and thus system is more unstable.

PERTURBATION SOLUTIONS

Let the initial basic state described by (9.11) is slightly perturbed so that perturbed state is given by

$$\left(u', v', w'\right) = 0 + \left(u'', v'', w''\right),$$

Figure 2. Basic state temperature and basic state nanoparticles volumetric distributions for different values of internal heat source strength parameter Hs

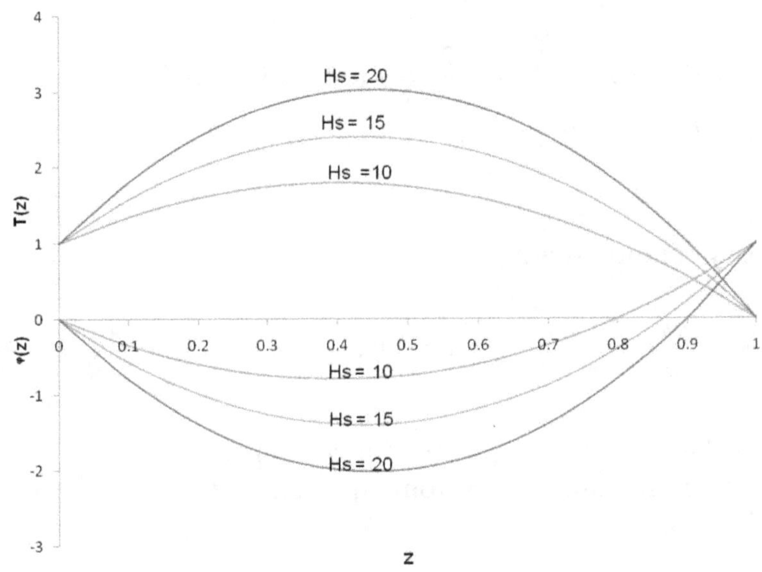

$$T' = T_b + T'', \tag{9.20}$$

$$\varphi' = \varphi_b + \varphi'',$$

$$p' = p_b + p'',$$

where $T_b = \frac{1}{2}\left(2 - 2z' + Hsz'\text{-}Hsz'\right)$, $\varphi_b = \phi_0 + N_A\left(z' - \frac{Hsz'}{2} + \frac{Hsz'}{2}\right)$ and $\left(u'',v'',w''\right)$, T'', φ'' and p'' respectively the perturbations in the initial velocity, temperature, volume fraction of the nanoparticles and pressure.

By substituting (9.20) in equations (9.6) – (9.9) and linearize by neglecting the product of the prime quantities, we obtained following equations

$$\nabla \cdot \mathbf{v} = 0, \tag{9.21}$$

$$\left(1 + F\frac{\partial}{\partial t}\right)\frac{1}{\Pr}\frac{\partial \mathbf{v}}{\partial t} = \left(1 + F\frac{\partial}{\partial t}\right)\left(-\nabla p + RaT\hat{e}_z - Rn\varphi\hat{e}_z\right) + \nabla^2\mathbf{v}, \tag{9.22}$$

$$\frac{\partial Æ}{\partial t}+wN_A=\frac{1}{Le}\nabla^2 Æ+\frac{N_A}{Le}\nabla^2 T, \tag{9.23}$$

$$\begin{aligned}&\frac{\partial T}{\partial t}-w\left(-2+Hs-2Hsz\right)=\nabla^2 T+\frac{2N_AN_B}{Le}\left(-2+Hs-2Hsz\right)\frac{\partial T}{\partial z}\\&+\frac{N_BN_A}{Le}\left(1-\frac{Hs}{2}+2Hsz\right)\frac{\partial T}{\partial z}+\frac{2N_B}{Le}\left(-2+Hs-2Hsz\right)\frac{\partial \phi}{\partial z}.\end{aligned} \tag{9.24}$$

Boundary conditions are

$$w=0,\ T=0,\quad \frac{\partial \phi}{\partial z}+N_A\frac{\partial T}{\partial z}=0 \ \text{ at } z=0,1. \tag{9.25}$$

[Dashes (") have been suppressed for convenience]

Eliminating pressure term p from equation (9.23) by operating the curl twice on it, we have

$$\left(1+F\frac{\partial}{\partial t}\right)\frac{1}{Pr}\frac{\partial}{\partial t}\nabla^2 w-\nabla^4 w-\left(1+F\frac{\partial}{\partial t}\right)\left(Ra\nabla_H^2 T\text{-}Rn\nabla_H^2\varphi\right)=0, \tag{9.26}$$

where $\nabla_H^2=\frac{\partial^2}{\partial x^2}+\frac{\partial^2}{\partial y^2}$ is the two-dimensional Laplacian operator on the horizontal plane.

NORMAL MODE ANALYSIS

Analyzing the disturbances into the normal modes and assuming that the perturbed quantities are of the form

$$\left[w,T,\varphi\right]=\left[W(z),\Theta(z),\Phi(z)\right]\exp\left(ik_x x+ik_y y+nt\right), \tag{9.27}$$

where k_x, k_y are wave numbers in x and y direction and n is growth rate of disturbances.

Using equation (9.27), equations (9.26), (9.23) and (9.24) become

$$\left[\left(D^2 - a^2\right)\left(D^2 - a^2 - \frac{n\left(1 + nF\right)}{Pr}\right)\right]W - \left(1 + nF\right)\left(a^2 Ra\Theta - a^2 Rn\Phi\right) = 0, \tag{9.28}$$

$$N_A W - \frac{N_A}{Le}\left(D^2 - a^2\right)\tilde{} - \left(\frac{1}{Le}\left(D^2 - a^2\right) - n\right)\Phi = 0, \tag{9.29}$$

$$-\frac{1}{2}\left(-2 + Hs - 2Hsz\right)W + \begin{pmatrix} D^2 - a^2 - n + \frac{N_A N_B}{Le}\left(-2 + Hs - 2Hsz\right)D \\ -\frac{N_A N_B}{Le}\left(1 - \frac{Hs}{2} + Hsz\right)D \end{pmatrix}\tilde{} \tag{9.30}$$

$$-\frac{N_B}{2Le}\left(-2 + Hs - 2Hsz\right)D\,| = 0,$$

where $D \equiv \frac{d}{dz}$ and $a = \sqrt{k_x^2 + k_y^2}$ is the dimensionless resultant wave number.

The boundary conditions of the problem in view of normal mode analysis are

$$W = 0,\ D^2W = 0,\tilde{}\ = 0,\ D\Phi + N_A D\Theta = 0 \ \text{ at } \ z = 0,1. \tag{9.31}$$

METHOD OF SOLUTION

The Galerkin weighted residuals method is used to obtain an approximate solution to the system of equations (9.28) – (9.30) with the corresponding boundary conditions (9.31). In this method, the test functions are the same as the base (trial) functions. Accordingly W, Θ and Φ are taken as

$$W = \sum_{p=1}^{N} A_p W_p,\ \Theta = \sum_{p=1}^{N} B_p \Theta_p,\ \Phi = \sum_{p=1}^{N} C_p \Phi_p, \tag{9.32}$$

where $W_p = \Theta_p = \sin p\pi z, \Phi_p = -N_A \sin p\pi z$, A_p, B_p and C_p are unknown coefficients, p = 1, 2, 3,..., N and the base functions W_p, Θ_p, and Φ_p satisfying the boundary conditions (9.31). Using expression for W, Θ and Φ in equations (9.28) – (9.30) and multiplying the first equation by W_p the second equation by Θ_p and third equation by Φ_p and then integrating in the limits from zero to unity, we obtain a set of 3N linear homogeneous equations with

3N unknown A_p, B_p and C_p; p = 1, 2, 3,..., N. For existing of nontrivial solution, the vanishing of the determinant of coefficients produces the characteristics equation of the system in term of Rayleigh number Ra.

LINEAR STABILITY ANALYSIS

For the present formulation, we have considered the which system of equations (9.28) – (9.30) together with the boundary conditions (9.31) constitute a linear eigenvalue problem with variable coefficient for the growth rate of disturbance n of the system. Substituting equation (9.32) into the system of equations (9.28) –(9.30) and multiplying the first equation by W_p the second equation by Θ_p and third equation by Φ_p and then integrating in the limits from zero to unity and performing some integration by parts, one obtains the following matrix equation

$$\begin{bmatrix} \frac{(\pi^2+a^2)^2}{(1+nF)} + \frac{n}{\Pr}(\pi^2+a^2) & -a^2 Ra & -a^2 N_A Rn \\ 1-\frac{Hs}{2}+Hsz & -(\pi^2+a^2+n) & 0 \\ N_A & \frac{N_A}{Le}(\pi^2+a^2) & -N_A\left(\frac{1}{Le}(\pi^2+a^2)+n\right) \end{bmatrix} \begin{bmatrix} W_0 \\ \Theta_0 \\ \Phi_0 \end{bmatrix} = \begin{bmatrix} 0 \\ 0 \\ 0 \end{bmatrix}. \tag{9.33}$$

The non-trivial solution of the above matrix requires that

$$Ra = \frac{1}{a^2\left(1-\frac{Hs}{2}+Hsz\right)}\left(\frac{(\pi^2+a^2)^2}{(1+nF)}+\frac{n(\pi^2+a^2)}{\Pr}\right)(\pi^2+a^2+n) - \frac{\left(1-\frac{Hs}{2}+Hsz\right)(\pi^2+a^2+n)+Le(\pi^2+a^2+n)}{(\pi^2+a^2)+nLe} N_A Rn. \tag{9.34}$$

The growth rate n is in general a complex quantity such that $n = \omega_r + i\omega$, the system with $\omega_r < 0$ is always stable, while for $\omega_r > 0$ it will become unstable. For neutral stability, the real part of n is zero. Hence, we now write $n = i\omega$, (where ω is real and is a dimensionless frequency) in equation (9.34), we have

$$\mathrm{Ra} = \Delta_1 + i\omega\Delta_2, \tag{9.35}$$

where

$$\Delta_1 = \frac{1}{a^2}\frac{\left(\pi^2+a^2\right)^3}{1+\omega^2F^2} - \omega^2\left(\frac{\left(\pi^2+a^2\right)}{\mathrm{Pr}} - \frac{F}{1+\omega^2F^2}\right) - \frac{\left(\pi^2+a^2\right)^2\left(1-\dfrac{Hs}{2}+Hsz\right)+\left(\pi^2+a^2\right)^2}{\left(\pi^2+a^2\right)^2+\omega^2Le^2}\mathrm{N_A Rn} \tag{9.36}$$

and

$$\Delta_2 = \frac{1}{\mathrm{a}^2}\frac{\left(\pi^2+a^2\right)^2}{\mathrm{Pr}} - \frac{F\left(\pi^2+a^2\right)}{1+\omega^2F^2} + \frac{\left(\pi^2+a^2\right)^2}{1+\omega^2F^2} - \frac{\left(\pi^2+a^2\right)-\left(1-\dfrac{Hs}{2}+Hsz\right)\left(\pi^2+a^2\right)+\left(\pi^2+a^2\right)Le}{\left(\pi^2+a^2\right)^2+\omega^2Le^2}\mathrm{N_A Rn}. \tag{9.37}$$

Since Ra is a physical quantity, so it must be real. Hence, it follows from the equation (9.35) that either $\omega = 0$ (exchange of stability, steady state) or $\Delta_2 = 0$ ($\omega \neq 0$ overstability or oscillatory onset).

STATIONARY CONVECTION

For the case of stationary (non- oscillatory) convection $[\mathrm{n} = \omega = 0]$, equation (9.34) reduces to

$$\left(\mathrm{Ra}\right)_s = \frac{\left(\pi^2+a^2\right)^3}{\mathrm{a}^2\left(1-\dfrac{Hs}{2}+Hsz\right)} - \left(1+\frac{Le}{\left(1-\dfrac{Hs}{2}+Hsz\right)}\right)\mathrm{N_A Rn}. \tag{9.38}$$

It is observed that stationary Rayleigh number Ra is function of the Lewis number Le, the modified diffusivity ratio $\mathrm{N_A}$, the nanoparticles Rayleigh Rn and heat source strength parameter Hs but independent of visco- elastic parameter F, Prandtl number Pr and modified particle- density increment N_B. Thus Maxwellian visco-elastic nanofluid behaves like an ordinary Newtonian nanofluid and instability is purely a phenomenon due to buoyancy coupled with the conservation of nanoparticles.

In the absence of heat source strength parameter (Hs =0) equation (9.38) reduces to

$$(\mathrm{Ra})_s = \frac{\left(\pi^2 + a^2\right)^3}{\mathrm{a}^2} - \left(1 + Le\right)\mathrm{N_A Rn}. \tag{9.39}$$

This is the good agreement of the result (1.47) obtained in Chapter 1.

To find the critical value of $(\mathrm{Ra})_s$, equation (9.38) is differentiated with respect to 'a^2' and then equated to zero. The minimum of first term of right-hand side of equation (9.38) is attained at $\mathrm{a_c} = \frac{\pi}{\sqrt{2}}$ and minimum value found to $\frac{27\pi^4}{4}$ so the corresponding critical Rayleigh number given by

$$(\mathrm{Ra})_c = \frac{27\pi^4}{4} - \left(1 + Le\right)\mathrm{N_A Rn}. \tag{9.40}$$

In the absence of heat source strength parameter and nanoparticles (Hs = Rn = Le = N_A = 0), one recovers the well- known results that the critical Rayleigh number is equal to $(\mathrm{Ra})_c = \frac{27\pi^4}{4}$.

This is good agreement of the result obtained by Chandrasekhar (1961).

Thus presence of the nanoparticles lowers the value of the critical Rayleigh number by usually by substantial amount. Also parameter N_B does not appear in the equation (9.40); thus instability is purely phenomena due to buoyancy coupled with conservation of nanoparticles.

RESULTS AND DISCUSSION

To study the effect of internal heat source strength parameter Hs, Lewis number Le, modified diffusivity ratio N_A and nanoparticles Rayleigh number Rn on stationary convection, we examine the behavior of

$\frac{\partial(\mathrm{R}a)_s}{\partial Hs}, \frac{\partial(\mathrm{R}a)_s}{\partial Le}, \frac{\partial(\mathrm{R}a)_s}{\partial N_A}$ and $\frac{\partial(\mathrm{R}a)_s}{\partial Rn}$ analytically.

From equation (9.38), we have

1. $\frac{(\partial \mathrm{R}a)_s}{\partial Hs} < 0,$
2. $\frac{(\partial \mathrm{R}a)_s}{\partial Le} < 0,$
3. $\frac{(\partial \mathrm{R}a)_s}{\partial N_A} < 0,$
4. $\frac{(\partial \mathrm{R}a)_s}{\partial Rn} < 0.$

These inequalities shows that heat source strength parameter Hs, Lewis number Le, modified diffusivity ratio N_A and nanoparticles Rayleigh number Rn have destabilizing effect on the stationary convection.

Expression for stationary Rayleigh number, which characterizes the stability of the system, is found to be depend upon heat source strength parameter, Lewis number, modified diffusivity ratio and nanoparticles Rayleigh number. The computations are carried out for different values of parameters considered in the range $10^2 \leq Ra \leq 10^5$ (thermal Rayleigh number), $10 \leq Hs \leq 10^2$ (heat source strength parameter), $1 \leq N_A \leq 10$ (modified diffusivity ratio), $10^2 \leq Le \leq 10^4$ (Lewis number), $10^{-1} \leq Rn \leq 10^1$ (nanoparticles Rayleigh number).

The variations of the stationary thermal Rayleigh number with wave number have been plotted graphically as shown in Figs. 3-6.

Figure 3 shows the variation of thermal Rayleigh number for different value of heat source strength parameter Hs and for the fixed value of other parameters. It is found that stationary Rayliegh number decreases as the value of heat source strength parameter Hs increases, indicating that heat source strength parameter Hs destabilizes the stationary convection. The temperature and volumetric fraction of nanoparticle distributions are having a destabilizing factor to make the system more unstable.

Figure 4 shows the variation of thermal Rayleigh number for different value of Lewis number Le and for the fixed value of other parameters. It is found that stationary Rayliegh number decreases as the value of Lewis number increases, indicating that Lewis number destabilizes the stationary convection. It is due to the fact that thermophoresis at a higher value of ther-

Figure 3. Variation of the stationary Rayleigh number with wave number for different value of heat source strength parameter

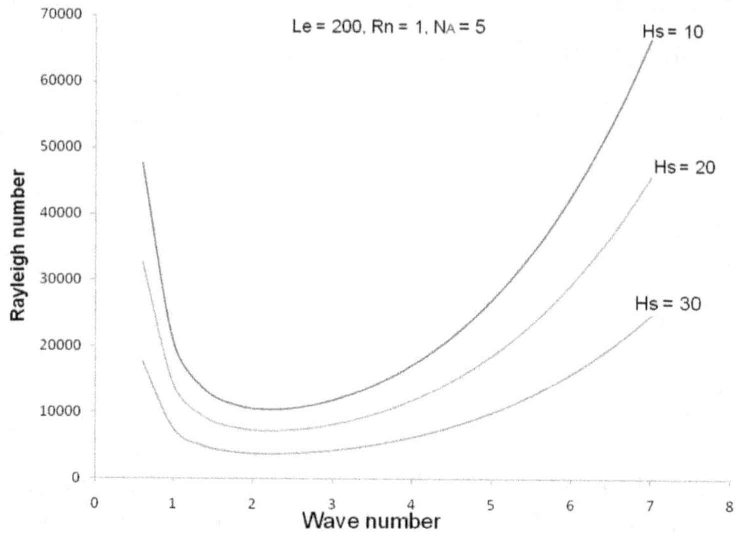

Figure 4. Variation of the stationary Rayleigh number with wave number for different value of Lewis number

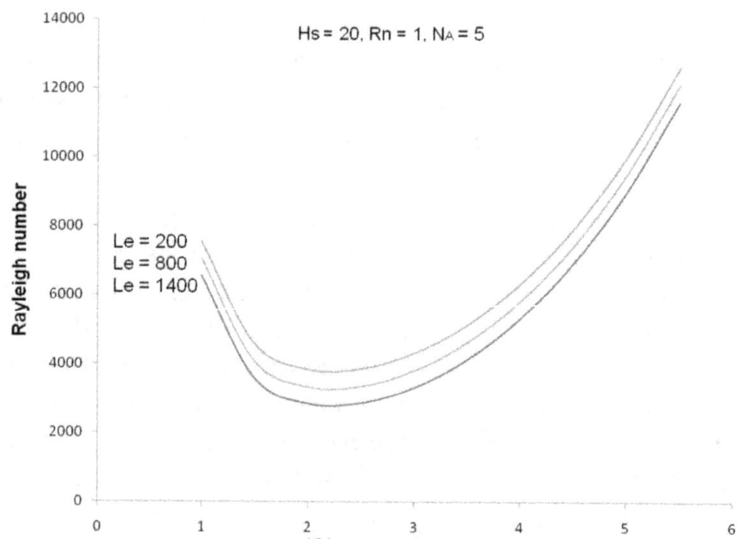

mophoretic diffusivity is more supportable to the disturbance in nanofluids, while both thermophoresis and Brownian motion are driving forces in favor of the motion of nanoparticles. This is good agrement of the result obtained by Chand and Rana (2014d, 2015a).

Figure 5 shows the variation of stationary Rayleigh number for different value of the modified diffusivity ratio N_A and fixed value of other parameters and it is found that Rayleigh number decreases with an increase in the value of the modified diffusivity ratio $N_{A;}$ which means that the modified diffusivity ratio N_A destabilizes on the fluid layer. This may lead to an increase in volumetric fraction, which shows that Brownian motion of the nanoparticles will also increase, which may cause destabilizing effect. This is good agrement of the result obtained by Chand and Rana (2014d, 2015a).

Figure 6 shows the variation of stationary Rayleigh number with wave number for different value of the nanoparticles Rayleigh number Rn and fixed value of other parameters and it is found that thermal Rayleigh number decreases with an increase in value of the nanoparticles Rayleigh number Rn, which means that the nanoparticles Rayleigh number Rn has destabilizing effect on fluid layer. It has destabilizing effect because the heavier nanoparticles moving through the base fluid makes more strong disturbances as compared with the lighter nanoparticles. This is good agrement of the result obtained by Chand and Rana (2014d, 2015a).

Figure 5. Variation of the stationary Rayleigh number with wave number for different value of modified diffusivity ratio

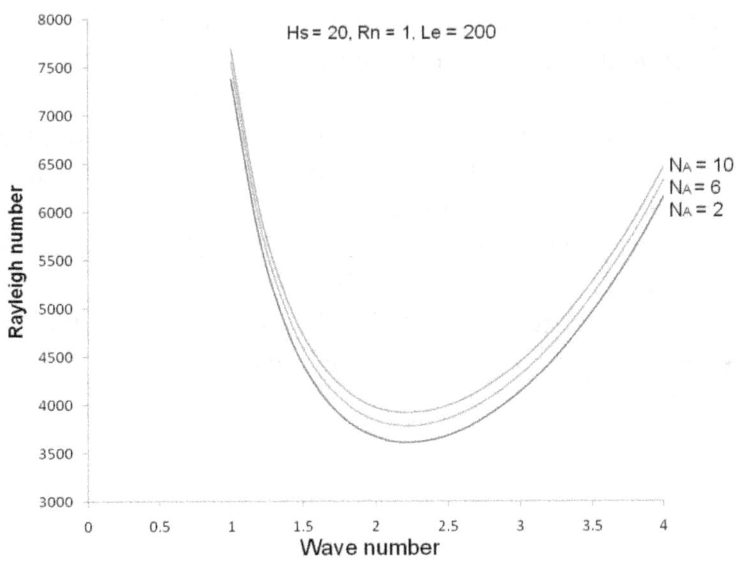

Figure 6. Variation of the stationary Rayleigh number with wave number for different value of nanoparticles Rayleigh number

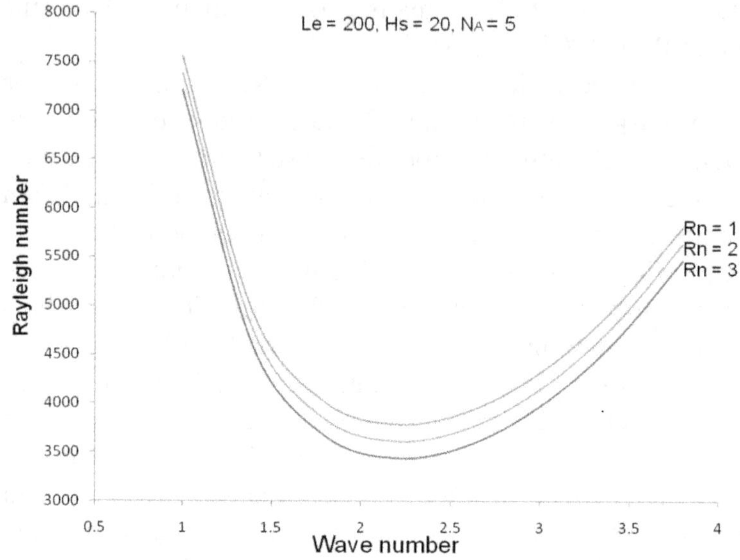

CONCLUSION

Effect of Internal heat source on the onset of thermal convection in a horizontal layer of Maxwellian visco-elastic nanofluid is studied. The flux of volume fraction of nanoparticles is taken to be zero on the isothermal boundaries and the eigen value problem is solved using the Galerkin residual method. The results have been presented both analytically and graphically.

The main conclusions derived from the present chapter are as follows:

1. The instability purely phenomenon due to buoyancy coupled with the conservation of nanoparticle and is independent of the contribution of Brownian motion and thermophoresis.
2. For stationary convection Maxwellian visco-elastic nanofluid behaves like an ordinary Newtonian nanofluid.
3. Heat source strength parameter Hs, Lewis number Le, modified diffusivity ratio N_A and nanoparticles Rayleigh number Rn destabilizes the stationary convection.

REFERENCES

Alex, S. M., Prabhamani, R. P., & Vankatakrishan, K. S. (2001). Variable gravity effects on thermal instability in a porous medium with internal heat source and inclined temperature gradient. *Fluid Dynamics Research*, *29*(1), 1–6. doi:10.1016/S0169-5983(01)00016-8

Chand, R., & Rana, G. C. (2014d). Thermal instability in a Brinkman porous medium saturated by nanofluid with no nanoparticle flux on boundaries. *Special Topics & Reviews in Porous Media: An International Journal*, *5*(4), 277–286. doi:10.1615/SpecialTopicsRevPorousMedia.v5.i4.10

Chand, R., & Rana, G. C. (2015a). Magneto convection in a layer of nanofluid in porous medium- a more realistic approach. *Journal of Nanofluids*, *4*(2), 196–202. doi:10.1166/jon.2015.1142

Char, M., & Chiang, K. (1994). Stability analysis of Bénard Marangoni convection in fluids with internal heat source. *Journal of Physics. D, Applied Physics*, *27*(4), 748–755. doi:10.1088/0022-3727/27/4/012

Kiran, P., Bhadauria, B. S., & Kumar, V. (2016). Thermal convection in a nanofluid saturated porous medium with internal heating and gravity modulation. *Journal of Nanofluids*, *5*(3), 1–12. doi:10.1166/jon.2016.1220

Nanjundappa, C. E., Shivakumara, I. S., Lee, J., & Ravisha, M. (2011). Effect of internal heat generation on the onset of Brinkman-Bénard convection in a ferrofluid saturated porous layer. *International Journal of Thermal Sciences*, *50*(2), 160–168. doi:10.1016/j.ijthermalsci.2010.10.003

Nield, D. A., & Kuznetsov, A. V. (2013). Onset of convection with internal heating in a porous medium saturated by a nanofluid. *Transport in Porous Media*, *99*(1), 73–83. doi:10.1007/s11242-013-0174-6

Rionero, S., & Straughan, B. (1990). Convection in a porous medium with internal heat source and variable gravity effects. *International Journal of Engineering Science*, *28*(6), 497–503. doi:10.1016/0020-7225(90)90052-K

Shivakumara, I. S., & Suma, S. P. (2000). Effect of through flow and internal heat generation on the onset of convection in a fluid layer. *Acta Mechanica*, *140*(3-4), 207–217. doi:10.1007/BF01182511

Yadav, D., Bhargava, R., & Agrawal, G. S. (2012b). Boundary and internal heat source effects on the onset of Darcy- Brinkman convection in a porous layer saturated by nanofluid. *International Journal of Thermal Sciences*, *60*, 244–254. doi:10.1016/j.ijthermalsci.2012.05.011

Yadav, D., Lee, J., & Cho, H. H. (2015). Brinkman convection induced by purely internal heating in a rotating porous medium layer saturated by a nanofluid. *Powder Technology*, *286*, 592–601. doi:10.1016/j.powtec.2015.08.048

Chapter 10 Double Diffusive Convection in a Layer of Maxwellian Visco-Elastic Nanofluid

INTRODUCTION

Double-diffusive convection or thermosolutal instability in a nanofluid occurs when the base fluid of the nanofluid is itself a binary fluid. Binary nanofluids such as when titanium dioxide nanoparticles (1% by mass) are dispersed in the mixture of water and eutectic of chloride salts (KCl-$CaCl_2$-$LiCl$). Double-diffusive convection is an important phenomenon that has various applications in the fields of chemical science, food processing, engineering and nuclear industries, geophysics, bioengineering and cancer therapy, movement of biological fluid, oceanography and also used as solar thermal applications. The onset of double-diffusive nanofluid convection in a layer of a saturated porous medium (the Horton-Rogers-Lapwood problem) was studied by Kuznetsov and Nield (2010a, 2010b, 2010c, 2011) and found that the stability boundaries for both non-oscillatory and oscillatory cases. The Cheng-Minkowycz problem for the double-diffusive natural convective

DOI: 10.4018/978-1-68318-006-7.ch010

boundary layer flow in a porous medium saturated by a nanofluid was studied by Nield and Kuznetsov (2009b). Yadav et al. (2012b) was examined the effects of boundary and internal heat source on the onset of Darcy-Brinkman convection in a porous layer saturated by nanofluid and they have obtained the critical Rayleigh number as well as critical wave number by using Galerkin-type weighted residuals method.

Hence due to the importance of binary nanofluid, the main objective in this chapter is to examine theoretically the double-diffusive convection of Maxwellian visco-elastic nanofluid for more realistic boundary conditions.

MATHEMATICAL FORMULATIONS OF THE PROBLEM

The Physical Problem

In this chapter we shall investigate the double diffusive convection in a horizontal layer Maxwellian visco-elastic nanofluid. The physical configuration of the problem to be considered as:

An infinite horizontal layer of Maxwellian visco-elastic nanofluid of thickness 'd' bounded by horizontal boundaries $z = 0$ and $z = d$. A Cartesian coordinate system (x, y, z) is chosen with the origin at the bottom of the fluid layer and the z- axis normal to the fluid layer. Fluid layer is acted upon by gravity force $\mathbf{g}(0, 0, -g)$ and heated from below in such a way that horizontal boundaries $z = 0$ and $z = d$ respectively maintained at a uniform temperature T_0 and T_1 ($T_0 > T_1$). The normal component of the nanoparticles flux has to vanish at an impermeable boundary and the temperature T is taken to be T_0 at $z = 0$ and T_1 at $z = d$, ($T_0 > T_1$) as shown in Figure 1. The reference scale for temperature and nanoparticles fraction is taken to be T_1 and φ_0 respectively.

Assumptions

The mathematical equations describing the physical model are based upon the following assumptions:

1. Thermophysical properties of fluid expect for density in the buoyancy force (Boussinesq Hypothesis) are constant,
2. The fluid phase and nanoparticles are in thermal equilibrium state and thus, the heat flow has been described using one equation model,
3. Dilute mixture,
4. Nanoparticles are spherical,

Figure 1. Physical configuration of the problem

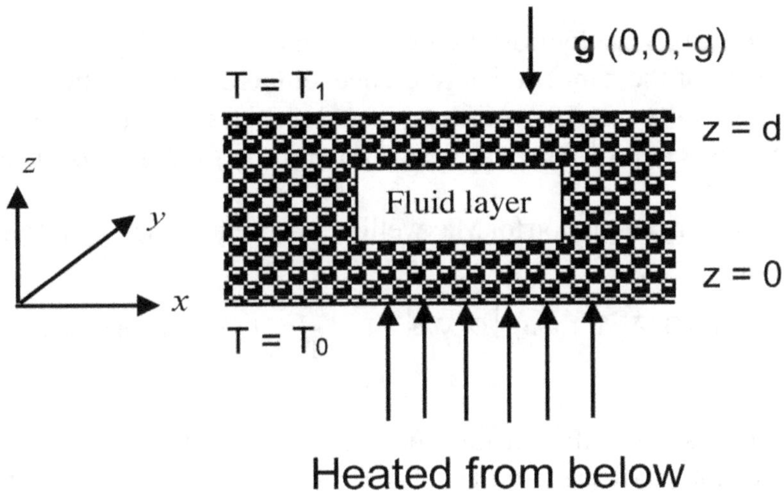

5. Nanoparticles are non-magnetic,
6. No chemical reactions take place in fluid layer,
7. Negligible viscous dissipation,
8. Radiative heat transfer between the sides of wall is negligible when compared with other modes of the heat transfer,
9. Nanofluid is incompressible, electrically conducting, Newtonian and laminar flow,
10. Each boundary wall is assumed to be impermeable and perfectly thermal conducting,
11. The nanoparticles do not affect the transport of the solute.

GOVERNING EQUATIONS

The governing equations for Maxwellian visco-elastic nanofluid in the presence of solute particle under the Boussinesq approximation are given as

$$\nabla \cdot \mathbf{v} = 0, \tag{10.1}$$

$$\rho\left(1+\lambda\frac{\partial}{\partial t}\right)\frac{d\mathbf{v}}{dt} = \left(1+\lambda\frac{\partial}{\partial t}\right)\left(-\nabla p + \left(\varphi\rho_p + (1-\varphi)\left\{\rho_f\left(1-\alpha_T\left(T-T_0\right)-\alpha_C\left(C-C_0\right)\right)\right\}\right)\mathbf{g}\right) + \mu\nabla^2\mathbf{v}, \tag{10.2}$$

where $\frac{d}{dt} = \frac{\partial}{\partial t} + (\mathbf{v} \cdot \nabla)$ is stands for convection derivative while $\mathbf{v}$ is the fluid velocity, p is the hydrostatic pressure, ρ is the density of nanofluid, ρ_0 is the density of the nanofluid at reference temperature, μ is the viscosity of the fluid, λ is the relaxation time, **g** is the acceleration due to gravity, and α_T is the coefficient of thermal expansion and α_C is the analogous to solute concentration.

Thermal energy equation for Maxwellian visco-elastic nanofluid is given by

$$\rho c \frac{\partial T}{\partial t} + \mathbf{v} \cdot \nabla T = k_m \nabla^2 T + (\rho c)_p \left(D_B \nabla \varphi \cdot \nabla T + \frac{D_T}{T_1} \nabla T \cdot \nabla T \right) + \rho c D_{TC} \nabla^2 C , \quad (10.3)$$

where ρc is heat capacity of fluid, $(\rho c)_p$ is heat capacity of nanoparticles, T_1 is the temperature of the fluid layer at z = d, D_{TC} is a diffusivity of Dufour type and k_m is thermal conductivity.

The conservation equation for solute concentration Kuznetsov and Nield (2010c) is of the form

$$\frac{\partial C}{\partial t} + \mathbf{v} \cdot \nabla T = \kappa' \nabla^2 C + D_{CT} \nabla^2 T ,$$

where κ' is the solutal diffusivity and D_{CT} is the diffusivity of Soret type.

Equation of continuity for the nanoparticles is given by

$$\frac{\partial \varphi}{\partial t} + \mathbf{v} \cdot \nabla \varphi = D_B \nabla^2 \varphi + \frac{D_T}{T_1} \nabla^2 T , \quad (10.4)$$

where D_B is the Brownian diffusion coefficient, given by Einstein-Stokes equation and D_T is the thermoporetic diffusion coefficient of the nanoparticles.

We assume that the temperature is constant and nanoparticles flux is zero on the boundaries. Thus boundary conditions are

$$w = 0, T = T_0, \; D_B \frac{\partial \phi}{\partial z} + \frac{D_T}{T_1} \frac{\partial T}{\partial z} = 0 \text{ at } z = 0$$

and

$$w = 0, T = T_1, \; D_B \frac{\partial \phi}{\partial z} + \frac{D_T}{T_1}\frac{\partial T}{\partial z} = 0 \; \text{ at } \; z = d. \tag{10.5}$$

Introducing non-dimensional variables as

$$(x',y',z') = \left(\frac{x,y,z}{d}\right), \; v'(u',v',w') = V\left(\frac{u,v,w}{\kappa}\right)d, \; t' = \frac{t^o}{d^2}, \; p' = \frac{pd^2}{\mu\kappa}, \; \varphi' = \frac{(\varphi - \varphi_0)}{\varphi_0},$$
$$T' = \frac{(T - T_1)}{(T_0 - T_1)}, \; C' = \frac{(C - C_1)}{(C_0 - C_1)},$$

where

$\kappa = \frac{k_m}{\rho c}$ is thermal diffusivity of the fluid.

Equations (10.1) - (10.4) in non-dimensional form can be written as

$$\nabla' \cdot \mathbf{v}' = 0, \tag{10.6}$$

$$\left(1 + F\frac{\partial}{\partial t'}\right)\frac{1}{\Pr}\frac{\partial \mathbf{v}'}{\partial t'} = (1 + F)\left(-\nabla' p' - Rm\hat{e}_z + RaT'\hat{e}_z - Rn\varphi'\hat{e}_z + \frac{Rs}{Ls}C'\hat{e}_z\right) + \nabla'^2\mathbf{v}', \tag{10.7}$$

$$\frac{\partial \varphi'}{\partial t'} + \mathbf{v}' \cdot \nabla'\varphi' = \frac{1}{Le}\nabla'^2\varphi' + \frac{N_A}{Le}\nabla'^2 T', \tag{10.8}$$

$$\frac{\partial T'}{\partial t'} + \mathbf{v}' \cdot \nabla' T' = \nabla'^2 T' + \frac{N_B}{Le}\nabla'\varphi' \cdot \nabla' T' + \frac{N_A N_B}{Le}\nabla' T' \cdot \nabla' T' + N_{TC}\nabla^2 C', \tag{10.9}$$

$$\frac{\partial C'}{\partial t'} + \mathbf{v}' \cdot \nabla' C' = \frac{1}{Ls}\nabla'^2 C' + N_{CT}\nabla'^2 T'.$$

Here the non-dimensional parameters are given as follows:

$\Pr = \frac{\mu}{\rho\kappa}$ is the Prandtl number,

$Le = \frac{\kappa}{D_B}$ is the thermo-nanofluid Lewis number

$Ls = \frac{o}{D_S}$ is the solute Lewis number,

$N_{TC} = \frac{D_{TC}\left(C_0 - C_1\right)}{\kappa\left(T_0 - T_1\right)}$ is the Dufour parameter

$N_{CT} = \frac{D_{CT}\left(T_0 - T_1\right)}{\kappa\left(C_0 - C_1\right)}$ is the Soret parameter

$F = \frac{\kappa\lambda}{d^2}$ is the stress relaxation parameter,

$Ra = \frac{\rho g \alpha_T \left(T_0 - T_1\right) d^3}{\text{¼}}$ is the Rayleigh number,

$Rs = \frac{\rho g \alpha_C \left(C_0 - C_1\right) d^3}{\text{¼}'}$ is the solutal Rayleigh number,

$Rm = \frac{\left(\rho_p \varphi_0 + \rho\left(1 - \varphi_0\right)\right) g d^3}{\text{¼}}$ is the density Rayleigh number,

$Rn = \frac{\left(\rho_p - \rho\right)\varphi_0 g d^3}{\text{¼}}$ is the nanoparticles Rayleigh number,

$N_A = \frac{D_T\left(T_0 - T_1\right)}{D_B T_1 \varphi_0}$ is the modified diffusivity ratio,

$N_B = \frac{(\rho c)_p \varphi_0}{(\rho c)_f}$ is the modified particle-density increment.

Equation (10.7) has been linearized by the neglect of a term proportional to the product of φ_0 and T. This approximation is valid in the case of small temperature gradients in a dilute suspension of nanoparticles.

The dimensionless boundary conditions are

$$w' = 0, \quad T' = 1,\ C' = 1,\ \frac{\partial \phi'}{\partial z'} + N_A \frac{\partial T'}{\partial z'} = 0 \ \text{at} \quad z' = 0$$

and

$$w' = 0,\ T' = 0,\ C' = 0,\ \frac{\partial \phi'}{\partial z'} + N_A \frac{\partial T'}{\partial z'} = 0 \ \text{at}\ z' = 1. \tag{10.10}$$

THE BASIC STATE AND ITS SOLUTIONS

The basic state was assumed to be quiescent and is given by

$$v_i' \left(u', v', w'\right) = 0,$$

$$p' = p_b(z),$$

$$T' = T_b(z),$$

$$C' = C_b(z), \tag{10.11}$$

$$\varphi' = \varphi_b(z) \ \text{and}$$

$$\rho = \rho_0 \left(1 + \alpha \left(T\text{-}T_0\right)\right).$$

Equations (10.6) – (10.9) reduce to

$$0 = -\frac{dp_b}{dz'} \text{-Rm} + \text{Ra}T_b + \frac{\text{Ra}}{Ls} C_b - \text{Rn}\varphi_b, \tag{10.12}$$

$$\frac{d^2 T_b}{dz'^2} + \frac{N_B}{\text{Le}} \frac{d\varphi_b}{dz'} \frac{dT_b}{dz'} + \frac{N_A N_B}{\text{Le}} \left(\frac{dT_b}{dz'}\right)^2 + N_{TC} \frac{d^2 C_b}{dz'^2} = 0, \tag{10.13}$$

$$\frac{d^2 \varphi_b}{dz'^2} + N_A \frac{d^2 T_b}{dz'^2} = 0, \tag{10.14}$$

Using boundary conditions in (10.10), equation (10.14) gives

$$\varphi_b = -N_A T_b + (1 - N_A)z' + N_A \tag{10.15}$$

On substituting the value of the φ_b from equation (10.15) in equation (10.13), we get

$$\frac{d^2T_b}{dz'} + \frac{(1\text{-}N_A)N_B}{Le}\frac{dT_b}{dz'} = 0 \tag{10.16}$$

On integrating equation (10.16) with respect to z' and using boundary conditions (10.10), we get

$$T_b = \frac{1 - e^{-(1-N_A)N_B(1-z')/Le}}{1 - e^{-(1-N_A)N_B/Le}} \tag{10.17}$$

For most nanofluid investigated so far Le is large, is of order 10^2-10^3, while N_A is no greater than about 10. Then, the exponents in equation (10.17) are small. By expanding the exponential function into the power series and retaining up to the first order is negligible. Thus an approximate solution for basic state is given by

$$T_b = 1 - z',$$

$$C_b = 1 - z'$$

and

$$\varphi_b = \phi_0 + N_A z'$$

PERTURBATION SOLUTIONS

Let the initial basic state described by (10.11) is slightly perturbed so that perturbed state is given by

$$(u',v',w') = 0 + (u'',v'',w''),$$

$$T' = T_b + T'', \tag{10.18}$$

$$C' = C_b + C'',$$

$$\varphi' = \varphi_b + \varphi'',$$

$$p' = p_b + p'',$$

where $T_b = 1 - z'$, $C_b = 1 - z'$, $\varphi_b = \phi_0 + N_A z'$ and (u'', v'', w''), T'', φ'' and p'' respectively the perturbations in the initial velocity, temperature, volume fraction of the nanoparticles and pressure.

By substituting (10.18) in equations (10.6) – (10.9) and linearize by neglecting the product of the prime quantities, we obtained following equations

$$\nabla \cdot \mathbf{v} = 0, \tag{10.19}$$

$$\left(1 + F\frac{\partial}{\partial t}\right)\frac{1}{\Pr}\frac{\partial \mathbf{v}}{\partial t} = \left(1 + F\frac{\partial}{\partial t}\right)\left(-\nabla p + \mathrm{Ra}T\hat{e}_z - \mathrm{Rn}\varphi\hat{e}_z + \frac{\mathrm{Rs}}{Ls}C\hat{e}_z\right) + \nabla^2\mathbf{v}, \tag{10.20}$$

$$\frac{\partial \varphi}{\partial t} + wN_A = \frac{1}{\mathrm{Le}}\nabla^2\varphi + \frac{N_A}{\mathrm{Le}}\nabla^2 T, \tag{10.21}$$

$$\frac{\partial T}{\partial t} - w = \nabla^2 T + \frac{N_B}{\mathrm{Le}}\left(N_A\frac{\partial T}{\partial z} - \frac{\partial \phi}{\partial z}\right) - \frac{2N_A N_B}{\mathrm{Le}}\frac{\partial T}{\partial z} + N_{TC}\nabla^2 C, \tag{10.22}$$

$$\frac{\partial C}{\partial t} - w = \frac{1}{\mathrm{Ls}}\nabla^2 C + N_{CT}\nabla^2 T. \tag{10.23}$$

Boundary conditions are

$$w = 0,\ T = 0,\ \ C = 0,\ \frac{\partial \phi}{\partial z} + N_A\frac{\partial T}{\partial z} = 0 \ \text{ at } z = 0,1. \tag{10.24}$$

[Dashes (") have been suppressed for convenience]

Eliminating pressure term p from equation (10.20) by operating the curl twice on it, we have

$$\left(1 + F\frac{\partial}{\partial t}\right)\frac{1}{\Pr}\frac{\partial}{\partial t}\nabla^2 w - \nabla^4 w - \left(1 + F\frac{\partial}{\partial t}\right)\left(\mathrm{Ra}\nabla_H^2 T + \frac{\mathrm{Rs}}{Ls}\nabla_H^2 C - \mathrm{Rn}\nabla_H^2\varphi\right) = 0, \tag{10.25}$$

where $\nabla_H^2 = \frac{\partial^2}{\partial x^2} + \frac{\partial^2}{\partial y^2}$ is the two-dimensional Laplacian operator on the horizontal plane.

NORMAL MODE ANALYSIS

Analyzing the disturbances into the normal modes and assuming that the perturbed quantities are of the form

$$\left[\mathrm{w,T,C,}\,\varphi\right] = \left[\mathrm{W(z),\Theta(z),\Gamma\left(z\right),\Phi(z)}\right]\exp\left(\mathrm{ik_x x + ik_y y + nt}\right), \tag{10.26}$$

where k_x, k_y are wave numbers in x and y direction and n is growth rate of disturbances.

Using equation (10.26), equations (10.25), (10.21) – (10.23) become

$$\left(\left(D^2 - a^2\right)\left(D^2 - a^2 - \frac{n\left(1+nF\right)}{\Pr}\right)\right)W - \left(1+nF\right)\left(a^2\mathrm{Ra}\Theta + a^2\frac{\mathrm{Rs}}{Ls}\Gamma - a^2\mathrm{Rn}\Phi\right) = 0, \tag{10.27}$$

$$W + \left(D^2 - a^2 - n - \frac{N_A N_B}{Le}D\right)\Theta - \frac{N_B}{Le}D\Phi + N_{TC}\left(D^2 - a^2\right)\Gamma = 0, \tag{10.28}$$

$$W + \frac{1}{Ls}\left(D^2 - a^2 - n\right)\Gamma + N_{CT}\left(D^2 - a^2\right)\Theta = 0, \tag{10.29}$$

$$N_A W - \frac{N_A}{Le}\left(D^2 - a^2\right)\Theta - \left(\frac{1}{Le}\left(D^2 - a^2\right) - n\right)\Phi = 0, \tag{10.30}$$

where $D \equiv \frac{d}{dz}$ and $a = \sqrt{k_x^2 + k_y^2}$ is the dimensionless resultant wave number.

The boundary conditions of the problem in view of normal mode analysis are

$$W = 0,\ D^2W = 0, \Theta = 0,\ \Gamma = 0, D\Phi + N_A D\Theta = 0 \ \text{at} \quad z = 0,1. \tag{10.31}$$

METHOD OF SOLUTION

The Galerkin weighted residuals method is used to obtain an approximate solution to the system of equations (10.27) – (10.30) with the corresponding boundary conditions (10.31). In this method, the test functions are the same as the base (trial) functions. Accordingly W, Θ and Φ are taken as

$$W = \sum_{p=1}^{N} A_p W_p,\ \Theta = \sum_{p=1}^{N} B_p \Theta_p, \Gamma = \sum_{p=1}^{N} C_p \Gamma_p, \Phi = \sum_{p=1}^{N} D_p \Phi_p, \tag{10.32}$$

where $W_p = \Theta_p = \sin p\pi z, \Phi_p = -N_A \sin p\pi z$, A_p, B_p, C_p and D_p are unknown coefficients, p = 1, 2, 3,..., N and the base functions W_p, Θ_p, Γ_p and Φ_p satisfying the boundary conditions (10.31). Using expression for W, Θ and Φ in equations (10.27) – (10.30) and multiplying the first equation by W_p the second equation by Θ_p, third equation by Γ_p and fourth by Φ_p and then integrating in the limits from zero to unity, we obtain a set of 4N linear homogeneous equations with 4N unknown A_p, B_p and C_p; p = 1, 2, 3,..., N. For existing of nontrivial solution, the vanishing of the determinant of coefficients produces the characteristics equation of the system in term of Rayleigh number Ra.

LINEAR STABILITY ANALYSIS

For the present formulation, we have considered the which system of equations (10.27) – (10.30) together with the boundary conditions (10.31) constitute a linear eigenvalue problem with variable coefficient for the growth rate of disturbance n of the system. Substituting equation (10.32) into the system of equations (10.27) – (10.30), one obtains the following matrix equation

$$\begin{bmatrix} \frac{(\pi^2+a^2)}{(1+nF)}\left(\pi^2+a^2+\frac{n(1+nF)}{\Pr}\right) & -a^2 Ra & -a^2\frac{Rs}{Ln} & -a^2 N_A Rn \\ 1 & -(\pi^2+a^2+n) & -N_{TC}(\pi^2+a^2) & 0 \\ 1 & -N_{CT}(\pi^2+a^2) & -\frac{1}{Ls}(\pi^2+a^2+n) & 0 \\ N_A & \frac{N_A}{Le}(\pi^2+a^2) & 0 & -N_A\left(\frac{1}{Le}(\pi^2+a^2)+n\right) \end{bmatrix} \begin{bmatrix} W_0 \\ \Theta_0 \\ \Gamma_0 \\ \Phi_0 \end{bmatrix} = \begin{bmatrix} 0 \\ 0 \\ 0 \\ 0 \end{bmatrix}. \tag{10.33}$$

The non-trivial solution of the above matrix requires that

$$\mathrm{Ra} = \frac{\left(\left(\pi^2+a^2+n\right)^2-\left(\pi^2+a^2\right)^2 N_{CT}N_{TC}Ls\right)}{\mathrm{a}^2\left(\left(\pi^2+a^2+n\right)-\left(\pi^2+a^2\right)N_{TC}Ls\right)}\left(\frac{\left(\pi^2+a^2\right)}{\left(1+nF\right)}\left(\pi^2+a^2+\frac{n\left(1+nF\right)}{\mathrm{Pr}}\right)\right)$$
$$+\frac{\left(\pi^2+a^2\right)N_{CT}-\left(\pi^2+a^2+n\right)}{\left(\left(\pi^2+a^2+n\right)-\left(\pi^2+a^2\right)N_{TC}Ls\right)}Rs$$
$$-\frac{\left(\pi^2+a^2+n\right)\left(\frac{\left(\pi^2+a^2\right)}{Le}+\left(\pi^2+a^2+n\right)\right)}{\left(\frac{\left(\pi^2+a^2\right)}{Le}+\mathrm{n}\right)\left(\left(\pi^2+a^2+n\right)-\left(\pi^2+a^2\right)N_{TC}Ls\right)}\mathrm{N_A Rn}. \tag{10.34}$$

The growth rate n is in general a complex quantity such that $\mathrm{n} = \omega_r + i\omega$, the system with $\omega_r < 0$ is always stable, while for $\omega_r > 0$ it will become unstable. For neutral stability, the real part of n is zero. Hence, for the case of stationary (non- oscillatory) convection [n = 0], equation (10.34) reduces to

$$\left(\mathrm{Ra}\right)_s = \frac{\left(\pi^2+a^2\right)^3}{\mathrm{a}^2}\frac{\left(1-N_{CT}N_{TC}Ls\right)}{1-LsN_{TC}}+\frac{Rs\left(N_{CT}-1\right)}{1-LsN_{TC}}-\frac{\left(LnN_{TC}+N_{CT}N_{TC}LsLe+1+Le\right)}{1-LsN_{TC}}\mathrm{N_A Rn}. \tag{10.35}$$

It is observed that stationary Rayleigh number Ra is function of the Soret parameter, Dufour parameter, solute Lewis number Ls, nanofluid Lewis number Le, Solute Rayleigh number Rs, the modified diffusivity ratio N_A and the nanoparticles Rayleigh Rn but independent of visco- elastic parameter F, Prandtl number Pr and modified particle- density increment N_B. Thus Maxwellian visco-elastic nanofluid behaves like an ordinary Newtonian nanofluid and instability is purely a phenomenon due to buoyancy coupled with the conservation of nanoparticles.

In the absence of the Dufour and Soret parameters ($N_{TC} = N_{CT} = 0$), equation (10.35) reduces to

$$\left(\mathrm{Ra}\right)_s = \frac{\left(\pi^2+a^2\right)^3}{\mathrm{a}^2} - Rs - \left(1+Le\right)\mathrm{N_A Rn}.$$

In the absence of solute particle ($N_{CT} = N_{TC} = Rs = 0$) equation (10.35) reduces to

$$\left(\text{Ra}\right)_s = \frac{\left(\pi^2 + a^2\right)^3}{\text{a}^2} - \left(1 + Le\right)\text{N}_\text{A}\text{Rn}.$$

This is the good agreement of the result (1.47) obtained in Chapter 1.

In the absence of solute and nanoparticles ($N_{CT} = N_{TC} = Rs = Rn = Le = N_A = 0$), equation (10.35) reduces to

$$\left(\text{Ra}\right)_s = \frac{\left(\pi^2 + a^2\right)^3}{\text{a}^2}. \tag{10.36}$$

To find the critical value of $\left(\text{Ra}\right)_s$, equation (10.36) is differentiated with respect to 'a^2' and then equated to zero. The minimum of equation (10.36) is attained at $\text{a}_c = \frac{\pi}{\sqrt{2}}$ and one recovers the well- known results that the critical Rayleigh number is equal to

$$\left(\text{Ra}\right)_c = \frac{27\pi^4}{4}. \tag{10.37}$$

This is good agreement of the result obtained by Chandrasekhar (1961).

Thus presence of the nanoparticles lowers the value of the critical Rayleigh number by usually by substantial amount. Also parameter N_B does not appear in the equation (10.35); thus instability is purely phenomena due to buoyancy coupled with conservation of nanoparticles. Thus average contribution of nanoparticles flux in the thermal energy equation is zero with one-term Galerkin approximation.

RESULTS AND DISCUSSION

Expression for stationary Rayleigh number, which characterizes the stability of the system, is found to be depend upon solutal Rayleigh number, solute Lewis number, Lewis number, Soret parameter, Dufour parameter, modified diffusivity ratio and nanoparticles Rayleigh number.

The computations are carried out for different values of parameters considered in the range $10^2 \le Ra \le 10^5$ (thermal Rayleigh number), $1 \le N_{CT} \le 10$ (Soret parameter), $10 \le N_{TC} \le 10^2$ (Dufour parameter), $1 \le N_A \le 10$ (modified diffusivity ratio), $10^2 \le Le \le 10^4$ (nanofluid Lewis number), $1 \le Ls \le 10$ (solute Lewis number), $10^{-1} \le Rn \le 10^1$ (nanofluid Rayleigh number).

The variations of the stationary thermal Rayleigh number with wave number have been plotted graphically as shown in Figures 2-8.

Figure 2 shows the variation of thermal Rayleigh number with wave number for different value of solutal Rayleigh number and for the fixed value of other parameters. It is found that stationary Rayliegh number increases as the value of solutal Rayleigh number increases, indicating that solutal Rayleigh number stabilizes the stationary convection.

Figure 3 shows the variation of thermal Rayleigh number with wave number for different value of solute Lewis number and for the fixed value of other parameters. It is found that stationary Rayliegh number increases as the value of solute Lewis number increases, indicating that solute Lewis number stabilizes the stationary convection.

Figure 2. Variation of the stationary Rayleigh number with wave number for different value of solutal Rayleigh number

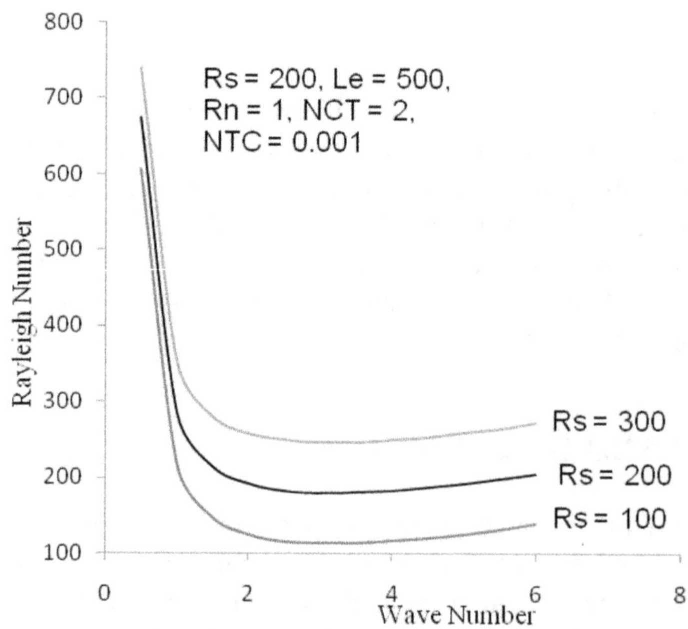

Figure 3. Variation of the stationary Rayleigh number with wave number for different value of solute Lewis number

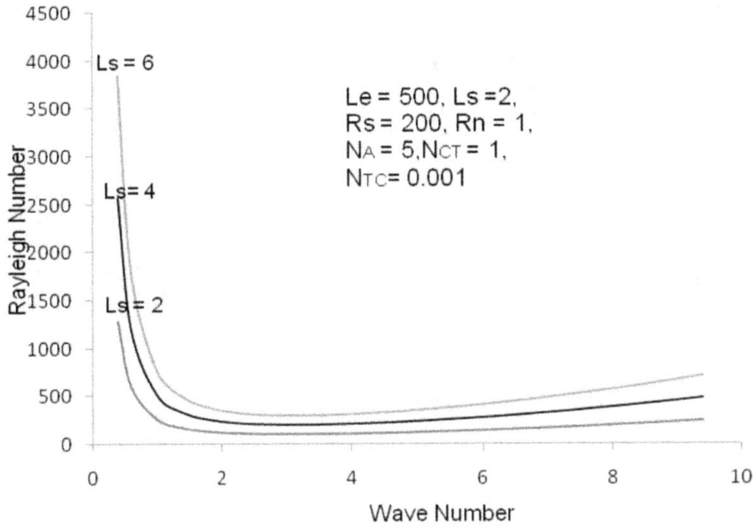

Figure 4 shows the variation of thermal Rayleigh number with wave number for different value of Lewis number Le and for the fixed value of other parameters. It is found that stationary Rayliegh number decreases as the value of Lewis number increases, indicating that Lewis number destabilizes the stationary convection. It is due to the fact that thermophoresis at a higher value of thermophoretic diffusivity is more supportable to the disturbance in nanofluids, while both thermophoresis and Brownian motion are driving forces in favor of the motion of nanoparticles.

Figure 5 shows the variation of thermal Rayleigh number with wave number for different value of Soret parameter and for the fixed value of other parameters. It is found that stationary Rayliegh number increases the value of Soret parameter increases, indicating that Soret parameter stabilizes the stationary convection.

Figure 6 shows the variation of thermal Rayleigh number with wave number for different value of Dufour parameter and for the fixed value of other parameters. It is found that stationary Rayliegh number first increases, then decreases and finally increases with an incresae in the values of Dufour parameter, indicating that Dufour parameter has stabilizes and destabilizing effect on stationary convection depending upon certain conditions.

Figure 7 shows the variation of stationary Rayleigh number with wave number for different value of the modified diffusivity ratio and fixed value

Figure 4. Variation of the stationary Rayleigh number with wave number for different value of nanofluid Lewis number

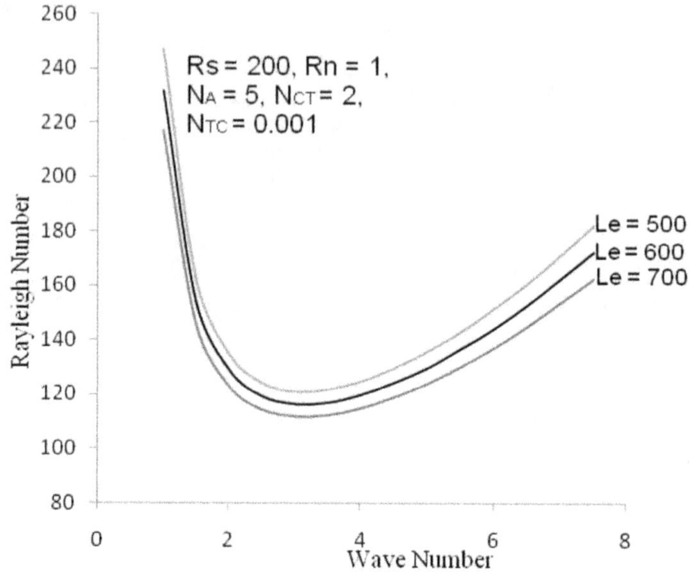

Figure 5. Variation of the stationary Rayleigh number with wave number for different value of Soret parameter

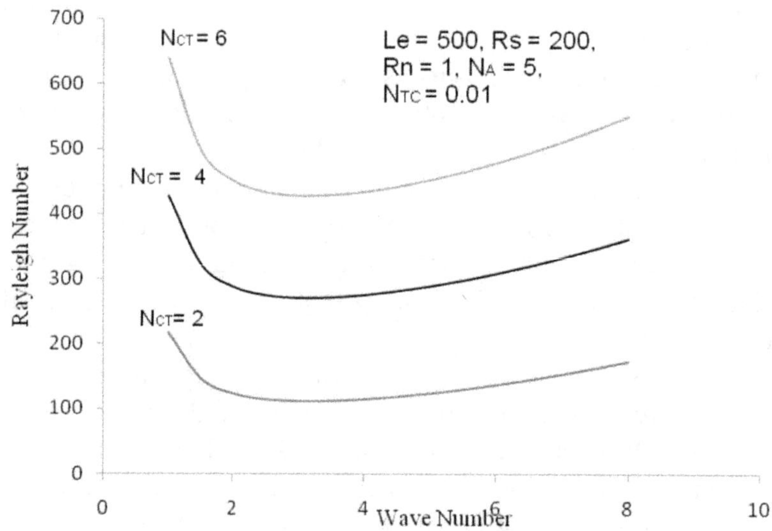

Figure 6. Variation of the stationary Rayleigh number with wave number for different value of Dufour parameter

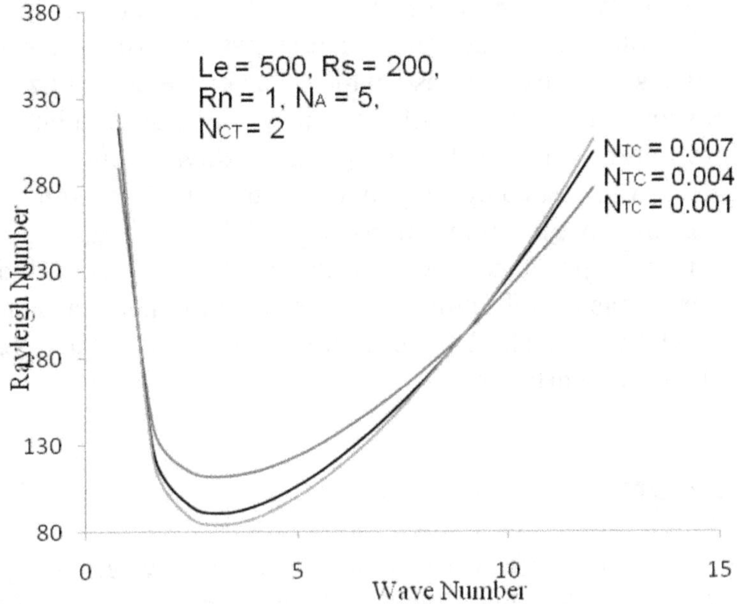

Figure 7. Variation of the stationary Rayleigh number with wave number for different value of modified diffusivity ratio

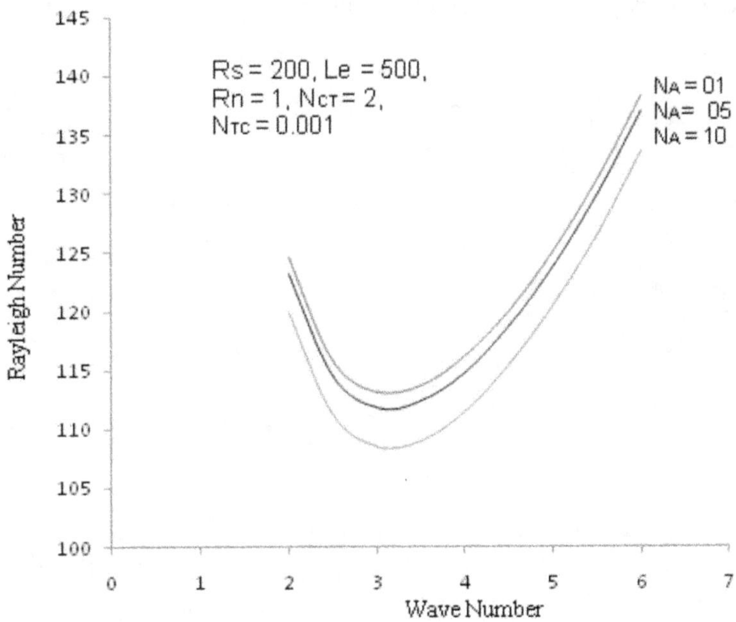

of other parameters and it is found that Rayleigh number decreases with an increase in the value of the modified diffusivity ratio which means that the modified diffusivity ratio destabilizes on the fluid layer. This may lead to an increase in volumetric fraction, which shows that Brownian motion of the nanoparticles will also increase, which may cause destabilizing effect.

Figure 8 shows the variation of stationary Rayleigh number with wave number for different value of the nanoparticles Rayleigh number and fixed value of other parameters and it is found that thermal Rayleigh number decreases with an increase in value of the nanoparticles Rayleigh number, which means that the nanoparticles Rayleigh number Rn has destabilizing effect on fluid layer. It has destabilizing effect because the heavier nanoparticles moving through the base fluid makes more strong disturbances as compared with the lighter nanoparticles.

CONCLUSION

Double diffusive convection in a horizontal layer of Maxwellian visco-elastic nanofluid is studied. The flux of volume fraction of nanoparticles is taken to be zero on the isothermal boundaries and the eigen value problem is solved

Figure 8. Variation of the stationary Rayleigh number with wave number for different value of nanoparticles Rayleigh number

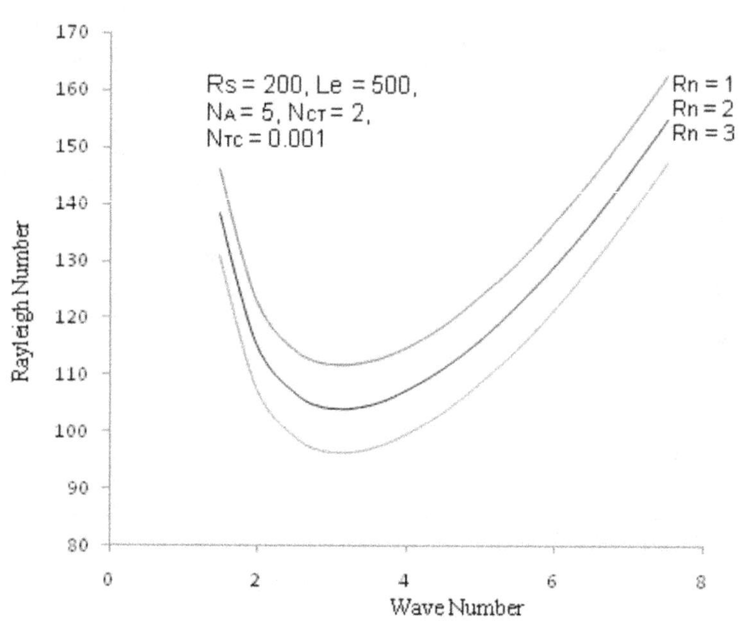

using the Galerkin residual method. The results have been presented both analytically and graphically.

The main conclusions derived from the present analysis are as follows:

1. The instability purely phenomenon due to buoyancy coupled with the conservation of nanoparticle and is independent of the contribution of Brownian motion and thermophoresis.
2. For stationary convection Maxwellian visco-elastic nanofluid behaves like an ordinary Newtonian nanofluid.
3. Solutal Rayleigh number, solute Lewis number, Soret parameter have stabilizing effect on stationary convection while Lewis number, modified diffusivity ratio and nanoparticles Rayleigh number have destabilizing effect on stationary convection.
4. Dufour parameter has both stabilizing and destabilizing effect on stationary convection depending upon certain conditions.

REFERENCES

Chandrasekhar, S. (1961). *Hydrodynamic and Hydromagnetic Stability*. New York: Dover Publication.

Kuznetsov, A. V., & Nield, D. A. (2010a). Effect of local thermal non-equilibrium on the onset of convection in a porous medium layer saturated by a nanofluid. *Transport in Porous Media*, *83*(2), 425–436. doi:10.1007/s11242-009-9452-8

Kuznetsov, A. V., & Nield, D. A. (2010b). Thermal instability in a porous medium layer saturated by a nanofluid: Brinkman Model. *Transport in Porous Media*, *81*(3), 409–422. doi:10.1007/s11242-009-9413-2

Kuznetsov, A. V., & Nield, D. A. (2010c). The onset of double-diffusive nanofluid convection in a layer of a saturated porous medium. *Transport in Porous Media*, *85*(3), 941–951. doi:10.1007/s11242-010-9600-1

Kuznetsov, A. V., & Nield, D. A. (2011). The effect of local thermal non-equilibrium on the onset of convection in porous medium layer saturated by a nanofluid: Brinkman model. *Journal of Porous Media*, *14*(4), 285–293. doi:10.1615/JPorMedia.v14.i4.10

Nield, D. A., & Kuznetsov, A. V. (2009b). The Cheng–Minkowycz problem for natural convective boundary-layer flow in a porous medium saturated by a nanofluid. *International Journal of Heat and Mass Transfer*, *52*(25-26), 5792–5795. doi:10.1016/j.ijheatmasstransfer.2009.07.024

Yadav, D., Bhargava, R., & Agrawal, G. S. (2012b). Boundary and internal heat source effects on the onset of Darcy- Brinkman convection in a porous layer saturated by nanofluid. *International Journal of Thermal Sciences*, *60*, 244–254. doi:10.1016/j.ijthermalsci.2012.05.011

Chapter 11
Double Diffusive Convection in a Layer of Maxwellian Visco-Elastic Nanofluid in a Porous Medium

INTRODUCTION

Double-diffusive convection is referred to buoyancy-driven flows induced by combined temperature and concentration gradients. Double diffusive convection in a fluid layer in a porous medium heated from below is regarded as a classical problem due to its wide range of applications in many engineering fields such as evaporative cooling of high temperature systems, soil sciences, enhanced oil recovery, agricultural product storage, packed-bed catalytic reactors and the pollutant transport in underground. Thermal convection in binary fluid driven by the Soret and Dufour effects has been investigated by Knobloch (1980) and showed that the equations are identical to the thermo-solutal problem except relation between the thermal and solutal Rayleigh numbers. The study of flow, heat and mass transfer about natural convection of non-Newtonian nanofluid fluids in a porous media has gained much attention from the researchers because of its engineering and industrial applica-

DOI: 10.4018/978-1-68318-006-7.ch011

tions. These applications include design of chemical processing equipment, formation and dispersion of fog, distributions of temperature and moisture over agricultural fields and groves of fruit trees and damage of crops due to freezing and pollution of the environment etc. Double diffusive convection in binary fluid layer in a porous medium was investigated by Malashetty and Kollur (2011), Nield and Bejan (2013). The onset of double-diffusive nanofluid convection in a layer of a saturated porous medium was studied by Kuznetsov and Nield (2010c), Yadev et al. (2013a, 2016a) and Rana et al. (2012, 2014b, 2014c).

In this chapter an attempt has been made to study the double diffusive convection in a horizontal layer of Maxwellian visco-elastic nanofluid layer in a Brinkman porous medium. Stability is discussed analytically as well as numerically using Galerkin-type weighted residuals method.

MATHEMATICAL FORMULATIONS OF THE PROBLEM

The Physical Problem

The physical configuration of the problem to be considered as An infinite horizontal layer of Maxwellian visco-elastic nanofluid in a porous medium of medium permeability k_1 and porosity ε and of thickness 'd' bounded by horizontal boundaries $z = 0$ and $z = d$. A Cartesian coordinate system (x, y, z) is chosen with the origin at the bottom of the fluid layer and the z- axis normal to the fluid layer. Fluid layer is acted upon by gravity force $\mathbf{g}(0, 0,-g)$. Fluid layer is heated from below in such a way that horizontal boundaries $z = 0$ and $z = d$ respectively maintained at a uniform temperature T_0 and T_1 $(T_0 > T_1)$. The normal component of the nanoparticles flux has to vanish at an impermeable boundary and the temperature T is taken to be T_0 at $z = 0$ and T_1 at $z = d$, $(T_0 > T_1)$ as shown in Figure 1. The reference scale for temperature and nanoparticles fraction is taken to be T_1 and φ_0 respectively.

Assumptions

The mathematical equations describing the physical model are based upon the following assumptions:

1. Thermophysical properties of fluid expect for density in the buoyancy force (Boussinesq Hypothesis) are constant,

Figure 1. Physical configuration of the problem

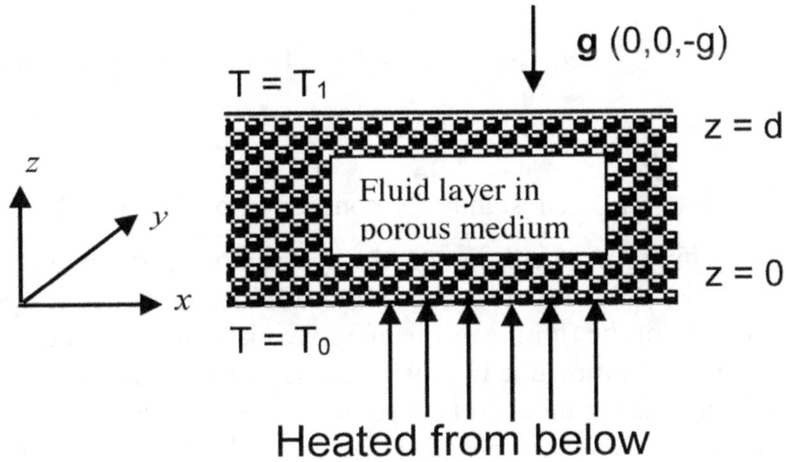

2. The fluid phase and nanoparticles are in thermal equilibrium state and thus, the heat flow has been described using one equation model,
3. Dilute mixture,
4. Nanoparticles are spherical,
5. Nanoparticles are non-magnetic,
6. No chemical reactions take place in fluid layer,
7. Negligible viscous dissipation,
8. Radiative heat transfer between the sides of wall is negligible when compared with other modes of the heat transfer,
9. Nanofluid is incompressible, electrically conducting, Newtonian and laminar flow,
10. Each boundary wall is assumed to be impermeable and perfectly thermal conducting,
11. The nanoparticles do not affect the transport of the solute,
12. Nanoparticles are suspended in the nanofluid using either surfactant or surface charge technology.

GOVERNING EQUATIONS

The governing equations for double-diffusive convection of Maxwellian visco-elastic nanofluid in a Brinkman porous medium under the Boussinesq approximation are

$$\nabla \cdot \mathbf{q} = 0, \tag{11.1}$$

$$\frac{\rho}{\varepsilon}\left(1+\lambda\frac{\partial}{\partial t}\right)\frac{d\mathbf{q}}{dt} = \left(1+\lambda\frac{\partial}{\partial t}\right)\left(-\nabla p + \left(\varphi\rho_p + (1-\varphi)\left\{\rho_f\left(1-\alpha_T\left(T-T_0\right)-\alpha_C\left(C-C_0\right)\right)\right\}\right)\mathbf{g}\right) + \tilde{\mu}\nabla^2\mathbf{q} - \frac{\mu}{k_1}\mathbf{q}, \tag{11.2}$$

where $\frac{d}{dt} = \frac{\partial}{\partial t} + \frac{1}{\varepsilon}(\mathbf{q}\cdot\nabla)$ is stands for convection derivative while **q** is the fluid velocity in porous medium, p is the hydrostatic pressure, ρ is the density of nanofluid, ρ_0 is the density of the nanofluid at reference temperature, μ is the viscosity of the fluid, $\tilde{\mu}$ is the effective viscosity, λ is the relaxation time, **g** is the acceleration due to gravity and α_T is the coefficient of thermal expansion and α_C is the analogous to solute concentration.

Thermal energy equation for Maxwellian visco-elastic nanofluid is given by

$$(\rho c)_m \frac{\partial T}{\partial t} + (\acute{A}c)_f\, \mathbf{q}\cdot\nabla T = k_m \nabla^2 T + (\rho c)_p \left(D_B \nabla\varphi\cdot\nabla T + \frac{D_T}{T_1}\nabla T\cdot\nabla T\right) + (\rho c)_f D_{TC}\nabla^2 C, \tag{11.3}$$

where $(\rho c)_m$ is heat capacity of fluid in porous medium, $(\rho c)_p$ is heat capacity of nanoparticles, T_1 is the temperature of the fluid layer at z = d, D_{TC} is a diffusivity of Dufour type and k_m is thermal conductivity.

The conservation equation for solute concentration Kuznetsov and Nield (2010c) is of the form

$$\frac{\partial C}{\partial t} + \frac{1}{\varepsilon}(\mathbf{q}\cdot\nabla T) = \kappa'\nabla^2 C + D_{CT}\nabla^2 T, \tag{11.4}$$

where κ' is the solutal diffusivity and D_{CT} is the diffusivity of Soret type.

Equation of continuity for the nanoparticles is given by

$$\frac{\partial\varphi}{\partial t} + \frac{1}{\varepsilon}(\mathbf{q}\cdot\nabla\varphi) = D_B\nabla^2\varphi + \frac{D_T}{T_1}\nabla^2 T, \tag{11.5}$$

where D_B is the Brownian diffusion coefficient, given by Einstein-Stokes equation and D_T is the thermoporetic diffusion coefficient of the nanoparticles.

We assume that the temperature is constant and nanoparticles flux is zero on the boundaries. Thus boundary conditions are

$$w = 0, T = T_0, \; D_B \frac{\partial \phi}{\partial z} + \frac{D_T}{T_1}\frac{\partial T}{\partial z} = 0 \text{ at } z = 0$$

and

$$w = 0, T = T_1, \; D_B \frac{\partial \phi}{\partial z} + \frac{D_T}{T_1}\frac{\partial T}{\partial z} = 0 \text{ at } z = d. \tag{11.6}$$

Introducing non-dimensional variables as

$$(x',y',z') = \left(\frac{x,y,z}{d}\right), \; q'(u',v',w') = q\left(\frac{u,v,w}{\kappa}\right)d, \; t' = \frac{t^{o}}{d^2}, \; p' = \frac{pk_1}{\mu\kappa},$$

$$\varphi' = \frac{(\varphi - \varphi_0)}{\varphi_0}, \; T' = \frac{(T - T_1)}{(T_0 - T_1)}, \; C' = \frac{(C - C_1)}{(C_0 - C_1)},$$

where

$$\sigma = \frac{(\rho c)_m}{(\rho c)_f}, {}^{o} = \frac{k_m}{(\rho c)_f} \text{ is thermal diffusivity of the fluid and.}$$

Equations (11.1) - (11.5) in non-dimensional form can be written as

$$\nabla' \cdot \mathbf{q}' = 0, \tag{11.7}$$

$$\left(1 + F\frac{\partial}{\partial t'}\right)\frac{1}{Va}\frac{\partial \mathbf{q}'}{\partial t'} = \left(1 + F\frac{\partial}{\partial t'}\right)\left(-\nabla' p' \text{-} Rm\hat{e}_z + RaT'\hat{e}_z - Rn\varphi'\hat{e}_z + \frac{Rs}{Ls}C'\hat{e}_z\right) + \tilde{D}a\nabla'^2\mathbf{q}' - \mathbf{q}', \tag{11.8}$$

$$\frac{\partial \varphi'}{\partial t'} + \frac{1}{\varepsilon}(\mathbf{q}' \cdot \nabla'\varphi') = \frac{1}{Le}\nabla'^2\varphi' + \frac{N_A}{Le}\nabla'^2 T', \tag{11.9}$$

$$\frac{\partial T'}{\partial t'} + \frac{1}{\varepsilon}(\mathbf{q}' \cdot \nabla' T') = \nabla'^2 T' + \frac{N_B}{Le}\nabla'\varphi' \cdot \nabla' T' + \frac{N_A N_B}{Le}\nabla' T' \cdot \nabla' T' + N_{TC}\nabla^2 C', \tag{11.10}$$

$$\frac{\partial C'}{\partial t'} + \frac{1}{\varepsilon}(\mathbf{q}' \cdot \nabla' C') = \frac{1}{Ls}\nabla'^2 C' + N_{CT}\nabla'^2 T'. \tag{11.11}$$

Here the non-dimensional parameters are given as follows

$\Pr = \frac{\mu}{\rho\kappa}$ is the Prandtl number,

$Da = \frac{k_1}{d^2}$ is the Darcy number,

$\tilde{D}a = \frac{\tilde{\mu} k_1}{\mu d^2}$ is the Brinkman-Darcy number,

$N_{TC} = \frac{D_{TC}\left(C_0 - C_1\right)}{\kappa\left(T_0 - T_1\right)}$ is the Dufour parameter,

$N_{CT} = \frac{D_{CT}\left(T_0 - T_1\right)}{\kappa\left(C_0 - C_1\right)}$ is the Soret parameter,

$Va = \frac{\varepsilon \Pr}{Da}$ is the Prandtl- Darcy Number (Vadasz Number),

$Le = \frac{\kappa}{D_B}$ is the thermo-nanofluid Lewis number,

$Ls = \frac{\kappa}{D_S}$ is the thermosolutal Lewis number,

$F = \frac{\kappa\lambda}{d^2}$ is the stress relaxation parameter,

$Ra = \frac{\rho g \alpha_T \left(T_0 - T_1\right) k_1 d}{¼}$ is the Rayleigh number,

$Rs = \frac{\rho g \alpha_C \left(C_0 - C_1\right) k_1 d}{\mu\kappa'}$ is the solutal Rayleigh number,

$Rm = \frac{\left(\rho_p \varphi_0 + \rho\left(1-\varphi_0\right)\right) g d k_1}{¼}$ is the density Rayleigh number,

$Rn = \frac{\left(\rho_p - \rho\right)\varphi_0 g k_1 d}{\mu\kappa}$ is the nanoparticles Rayleigh number,

$$N_A = \frac{D_T\left(T_0\text{-}T_1\right)}{D_B T_1 \varphi_0}$$ is the modified diffusivity ratio,

$$N_B = \frac{\left(\rho c\right)_p \varphi_0}{\left(\rho c\right)_f}$$ is the modified particle-density increment.

In spirit of Oberbeck-Boussinesq approximation, equation (11.8) has been linearized by the neglect of a term proportional to the product of $Æ_0$ and T. This approximation is valid in the case of small temperature gradients in a dilute suspension of nanoparticles.

The dimensionless boundary conditions are

$$w' = 0, \quad T' = 1, \; C' = 1, \; \frac{\partial \phi'}{\partial z'} + N_A \frac{\partial T'}{\partial z'} = 0 \text{ at } \quad z' = 0$$

and

$$w' = 0, \; T' = 0, \; C' = 0, \; \frac{\partial \phi'}{\partial z'} + N_A \frac{\partial T'}{\partial z'} = 0 \text{ at } z' = 1. \tag{11.12}$$

THE BASIC STATE AND ITS SOLUTIONS

The basic state was assumed to be quiescent and is given by

$$v_i' \left(u', v', w'\right) = 0,$$

$$p' = p_b(z),$$

$$T' = T_b(z),$$

$$C' = C_b(z), \tag{11.13}$$

$$\varphi' = \varphi_b(z) \text{ and}$$

$$\rho = \rho_0\left(1 + \alpha\left(T\text{-}T_0\right)\right).$$

Equations (11.7) – (11.11) reduce to

$$0 = -\frac{dp_b}{dz'}\text{-Rm} + \text{Ra}T_b + \frac{\text{Ra}}{Ls}C_b - \text{Rn}\varphi_b, \tag{11.14}$$

$$\frac{d^2T_b}{dz'^2} + \frac{N_B}{\text{Le}}\frac{dÆ_b}{dz'}\frac{dT_b}{dz'} + \frac{N_A N_B}{\text{Le}}\left(\frac{dT_b}{dz'}\right)^2 + N_{TC}\frac{d^2C_b}{dz'^2} = 0, \tag{11.15}$$

$$\frac{1}{\text{Ls}}\frac{d^2C_b}{dz'} + N_{CT}\frac{d^2T_b}{dz'} = 0, \tag{11.16}$$

$$\frac{d^2\varphi_b}{dz'^2} + N_A\frac{d^2T_b}{dz'^2} = 0. \tag{11.17}$$

According to Buongiorno (2006), for most nanofluid investigated so far Le is large, is of order 10^2-10^3, while N_A is no greater than about 10. Using this approximation and solving equations (11.14) - (11.17), we have the an approximate solution of basic state is given by

$$T_b = 1 - z', \; C_b = 1 - z' \text{ and } \varphi_b = \phi_0 + N_A z'$$

PERTURBATION SOLUTIONS

Let the initial basic state described by (11.13) is slightly perturbed so that perturbed state is given by

$$(u',v',w') = 0 + (u'',v'',w''),$$

$$T' = T_b + T'', \tag{11.18}$$

$$C' = C_b + C'',$$

$$\varphi' = \varphi_b + \varphi'',$$

$$p' = p_b + p'',$$

where $T_b = 1 - z'$, $C_b = 1 - z'$, $\varphi_b = \phi_0 + N_A z'$ and (u'', v'', w''), T'', φ'' and p'' respectively the perturbations in the initial velocity, temperature, volume fraction of the nanoparticles and pressure.

By substituting (11.18) in equations (11.7) – (11.11) and linearize by neglecting the product of the prime quantities, we obtained following equations

$$\nabla \cdot \mathbf{q} = 0, \tag{11.19}$$

$$\left(1 + F\frac{\partial}{\partial t}\right)\frac{1}{Va}\frac{\partial \mathbf{q}}{\partial t} = \left(1 + F\frac{\partial}{\partial t}\right)\left(-\nabla p + \mathrm{Ra}T\hat{e}_z - \mathrm{Rn}\varphi\hat{e}_z + \frac{\mathrm{Rs}}{Ls}C\hat{e}_z\right) + \tilde{D}a\nabla^2\mathbf{q} - \mathbf{q}, \tag{11.20}$$

$$\frac{\partial T}{\partial t} - w = \nabla^2 T + \frac{N_B}{\mathrm{Le}}\left(N_A\frac{\partial T}{\partial z} - \frac{\partial \phi}{\partial z}\right) - \frac{2N_A N_B}{\mathrm{Le}}\frac{\partial T}{\partial z} + N_{TC}\nabla^2 C, \tag{11.21}$$

$$\frac{1}{\sigma}\frac{\partial C}{\partial t} - \frac{1}{\varepsilon}w = \frac{1}{\mathrm{Ls}}\nabla^2 C + N_{CT}\nabla^2 T, \tag{11.22}$$

$$\frac{1}{\sigma}\frac{\partial \varphi}{\partial t} + \frac{1}{\varepsilon}wN_A = \frac{1}{\mathrm{Le}}\nabla^2\varphi + \frac{N_A}{\mathrm{Le}}\nabla^2 T. \tag{11.23}$$

Boundary conditions are

$$w = 0, T = 0, \; C = 0, \; \frac{\partial \phi}{\partial z} + N_A\frac{\partial T}{\partial z} = 0 \text{ at } z = 0, 1. \tag{11.24}$$

[Dashes (") have been suppressed for convenience]
Eliminating pressure term p from equation (11.20), we have

$$\left(1 + F\frac{\partial}{\partial t}\right)\frac{1}{\mathrm{Va}}\frac{\partial}{\partial t}\nabla^2 w - \tilde{D}a\nabla^4 w + \nabla^2 w - \left(1 + F\frac{\partial}{\partial t}\right)\left(\mathrm{Ra}\nabla_H^2 T + \frac{\mathrm{Rs}}{Ls}\nabla_H^2 C - \mathrm{Rn}\nabla_H^2\varphi\right) = 0, \tag{11.25}$$

where $\nabla_H^2 = \frac{\partial^2}{\partial x^2} + \frac{\partial^2}{\partial y^2}$ is the two-dimensional Laplacian operator.

NORMAL MODE ANALYSIS

We shall now analyze an arbitrary perturbation into a complete set of normal modes and then examine the stability of each of those modes individually. For the system of equations (11.25), (11.21), (11.22) and (11.23) the analysis can be made in terms of two dimensional periodic wave numbers. Thus, assuming that the perturbed quantities are of the form

$$[w, T, C, \varphi] = [W(z), \Theta(z), \Gamma(z), \Phi(z)]\exp\left(ik_x x + ik_y y + nt\right), \tag{11.26}$$

where k_x, k_y are wave numbers in x and y direction and n is growth rate of disturbances.

Using equation (11.26), equations (11.25), (11.21) - (11.23) become

$$\left((D^2 - a^2)\left(\tilde{D}a(D^2 - a^2) - 1 - \frac{n(1+nF)}{Va}\right)\right)W - (1+nF)\left(a^2 Ra\Theta + a^2\frac{Rs}{Ls}\Gamma - a^2 Rn\Phi\right) = 0, \tag{11.27}$$

$$W + \left(D^2 - a^2 - n - \frac{N_A N_B}{Le}D\right)\Theta - \frac{N_B}{Le}D\Phi + N_{TC}\left(D^2 - a^2\right)\Gamma = 0, \tag{11.28}$$

$$\frac{1}{\varepsilon}W + \frac{1}{Ls}\left(D^2 - a^2 - \frac{n}{\sigma}\right)\Gamma + N_{CT}\left(D^2 - a^2\right)\Theta = 0, \tag{11.29}$$

$$\frac{1}{\varepsilon}WN_A - \frac{N_A}{Le}\left(D^2 - a^2\right)\Theta - \left(\frac{1}{Le}\left(D^2 - a^2\right) - \frac{n}{\sigma}\right)\Phi = 0, \tag{11.30}$$

where $D \equiv \frac{d}{dz}$ and $a = \sqrt{k_x^2 + k_y^2}$ is the dimensionless resultant wave number.

The boundary conditions of the problem in view of normal mode analysis are

$$W = 0,\ D^2W = 0, \Theta = 0,\ \Gamma = 0, D\Phi + N_A D\Theta = 0 \quad \text{at} \quad z = 0, 1. \tag{11.31}$$

METHOD OF SOLUTION

The Galerkin weighted residuals method is used to obtain an approximate solution to the system of equations (11.27) – (11.30) with the corresponding boundary conditions (11.31). In this method, the test functions are the same as the base (trial) functions. Accordingly W, Θ and Φ are taken as

$$W = \sum_{p=1}^{N} A_p W_p,\ \Theta = \sum_{p=1}^{N} B_p \Theta_p, \Gamma = \sum_{p=1}^{N} C_p \Gamma_p, \Phi = \sum_{p=1}^{N} D_p \Phi_p, \tag{11.32}$$

where $W_p = \Theta_p = \sin p\pi z, \Phi_p = -N_A \sin p\pi z$, A_p, B_p, C_p and D_p are unknown coefficients, p = 1, 2, 3,..., N and the base functions W_p, Θ_p, Γ_p and Φ_p satisfying the boundary conditions (11.31). Using expression for W, Θ and Φ in equations (11.27) – (11.30) and multiplying the first equation by W_p the second equation by Θ_p third equation by Γ_p and fourth by Φ_p and then integrating in the limits from zero to unity, we obtain a set of 4N linear homogeneous equations with 4N unknown A_p, B_p and C_p; p = 1, 2, 3,..., N. For existing of nontrivial solution, the vanishing of the determinant of coefficients produces the characteristics equation of the system in term of Rayleigh number Ra.

LINEAR STABILITY ANALYSIS

For the present formulation, we have considered the which system of equations (11.27) – (11.30) together with the boundary conditions (11.31) constitute a linear eigenvalue problem with variable coefficient for the growth rate of disturbance n of the system. Substituting equation (11.32) in system of equations (11.27) – (11.30) and multiplying the first equation by Wp the second equation by Θp third equation by Γp and fourth by Φp and then integrating in the limits from zero to unityand performing some integration by parts, one obtains the following matrix equation

$$\begin{bmatrix} \frac{(\pi^2+a^2)}{(1+nF)}\left(\tilde{D}a(\pi^2+a^2)+1+\frac{n(1+nF)}{Va}\right) & -a^2 Ra & -a^2\frac{Rs}{Ls} & -a^2 N_A Rn \\ 1 & -(\pi^2+a^2+n) & -N_{TC}(\pi^2+a^2) & 0 \\ \frac{1}{\varepsilon} & -N_{CT}(\pi^2+a^2) & -\frac{1}{Ls}\left(\pi^2+a^2+\frac{n}{\sigma}\right) & 0 \\ \frac{N_A}{\varepsilon} & \frac{N_A}{Le}(\pi^2+a^2) & 0 & -N_A\left(\frac{1}{Le}(\pi^2+a^2)+\frac{n}{\sigma}\right) \end{bmatrix} \begin{bmatrix} W_0 \\ \Theta_0 \\ \Gamma_0 \\ \Phi_0 \end{bmatrix} = \begin{bmatrix} 0 \\ 0 \\ 0 \\ 0 \end{bmatrix}. \tag{11.33}$$

The non-trivial solution of the above matrix for the case of stationary convection (n = 0) requires that

$$(\mathrm{Ra})_s = \frac{(\pi^2+a^2)^2}{\mathrm{a}^2}\frac{\left(\tilde{D}a(\pi^2+a^2)+1\right)(1-N_{CT}N_{TC}Ln)}{(\varepsilon - LsN_{TC})} + \frac{Rs(N_{CT}-1)}{(\varepsilon - LsN_{TC})} - \frac{\left(1+\frac{Le}{\varepsilon}\right) - \frac{LsN_{TC}}{\varepsilon}(1+LeN_{CT})}{(\varepsilon - LsN_{TC})}\mathrm{N_A Rn}. \tag{11.34}$$

It is observed that stationary Rayleigh number Ra is function of the Soret parameter, Dufour parameter, solute Lewis number Ls, nanofluid Lewis number Le, Solute Rayleigh number Rs, Brinkman-Darcy number, the modified diffusivity ratio N_A and the nanoparticles Rayleigh Rn but independent of visco- elastic parameter F, Prandtl number Pr and modified particle- density increment N_B. Thus Maxwellian visco-elastic nanofluid behaves like an ordinary Newtonian nanofluid and instability is purely a phenomenon due to buoyancy coupled with the conservation of nanoparticles.

In the absence of solute particle ($N_{CT} = N_{TC} = Rs = 0$) equation (11.34) reduces to

$$(\mathrm{Ra})_s = \frac{\tilde{D}a(\pi^2+a^2)^3 + (\pi^2+a^2)^2}{\mathrm{a}^2} - \left(1+\frac{Le}{\varepsilon}\right)\mathrm{N_A Rn}. \tag{11.35}$$

This is the same expression for thermal Rayleigh-Darcy number as obtained by Kuznetsov and Nield (2010b).

In the absence of nanoparticle and if the base fluid is one component, then equation (11.35) gives

$$(\mathrm{Ra})_s = \frac{\tilde{D}a(\pi^2+a^2)^3 + (\pi^2+a^2)^2}{\mathrm{a}^2}. \tag{11.36}$$

This is the exactly same result for regular fluid Nield and Bejan (2013).

Thus presence of the nanoparticles lowers the value of the critical Rayleigh number by usually by substantial amount. Also parameter N_B does not appear in the equation (11.34); thus instability is purely phenomena due to buoyancy coupled with conservation of nanoparticles. Thus average contribution of nanoparticles flux in the thermal energy equation is zero with one-term Galerkin approximation.

RESULTS AND DISCUSSION

Expression for stationary Rayleigh number, which characterizes the stability of the system, is found to be depend upon Soret parameter, Dufour parameter, solute Lewis number Ls, nanofluid Lewis number Le, Solute Rayleigh number Rs, Brinkman-Darcy number, the Brinkman Darcy number, the modified diffusivity ratio N_A and the nanoparticles Rayleigh Rn.

The computations are carried out for different values of parameters considered in the range $10^2 \leq Ra \leq 10^5$ (thermal Rayleigh number), $1 \leq N_{CT} \leq 10$ (Soret parameter), $10^{-3} \leq N_{TC} \leq 10^{-1}$ (Dufour parameter), $1 \leq N_A \leq 10$ (modified diffusivity ratio), $10^2 \leq Le \leq 10^4$ (nanofluid Lewis number), $1 \leq Ls \leq 10$ (solute Lewis number), $10^{-1} \leq Rn \leq 10^1$ (nanofluid Rayleigh number), $10^{-2} < \varepsilon < 1$ (porosity parameter).

The variations of the stationary thermal Rayleigh number with wave number have been plotted graphically as shown in Figures 2-9.

Figure 2 shows the variation of thermal Rayleigh number with wave number for different value of solutal Rayleigh number and for the fixed value of other parameters. It is found that stationary Rayliegh number increases as the value of solutal Rayleigh number increases, indicating that solutal Rayleigh number stabilizes the stationary convection.

Figure 2. Variation of the stationary Rayleigh number with wave number for different value of solutal Rayleigh number

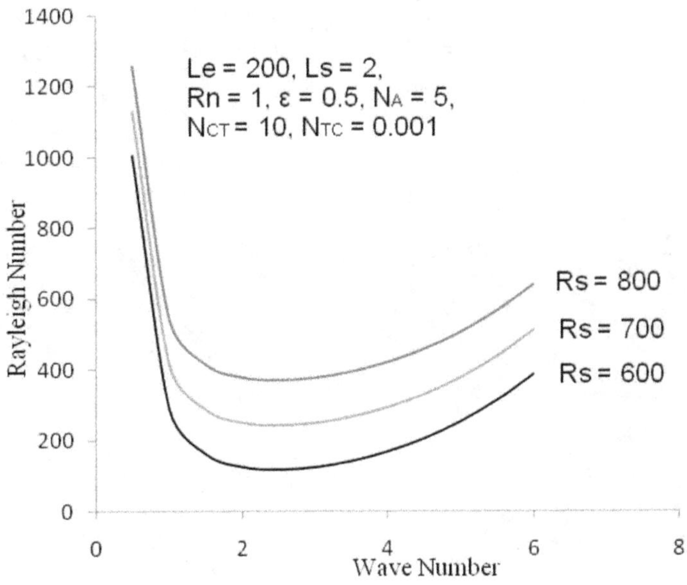

Figure 3 shows the variation of thermal Rayleigh number with wave number for different value of solute Lewis number and for the fixed value of other parameters. It is found that stationary Rayliegh number increases as the value of solute Lewis number increases, indicating that solute Lewis number stabilizes the stationary convection.

Figure 4 shows the variation of thermal Rayleigh number with wave number for different value of Lewis number and for the fixed value of other parameters. It is found that stationary Rayliegh number decreases with an increase in the value of Lewis number increases, indicating that Lewis number destabilizes the stationary convection. It is due to the fact that thermophoresis at a higher value of thermophoretic diffusivity is more supportable to the disturbance in nanofluids, while both thermophoresis and Brownian motion are driving forces in favor of the motion of nanoparticles.

Figure 5 shows the variation of thermal Rayleigh number with wave number for different value of Soret parameter and for the fixed value of other parameters. It is found that stationary Rayliegh number increases as the value of Soret parameter increases, indicating that Soret parameter stabilizes the stationary convection.

Figure 3. Variation of the stationary Rayleigh number with wave number for different value of solute Lewis number

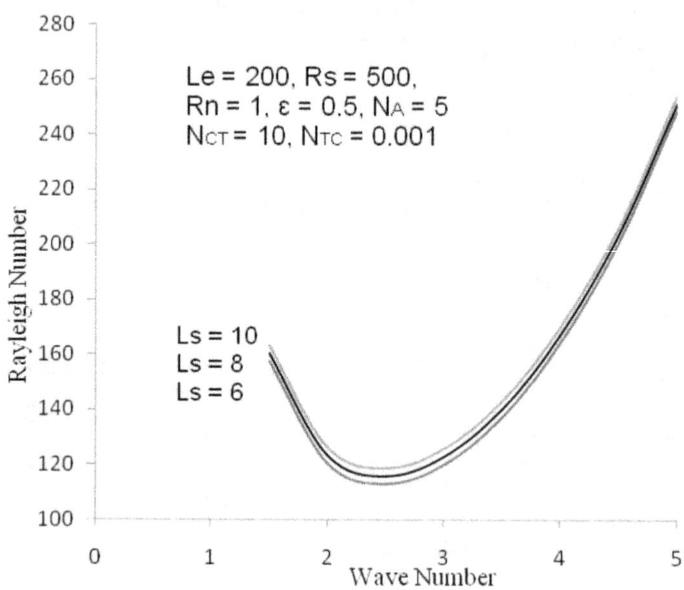

Figure 4. Variation of the stationary Rayleigh number with wave number for different value of nanofluid Lewis number

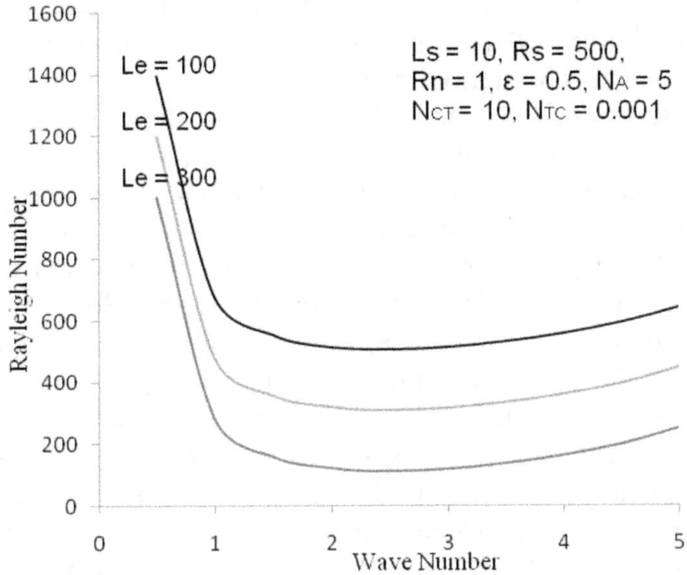

Figure 5. Variation of the stationary Rayleigh number with wave number for different value of Soret parameter

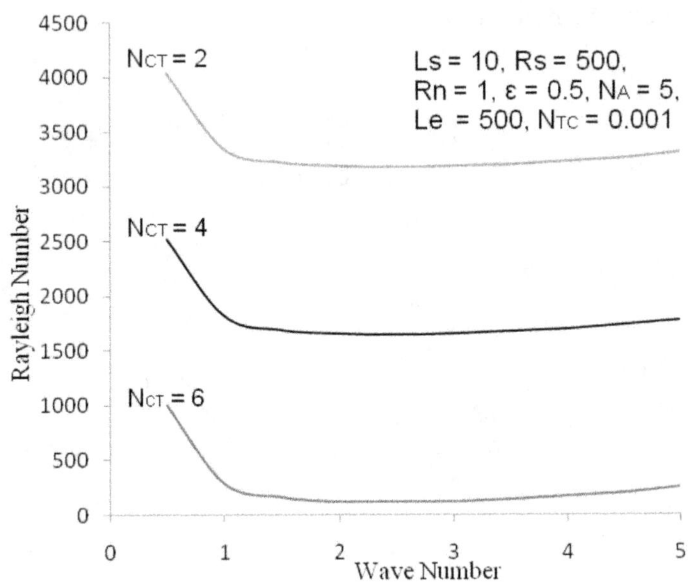

Figure 6 shows the variation of thermal Rayleigh number with wave number for different value of Dufour parameter and for the fixed value of other parameters. It is found that stationary Rayliegh number increases with an increase in the value of Dufour parameter, indicating that Dufour parameter has stabilizing effect on stationary convection.

Figure 7 shows the variation of stationary Rayleigh number with wave number for different value of the modified diffusivity ratio and fixed value of other parameters and it is found that Rayleigh number decreases with an increase in the value of the modified diffusivity ratio which means that the modified diffusivity ratio destabilizes on the fluid layer. This may lead to an increase in volumetric fraction, which shows that Brownian motion of the nanoparticles will also increase, which may cause destabilizing effect.

Figure 8 shows the variation of stationary Rayleigh number with wave number for different value of the nanoparticles Rayleigh number and fixed value of other parameters and it is found that thermal Rayleigh number decreases with an increase in value of the nanoparticles Rayleigh number, which means that the nanoparticles Rayleigh number Rn has destabilizing effect on fluid layer. It has destabilizing effect because the heavier nanoparticles moving through the base fluid makes more strong disturbances as compared with the lighter nanoparticles.

Figure 6. Variation of the stationary Rayleigh number with wave number for different value of Dufour parameter

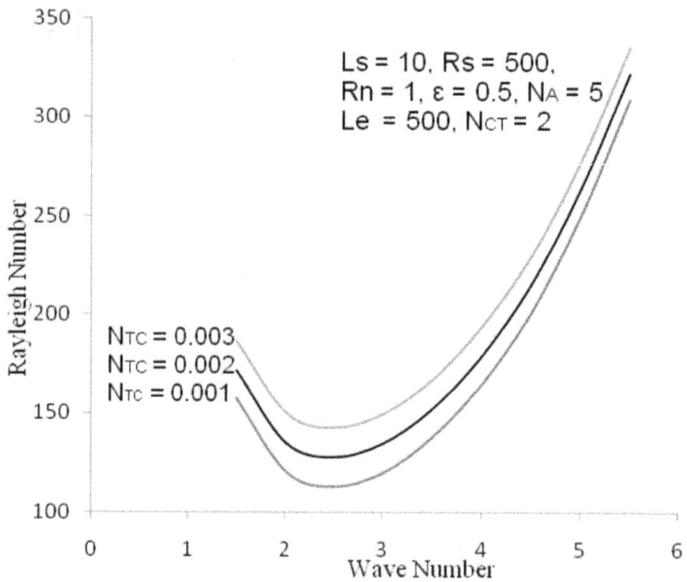

Figure 7. Variation of the stationary Rayleigh number with wave number for different value of modified diffusivity ratio

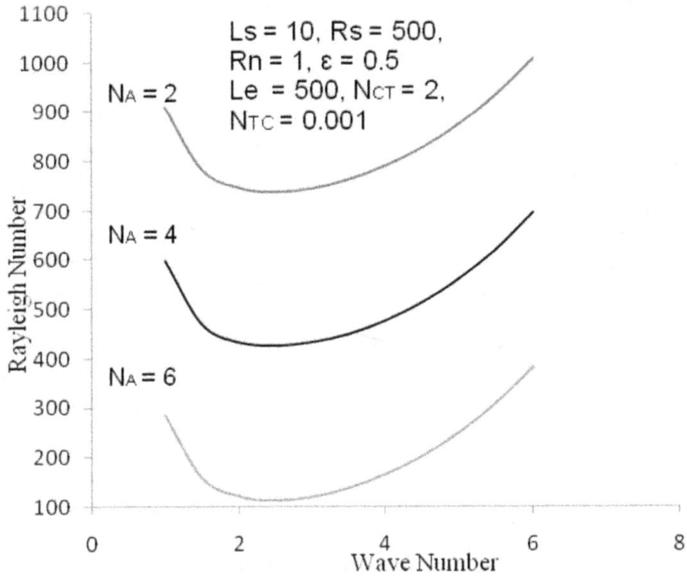

Figure 8. Variation of the stationary Rayleigh number with wave number for different value of nanoparticles Rayleigh number

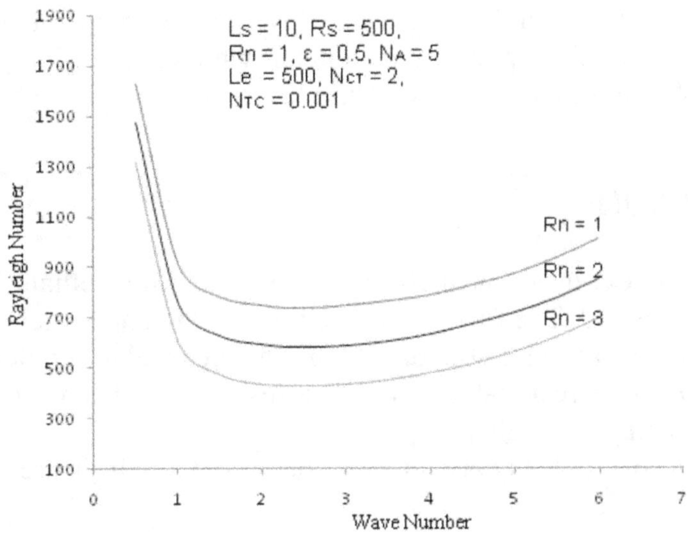

Figure 9. Variation of the stationary Rayleigh number with wave number for different value of porosity parameter

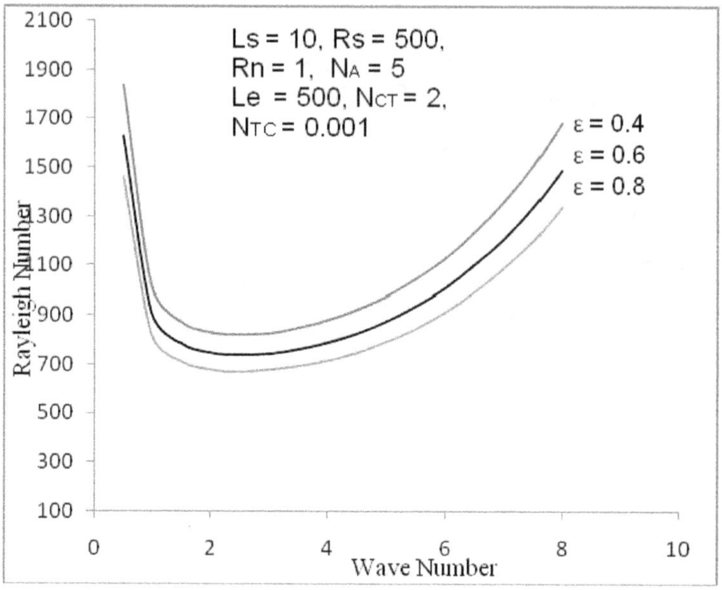

Figure 9 shows the variation of stationary Rayleigh numbers with wave number for different value of the porosity parameter and fixed value of other parameters and it is found that thermal Rayleigh number decreases with an increase in value of the porosity parameter, which means that porosity parameter has destabilizing effect on fluid layer.

CONCLUSION

Double diffusive convection in a horizontal layer of Maxwellian visco-elastic nanofluid is studied. The flux of volume fraction of nanoparticles is taken to be zero on the isothermal boundaries and the eigenvalue problem is solved using the Galerkin residual method. The results have been presented both analytically and graphically.

The main conclusions derived from the present analysis are as follows

1. The instability purely phenomenon due to buoyancy coupled with the conservation of nanoparticles and is independent of the contribution of Brownian motion and thermophoresis.

2. For stationary convection Maxwellian visco-elastic nanofluid behaves like an ordinary Newtonian nanofluid.
3. Solutal Rayleigh number, solute Lewis number, Dufour parameter and Soret parameter have stabilizing while Lewis number, modified diffusivity ratio, nanoparticles Rayleigh number and porosity parameter have destabilizing effect on the stationary convection.

REFERENCES

Knobloch, E. (1980). Convection in binary fluids. *Physics of Fluids*, *23*(9), 1918–1927. doi:10.1063/1.863220

Kuznetsov, A. V., & Nield, D. A. (2010c). The onset of double-diffusive nanofluid convection in a layer of a saturated porous medium. *Transport in Porous Media*, *85*(3), 941–951. doi:10.1007/s11242-010-9600-1

Malashetty, M. S., & Kollur, P. (2011). The onset of double diffusive convection in a couple stress fluid saturated anisotropic porous layer. *Transport in Porous Media*, *86*(2), 435–459. doi:10.1007/s11242-010-9630-8

Nield, D. A., & Bejan, A. (2013). *Convection in porous media*. New York: Springer-Verlag. doi:10.1007/978-1-4614-5541-7

Rana, G. C., Thakur, R. C., & Kango, S. K. (2014b). On the onset of thermosolutal instability in a layer of an elastico-viscous nanofluid in porous medium. *FME Transactions*, *42*(1), 1–9. doi:10.5937/fmet1401001R

Rana, G. C., Thakur, R. C., & Kango, S. K. (2014c). On the onset of double-diffusive convection in a layer of nanofluid under rotation saturating a porous medium. *Journal of Porous Media*, *17*(8), 657–667. doi:10.1615/JPorMedia.v17.i8.10

Rana, G. C., Thakur, R. C., & Kumar, S. (2012). Thermosolutal convection in compressible Walters (model B) fluid permeated with suspended particles in a Brinkman porous medium. *The Journal of Computational Multiphase Flows*, *4*(2), 211–224. doi:10.1260/1757-482X.4.2.211

Yadav, D., Bhargava, R., & Agrawal, G. S. (2013a). The onset of double diffusive nanofluid convection in a layer of a saturated porous medium with thermal conductivity and viscosity variation. *Journal of Porous Media*, *16*(2), 105–121. doi:10.1615/JPorMedia.v16.i2.30

Yadav, D., Lee, D., Cho, H. H., & Lee, J. (2016a). The onset of double-diffusive nanofluid convection in a rotating porous medium layer with thermal conductivity and viscosity variation: A revised model. *Journal of Porous Media*, *19*(1), 1–16. doi:10.1615/JPorMedia.v19.i1.30

Conclusion

Thermal instability in a horizontal layer of Maxwellian visco-elastic nanofluid is investigated theoretically. The model used for nanofluid incorporates the effect of Brownian diffusion, thermophoresis and electrophoresis. The flux of volume fraction of a nanoparticle is taken to be zero on the isothermal boundaries. Using linear stability theory, an expression for the Rayleigh number is obtained for both stationary and oscillatory convection in terms of various non-dimensional parameters. The stability criterions for stationary and oscillatory convection have been derived and graphs have been plotted to study the effects of various parameters on the on stationary convection. In the present study we theoretically investigate the effect of rotation, magnetic field, porous medium, internal heat source, Dufour and Soret parameter on the onset of thermal instability of Maxwellian visco-elastic nanofluid. The main conclusions derived from the present study are as follows

1. The instability purely phenomenon due to buoyancy coupled with the conservation of nanoparticle and is independent of the contribution of Brownian motion and thermophoresis.
2. For stationary convection Maxwellian visco-elastic nanofluid behaves like an ordinary Newtonian nanofluid.
3. Rotation, magnetic field, Brinkman-Darcy number, porosity, solutal Rayleigh number, solute Lewis number, Lewis number, Soret parameter stabilizing effect on the stationary convection.
4. Hall effect, Heat source strength parameter, modified diffusivity ratio and nanoparticles Rayleigh number have destabilizing effect on stationary convection.
5. Dufour parameter has both stabilizing and destabilizing effect on stationary convection depending upon certain conditions.
6. Stationary convection has stabilizing effect when the gravity parameter varies as $h(z) = z^2 - 2z$, $h(z) = -z$, $h(z) = -z^2$ and has destabilizing effect when gravity parameter varies as $h(z) = z$. In other word decreasing gravity parameter has stabilizing effect while increasing gravity parameter has destabilizing effect on the stationary convection.

FUTURE WORK SCOPE

In the present study the thermal instability in a horizontal layer of Maxwellian visco-elastic nanofluid is investigated theoretically. The model used for nanofluid incorporates the effect of Brownian diffusion, thermophoresis and electrophoresis. The flux of volume fraction of nanoparticles is taken to be zero on the isothermal boundaries and the eigen value problem is solved using the single-term Galerkin method. The same work can be extended up to the six terms Galerkin method to investigate the same study. Yadav et al. (2012b) studied the boundary and internal heat source effects on the onset of Darcy- Brinkman convection in a porous layer saturated by nanofluid by six terms Galerkin method. The work can be extended by taking different type of non-Newtonian fluids as base fluid. e.g. We can extend the work for Oldroydian visco elastic nanofluid, Rivlin-Ericksen elastico-viscous nanofluid, Walter's (B' model) visco-elastic nanofluid and couple stress nanofluids etc. for more realistic boundary conditions. Authors Sheu (2011a), Chand and Rana (2012c), Rana et al. (2014b) and Shivkumara et al. (2006, 2015) studied some thermal instability in a horizontal layer of non-Newtonian nanofluid by taking different type of non-Newtonian fluid as base fluid but work can be extended for for more realistic boundary conditions. Recently Khan et al. (2013), Kumar and Awasthi (2016), studied the triple- diffusive convection in a nanofluid layer so this work can be extended triple- diffusive convection in a Maxwellian visco elastic nanofluid layer or other non-Newtonian fluid for for more realistic boundary conditions. Nonlinear stability analysis by energy method and weakly nonlinear analysis of the problem can be explored.

REFERENCES

Chand, R., & Rana, G. C. (2012c). Thermal instability of Rivlin-Ericksen elastico-viscous nanofluid saturated by a porous medium. *Journal of Fluids Engineering*, *134*(12), 121203. doi:10.1115/1.4007901

Khan, Z. H., Khan, W. A., & Pop, I. (2013). Triple diffusive free convection along a horizontal plate in porous media saturated by a nanofluid with convective boundary condition. *International Journal of Heat and Mass Transfer*, *66*, 603–612. doi:10.1016/j.ijheatmasstransfer.2013.07.074

Kumar, V., & Awasthi, M. K. (2016). Onset of triple-diffusive convection in a nanofluid layer. *Journal of Nanofluids*, *5*(2), 284–291. doi:10.1166/jon.2016.1217

Rana, G. C., Thakur, R. C., & Kango, S. K. (2014b). On the onset of thermosolutal instability in a layer of an elastico-viscous nanofluid in porous medium. *FME Transactions*, *42*(1), 1–9. doi:10.5937/fmet1401001R

Sheu, L. J. (2011a). Thermal instability in a porous medium layer saturated with a visco-elastic nanofluid. *Transport in Porous Media*, *88*(3), 461–477. doi:10.1007/s11242-011-9749-2

Shivakumara, I. S., Malashetty, M. S., & Chavaraddi, K. B. (2006). Onset of convection in a viscoelastic-fluid-saturated sparsely packed porous layer using a thermal nonequilibrium model. *Canadian Journal of Physics*, *84*(11), 973–990. doi:10.1139/p06-085

Shivakumara, I. S., Dhananjaya, M., & Ng, C. O. (2015). Thermal convective instability in an Oldroyd-B nanofluid saturated porous layer. *International Journal of Heat and Mass Transfer*, *84*, 167–177. doi:10.1016/j.ijheatmasstransfer.2015.01.010

Yadav, D., Bhargava, R., & Agrawal, G. S. (2012b). Boundary and internal heat source effects on the onset of Darcy- Brinkman convection in a porous layer saturated by nanofluid. *International Journal of Thermal Sciences*, *60*, 244–254. doi:10.1016/j.ijthermalsci.2012.05.011

Appendix

LIST OF SYMBOLS

a	wave number
c	specific heat
c_v	specific heat at constant volume
C	speed of light
d	thickness of the horizontal layer
D_B	Brownian diffusion coefficient
D_T	thermophoretic diffusion coefficient
Da	Darcy number
$\acute{D}a$	Brinkman Darcy number
D_{TC}	diffusivity of Dufour type
D_{CT}	diffusivity of Soret type
e	charge of electron
F	stress relaxation parameter
$\mathbf{g}$	acceleration due to gravity
$\mathbf{H}$	magnetic field
Hs	constant of heat source strength
$\mathbf{j_p}$	mass flux
k_1	medium permeability
k_m	thermal conductivity of porous medium
k_B	Boltzmann constant
$k_{x,}$	wave numbers in x- direction
k_y	wave numbers in y- direction
Le	Lewis number
Ls	thermosolutal Lewis number
M	Hall effect parameter
n	growth rate of disturbances
N	electron number density
N_A	modified diffusivity ratio
N_B	modified particle -density increment
N_{CT}	Soret parameter

N_{TC} Dufour parameter
p pressure
Pr Prandtl number
Pr_M magnetic Prandtl number
q Darcy velocity vector
Q Chandrasekhar number
Q_0 distributed effective volumetric internal heat source
Ra Rayleigh number
Ra_c critical Rayleigh number
Rm density Rayleigh number
Rn concentration Rayleigh number
Rs solutal Rayleigh number
t time
T temperature
T_0 temperature at lower layer
T_1 temperature at upper layer
Ta Taylor number
v velocity of fluid
Va Vadasz number
(u,v,w) velocity components
(x,y,z) space co-ordinates

Greek Symbols

α thermal expansion coefficient
α_C analogous to solute concentration
μ viscosity
$\tilde{\mu}$ effective viscosity
μ' kinematic visco-elasticity
μ_e magnetic permeability
λ relaxation time
ε porosity
Ω angular velocity
ρ density of the fluid
ρ_f density of base fluid
ρc heat capacity of nanofluid
$(\rho c)_m$ heat capacity of nanofluid in porous medium
$(\rho c)_p$ heat capacity of nanoparticles
φ volume fraction of the nanoparticles
φ_0 volume fraction of the nanoparticles at reference scale

ρ_p density of nanoparticles

$\varsigma = \frac{\partial v}{\partial x} - \frac{\partial u}{\partial y}$ z-components of vorticity

$\xi = \frac{\partial h_y}{\partial x} - \frac{\partial h_x}{\partial y}$ thermal diffusivity

κ' solutal diffusivity

σ thermal capacity ratio

ω dimensionless frequency of oscillation

ς z-components of current density

η thermal anisotropy parameter

$\delta h(z)$ variable gravity parameter

Superscripts

' non- dimensional variables

'' perturbed quantity

Subscripts

p particle

f fluid

b basic state

0 lower boundary

1 upper boundary

s stationary

osc oscillatory

c critical

H horizontal plane

$D \equiv \frac{d}{dz}$.

About the Author

Ramesh Chand received his M.Sc. degree (Applied Mathematics) in 1995 and Ph.D. (Physical Sciences) in the year 2001 from Himachal Pradesh University Shimla, India. Presently he is working as Assistant Professor in Department of Mathematics at Govt. Arya Degree College Nurpur, HP, India. His research area is Hydrodynamic and Hydromagnetic Stability, Fluid dynamics, Heat and Mass Transfer, Flows in Porous Media, Instability in Newtonian and Non- Newtonian fluid and Nanofluid. He has guided 01 Ph.D Research Scholar and 11 M.Phil Research Scholars. He has published 65 papers in various national and international research journals. He has been conferred with 'Bharat Jyoti Award' for Meritorious Services for Outstanding Performance and Remarkable Role in the field of Education by India International Friendship Society, New Delhi in 2013 and 'Best Citizen of India' Award by Best Citizen Publishing House New Delhi in 2014. His short biography is published in "Marquis Who's Who in the World 2014" (31st Edition).

Index

B

D

F

L

M

N

P

R

T

V

W

www.ingramcontent.com/pod-product-compliance
Lightning Source LLC
LaVergne TN
LVHW061543290926
849162LV00004B/888